M000214111

ASSURANCE TECHNOLOGIES PRINCIPLES AND PRACTICES

ASSURANCE TECHNOLOGIES PRINCIPLES AND PRACTICES
A Product, Process, and System Safety Perspective

Second Edition

DEV G. RAHEJA
MICHAEL ALLOCCO

A JOHN WILEY & SONS, INC. PUBLICATION

Published by John Wiley & Sons, Inc., Hoboken, New Jersey
Published simultaneously in Canada

For general information on our other products and services or for technical support, please contact our
Customer Care Department within the United States at (800) 762-2974, outside the United States at
(317) 572-3993 or fax (317) 572-4002.

Wiley also publishes its books in a variety of electronic formats. Some content that appears in print
may not be available in electronic formats. For more information about Wiley products, visit our web site
at www.wiley.com.

Library of Congress Cataloging-in-Publication Data:

Raheja, Dev.
 Assurance technologies: principles and practices: a product, process, and system safety perspective/
Dev G. Raheja, Michael Allocco.—2nd ed.
  p. cm.
 Includes index.
 ISBN-13: 978-0-471-74491-7 (cloth)
 ISBN-10: 0-471-74491-3 (cloth)
  1. Quality assurance. 2. Design, Industrial. 3. Reliability (Engineering). I. Allocco, Michael. II. Title.

 TS156.6.R34 2006
 658.5′.62—dc22   2005033606

Printed in the United States of America

10 9 8 7 6 5 4 3 2 1

CONTENTS

**CHAPTER 7   LOGISTICS SUPPORT ENGINEERING AND
SYSTEM SAFETY CONSIDERATIONS   237**

CHAPTER 9   SOFTWARE PERFORMANCE ASSURANCE   289

PREFACE

Over 60% of the problems in complex systems develop from the incomplete, vague, and poorly written specifications. A good specification will mitigate at least 80% of risks. This is the purpose of our book. One must read the entire book to validate the quality of the specification. Each chapter reveals a different side of the specification. If you construct a new house, it is not sufficient to describe the front side. Detailed attention has to be paid to each side. Similarly, we must describe the product from various sides such as system safety, reliability, maintainability, human engineering, logistics, software integrity, and system integration to recognize the relationships of all the sides fitting well. Otherwise, one may wind up with a house where the sides are misaligned and the walls do not meet the floor.

We have a great passion for safety. We cannot stand the major accidents and about 125 automotive safety recalls every year. Too many people are dying in vain and families are suffering. That is why every chapter describes ways to integrate safety in everything we do.

This book emphasizes that safety is an excellent investment, not a necessary cost as most believe out of ignorance. In 2004, 1.2 million people died just in automotive accidents at a cost of $230 billion. Preventing accidents is certainly much cheaper.

The first edition of this book was published by McGraw-Hill in 1991 under the title *Assurance Technologies: Principles and Practices*. It was on the Technical Best Seller List for two years. Later, the printing was discontinued because McGraw-Hill stopped publishing engineering management books. The authors are thankful to John Wiley & Sons, for encouraging us to write this book with safety as a strong foundation, and to our readers, who made the first edition a best seller.

Dev G. Raheja
Michael Allocco

CHAPTER 1

ASSURANCE TECHNOLOGIES, PROFITS, AND MANAGING SAFETY-RELATED RISKS

1.1 INTRODUCTION

Assurance technologies are processes for ensuring that a product performs well during its expected lifetime. They call for proactive care in design, manufacture, and maintenance, and they must be integrated if they are to succeed. Manufacturers of large systems such as aircraft or nuclear plants call the integration process *system assurance*. Manufacturers of consumer or industrial products call it *product assurance*. The major difference lies in the degree of complexity.

The most important mistake companies often make is not putting enough effort into writing system performance specifications. Usually many design assurance functions are vague, missing, or incomplete. For example, a specification may say the reliability shall be 95%. This is a vague statement. It does not say for how long. A year? Is it for the duration of the warranty? Or is it for the expected life such as 100,0000 miles for an automobile? The confidence level is completely missing. The description of duty cycles is also missing. Anyone writing a specification must read this whole book. Each chapter points out the features that are required in specifications.

The basic assurance technologies are the following:

Reliability Analysis: Lowers product failure rate over the long term, reduces producer's warranty costs, and makes the customer want to come back—if the price is not unreasonable. Product reliability depends on robustness of

Assurance Technologies Principles and Practices: A Product, Process, and System Safety Perspective, Second Edition, by Dev G. Raheja and Michael Allocco
Copyright © 2006 John Wiley & Sons, Inc.

design, component quality, and manufacturing. Its design requirements come from preliminary analysis before any hardware or software is acquired.

Maintainability Analysis: Minimizes downtime, reduces repair time, and, as a result, reduces maintenance costs. Some of the inputs for maintainability come from reliability analysis.

System Safety Engineering and Management: Enables the identification, elimination, and control of safety-related risks throughout the life cycle of the product, process, or system; thereby providing a safe product, process, or system.

Quality Assurance Engineering: Incorporates customer satisfaction requirements in design and assures that specifications have been met and mitigates risks associated with manufacturing errors or defects.

Human Factors Engineering: Recognizes the role of humans in product, process, or system and provides effective integration. It helps designers prevent human-induced mishaps by making designs insensitive to human errors of use. It also mitigates risks to humans that interface with the product, process, or system.

Logistics Engineering: Reduces ongoing field support costs, most of which result from poor quality, reliability, maintainability, and safety. A design analysis at the outset helps avoid much of this unnecessary waste of resources.

1.2 CHEAPER, BETTER, AND FASTER PRODUCTS

The purpose of assurance technologies is to allow an organization to make products better, cheaper, and faster. These benefits are the best leverage an organization has in a competitive market.

The U.S. Department of Defense numbers on life-cycle costs [1] show that 85% of costs are incurred during production and in service and only 15% during design and development. The life-cycle cost breakdown is as follows:

Concept stage	3%
Development stage	12%
Production stage	35%
Operation and support stage	50%

If the life-cycle costs (LCCs) are $100 million for a product, $85 million are spent in the post-design efforts, most in firefighting such as reducing manufacturing waste, paying for warrantees or recalls, and fixing problems. A good product assurance organization should reverse this negative trend. There should be much more spending upfront and much less investment in the later life cycle. This will reduce life-cycle costs dramatically. Even though the customer pays most of these costs, they have tremendous influence on the supplier. The customer can always go to another supplier whose product requires less maintenance and repair (see Chapter 4 for an example; it tells how Canon took away a 90% share of a market from a competitor). Even an old, established supplier may not survive when this happens. Aside from inconvenience and downtime, a high maintenance and repair rate means higher life-cycle cost. A supplier whose total life cycle cost is too high inevitably starts losing customers.

A cost reduction effort requires some investment in engineering analyses and in improving the product, the process design, and process controls. There is enough evidence to show that investment in these technologies is not an add-on cost, but a smart investment. Companies that are not willing to invest $1 today to save $10 later (proprietary data show as much as $100) are likely to stagnate.

Figure 1.1 makes the benefits of assurance technologies clearer. It shows that life-cycle costs are highest when no investment in assurance technologies is made and become lower and lower as more technologies are integrated. For example, a LCC of $100 million can be reduced to $50 million with assurance technologies. The benefits go to both the supplier and the customer (see the discussion of LCCs with and without stress screening in Chapter 3). A company should not be afraid to invest $5 million to save $50 million in future costs—except if it is run only for short-term survival and quick profits. So far, the benefits of low cost and good performance have been stressed. Fortunately, these benefits are also the primary requisites for adhering to an efficient implementation schedule. The *Challenger* space shuttle disaster taught lessons in this regard. The accident showed that space shuttle schedules were totally out of hand because of technical problems. Schedules can be held to only if they are realistic and are backed by sufficient resources to identify and mitigate problems before they become problems. This is what Oppenheimer meant when he said "genius is one who has answers before questions." Problems can be mitigated using various proactive methods covered in this text.

A success story happened at Ford. When the 1995 Lincoln Continental was analyzed for internal and external interfaces and other issues, the engineers made over 700 changes in the specification. This resulted in a reduction of the manufacturing costs from $90 million down to $30 million. The project started late but finished 4 months early. Only those who have accomplished such results know that making a system cheaper, better, and faster is a science.

Integrating assurance technologies is very complex and time consuming. The answer is not to compromise on the intent but to use concurrent engineering to perform tasks in concert. This process allows many kinds of analyses simultaneously so that problems are prevented downstream. For example, an associate of one of the authors made a reliability analysis of a new product for a defense electronics firm in Minneapolis concurrently with product design. He showed that many components were operating in the upper 50th percentile range of their ratings in the current design. He convinced the company that this design was not in conformity with reliability engineering principles of derating and he helped redesign the product so that all the parts operated below the 50% rating. The result took design engineers by surprise. When they tested the product for qualification, the

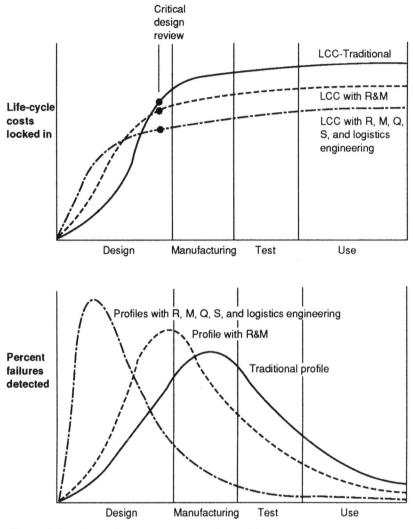

Figure 1.1   Effects of assurance technologies on life-cycle costs and performance.

product passed! They had never had this experience before; they had always gone back and forth several times, testing and fixing the problems. This time they were 3 months ahead of schedule because the new product did not require fixing. Better designed products can help reduce scheduled time and costs. This example shows a result of reliability improvement alone. If similar improvements are also made in quality assurance, system safety, maintainability, and logistics support engineering, new product introduction schedules could be shortened even more.

Caution: The key is concurrent engineering. If the analyses are done sequentially, as they are in many companies, they will always be late and have little value, and schedules will be almost always beyond control.

1.3 WHAT IS SYSTEM ASSURANCE?

To understand system assurance, one has to understand the definition of a failure and hazard. If a system does not meet the reasonable expectation of the user, then it has failed, even though it meets the specifications. When failures result in hazards, accidents can occur. This statement will surprise many readers, but millions of unsafe products are recalled annually for this reason. An example: a man wearing a programmable pacemaker walked through a department store's electromagnetic antitheft device. The device caused the pacemaker to malfunction; the man died of a heart attack as a result. The pacemaker and the antitheft device met their specifications but did not meet the expectations of their users. A customer expects a safe[1] and least troublesome product.

1.4 KEY MANAGEMENT RESPONSIBILITIES

1.4.1 Integration

For a customer, the system performance depends on the integration of at least five elements that only supplier management can integrate: hardware, software, people, environment, and methods. The system should be robust against errors, failures, and hazards in these areas in the user's environment. The user expects this robustness throughout the entire life cycle. These concepts are shown in Fig. 1.2.

The block diagram in Fig. 1.2a shows that, for a system to work, the hardware, the software, the manufacturing methods, and the human factor (operating and maintenance personnel) must be reliable simultaneously. Further consider that this model operates within an environment. Also implied in the model is that all of the interfaces among these five elements must be reliable.

Since human behavior is the most unpredictable among the elements, it may be the weakest link in the chain. Figure 1.2b shows an improved model in which the human user is a redundant element. With increasing use of software, the system needs less human interaction.

1.4.2 Budget Consistent with Objectives

Once a system is defined, management must observe the ground rules and manage resources effectively. Resources are needed to finance highly detailed analysis

---

[1]Nothing is completely safe: a safe system should indicate that the identified safety-related risks throughout the life cycle have been identified, eliminated, or controlled to an acceptable level.

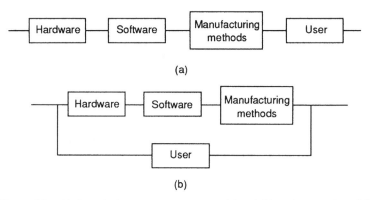

Figure 1.2   (a) A typical system assurance model and (b) an improved model.

during early design. An investment is most productive at this point. Unfortunately, the lack of a budget prevents managers from putting investment dollars on the table. To develop budgets for such activities, managers should learn to anticipate the investment needed to make several hundred design improvements.

1.4.3 Managing Risk

The process of risk management is very crudely defined in most management texts. Most of them use the projection of past data. They do not take into account that there is always a first-time disaster waiting to happen. This book approaches risk in a way designed to avoid many first-time disasters with the help of proactive designs.

The classical definition of risk as "a product of frequency and severity" is vague. It does not indicate what level of severity should be considered if the product has, say, ten levels of severity. It does not specify which out of, say, 5000 or so failures should be considered when frequency of failure is computed. Harold Geenan, former chairman of ITT Corp., writes: "Theory should be considered a starting point. Once you have broken that barrier you have to expand into the reality" [2]. This definitely applies to risk. The first reality is that a complex product, the *Columbia* shuttle, for example, can fail in several thousand ways. Each failure or hazard can have several levels of severity from no harm to loss of the system and loss of life. Indeed, the *Columbia* disaster may have been even more severe—it might have crashed in a crowded section of a city. A second reality is that a complex product, such as the shuttle, can fail again in a different manner.

1.4.3.1 Managing Safety-Related Risk   In the context of risk management, risk control includes concepts of risk acceptance, risk reduction, risk avoidance, spreading of the risk, and risk transfer. When considering the big picture, while initial design concepts are thought of these risk management concepts should be understood and applied.

Risk Acceptance and Risk Reduction   Risk acceptance and risk reduction are techniques that are used usually later within a system's life cycle. Decisions that involve risk avoidance can be made early during the inception of a design. A choice can be made to avoid a risk entirely. It may be concluded that the system, operation, or process may be too risky.

Risk Substitution   Substitution of a lower risk for a higher risk equates to risk reduction. This kind of decision can also be made early in concept development. For example, substitution of a low-power device for a high-power device could be made because of the risk of electrical shock.

Spreading of Risk   The method of spreading the risk involves insurance. Similar risks are insured and spread among the many. The insureds provide premiums, which are the cost of insurance. Insurance companies pay out for losses that are incurred. When the ratio of losses and premiums are less then 0.60, insurance companies will underwrite the business (risk). When total losses exceed premiums, insurance companies will not underwrite the risk.

Transfer of Risk   Risk transfer involves the actual transfer of a risk from one entity to another. High-risk operations may be farmed out to subcontractors who have more suitable risk control. For example, it would be less risky to have a car professionally painted rather than attempting to paint a car in the back yard. The professional contractor may be equipped with an appropriate paint booth, respirators, flammable liquids storage, and the electrical installation meeting the National Fire Protection Association Standards. There may have been apparent risks that were then transferred, such as risks of toxic inhalation, fire, and explosion—not to exclude the potential for poor quality.

1.4.3.2 *Risk Assessment*   Risk assessment provides a capability to rank system risks in order to allocate resources to fix higher risks in descending order to lower risks. It also provides a capability to compare risks within very similar systems. Comparisons can also be made between initial and residual risk. The overall system objective should be to design a complex system with acceptable risks.

The analyst should be concerned with safety-related risks. In evaluating complex systems, an initial concept to recognize is that risk is inherent in every system, subsystem, operation, task, or endeavor. Risk is an expression of the possibility of an accident in terms of severity and likelihood. Hazards in the form of initiators and contributors combine and present the potential risk of an accident.

1.4.3.3 *Risk Types*   There are many types of risk to consider: speculative, static, dynamic, and inherent. A speculative risk is one for which it is uncertain as to whether the final outcome will be a gain or loss. Gambling is a speculative risk. Usually, speculative risks cannot be insured. If a company introduces a new technology without sufficient understanding, the outcome is speculative. Speculative risks, for example, include loss of systems, process interruption, operational deviation, theft of valuables, vandalism, loss of key personnel, flood, water damage, pollution, environmental damage, fire, and explosion. Static risk is continuous and unchanging; the exposure may be constant and examples include constant forms of solar radiation, environmental pollution, nuclear power, and radon. Dynamic risk is fluctuating and it may be unpredictable such as weather anomalies, sunspots, or meteor strikes. Inherent risk may be unavoidable; it is the residual risk within a system after implementation of a design. It may be considered the risk of doing business.

1.4.3.4 Risk Terms   In conducting risk assessment, consideration is given to concepts that address identified versus unidentified risk, unacceptable versus acceptable risk, and residual risk. Identified risk is the risk that is recognized and comprehended as a result of assessment. It is not possible to be all-inclusive and indicate that all risk has been identified. There is an unknown aspect of any endeavor, which may present risk. Risk acceptability relates to sociology and behavioral science and whether the people being exposed to the risk will accept it. Acceptable risk is the part of identified and unidentified risk that is allowed to persist without further risk control, elimination, or mitigation. Residual risk is the risk within the system after all aspects of assurance technology, system safety, and risk management have been employed.

1.4.3.5 Risk Knowledge   An analyst should acquire as much knowledge of risk as possible. This knowledge may be gained by analysis, synthesis, and simulation or testing. In conducting these efforts, an attempt should be made to conduct analysis, synthesis, simulation, and testing as close as possible to the real world. The assumptions made, the calculations conducted, and the integration testing should all be suitable and reflect reality. Lack of knowledge of complex system actions, synergistic effects, and systemic anomalies all equate to the possibility that risk knowledge may be inadequate. When conducting analysis, synthesis, and simulation and testing, errors or oversights can be made that may skew results and reality, and consequently risk knowledge.

1.5 IS SYSTEM ASSURANCE A PROCESS?

All assurance technologies are processes. Thus, system assurance is a process. However, like any innovative process, it will only work if top management treats it as a business goal. Management must not only be committed to system assurance but should also be actively involved. Management that looks only at financial results will no longer be able to compete.

Example 1.1   A classic example of the above is Yokogawa–Hewlett Packard, where management as well as all the research and development (R&D) engineers learned and practiced reliability engineering. The data in Fig. 1.3a shows the annual failure rate versus the list price. As it should be, the reduction in price is

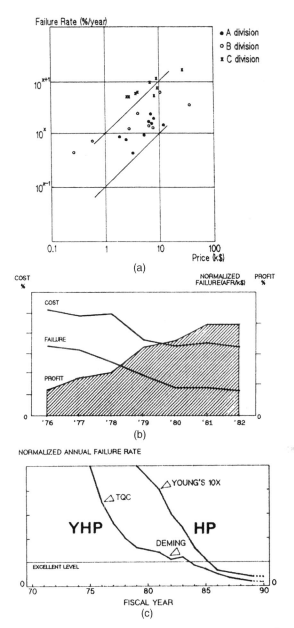

Figure 1.3 (a) Annual failure rate versus list price. (b) YHP product-line achievements. (c) Reliability improvement. (From a presentation by Kenzo Sasaoka, President, Yokogawa–Hewlett Packard, to the International Symposium on Reliability and Maintainability, organized by the Japanese Union of Scientists and Engineers, 1990. Reprinted with permission of Mr. Sasaoka.)

accompanied by the corresponding reduction in failure rate. If assurance technologies are used effectively, the low failure rate should result in higher profits, as shown in Fig. 1.3b. In the 1980s, John Young, President and Chief Executive Officer of Hewlett Packard, set the company-wide goal of tenfold improvement in reliability. Figure 1.3c is proof that such goals have been achieved. In summary, it is fair to say that assurance technologies reduce cost and increase profits.

1.6   SYSTEM ASSURANCE PROGRAMS

The tasks, activities, and functions that define system assurance need to be formally defined in order to plan efforts and allocate resources. Throughout this text, the various tasks, activities, and functions are described for the specific disciplines involved within system assurance. Specific program plans have evolved for each discipline. They are discussed within the following chapters. An integrated systems engineering or system assurance plan is used to define the concurrent engineering tasks, activities, and functions, such as formal concurrent reviews, working group meetings, scheduling, and coordination.

REFERENCES

1. M. B. Darch, Design Review. In: *Reliability and Maintainability of Electronic Systems*, J. E. Avsenault and J. A. Roberts (Eds.), Computer Science Press, Potomac, MD, 1980.
2. H. Geenan, *Managing*, Doubleday, New York, 1984.

FURTHER READING

Ireson, G. W., and C. F. Coombs, Jr., *Handbook of Reliability Engineering and Management*, McGraw-Hill, New York, 1988.

Ohmae, K., *The Mind of the Strategist*, McGraw-Hill, New York, 1982.

Raheja, D., There Is a Lot More to Reliability than Reliability, Society of Automotive Engineers World Congress, Paper 2004-866, 2004.

CHAPTER 2

INTRODUCTION TO STATISTICAL CONCEPTS

2.1 PROBABILISTIC DESIGNS

This chapter covers statistics only to the extent necessary for understanding the other chapters in the book. The reader is advised to read References 1, 2, and 3 for more details.

Performance of most designs is treated as deterministic, but it is always probabilistic. If an engineer uses Young's law to calculate the strain for the stress applied to each of a group of similar components, the law will give the same value for each. In real life, the strain for each component will not be the same. The range of strains can be described with statistical theory. The same variability principles apply to all engineering measurements and values. There is always a variation, and the range of the variation can be quantified.

To study variation, we draw a histogram of data and determine what percentage is below or above a target value. When an instrument supplier promises an accuracy of $\pm 5\%$, it means a very high percentage of measurements will fall within $\pm 5\%$ around a target measurement.

If histograms are plotted so that each bar represents the frequency as a percentage of all the values in the population, and if each interval is small, the heights of the bars can be joined to form a smooth curve. This curve represents the relative frequency with which the events represented by the interval occur. The relative frequency,

Assurance Technologies Principles and Practices: A Product, Process, and System Safety Perspective, Second Edition, by Dev G. Raheja and Michael Allocco
Copyright © 2006 John Wiley & Sons, Inc.

when used as an estimate of the population fraction, is commonly referred to as the *probability*. The mathematical model of this curve is called the *probability density function* (pdf) or $f(t)$ [1, 2].

This chapter covers some distributions that are applicable to engineering situations. Special distributions may be required if these models do not match the data. Section 2.2 covers the generic construction of a probability density function. Other sections describe the models frequently used in assurance technologies.

2.2 PROBABILITY COMPUTATIONS FOR RELIABILITY, SAFETY, AND MAINTAINABILITY

Probability of failure is the single most important concept encountered in reliability, maintainability, and safety assessment. The following generic concepts will be very helpful, no matter what the shape of the distribution.

2.2.1 Construction of a Histogram and the Empirical Distribution

A probability distribution is determined from a histogram of the data. Even though the future distribution may not be the same, a histogram is an estimate that can be modified with the accumulation of more knowledge in the future. Constructing the histogram involves grouping the data in mutually exclusive and collectively exhaustive intervals, computing the sample fraction for each interval, and plotting fraction versus the interval. *Mutually exclusive* means no two intervals will overlap. *Collectively exhaustive* means we must consider all the intervals that are possible even though there are no data in some intervals. For example, the following liability claims were made during a 10-year period:

Size of Claims (dollars)	Number of Claims
0–4,999	13
5,000–9,999	31
10,000–14,999	14
15,000–19,999	11
20,000–24,999	5
25,000–29,999	2
30,000–34,999	0
36,000–39,999	1

Each interval in the above data is already arranged so that there is no overlapping. The next step is to determine if the intervals are collectively exhaustive.

Since there can be no negative claims, the claims should start from 0. Up to $40,000, all possibilities are accounted for. Can a claim exceed $40,000 even though records do not show that any has in the past? It is possible; future claims might be very large. So one more interval is necessary. It can be labeled $40,000+ to include all possible claims beyond $40,000. Even though the sample shows the observed fraction for $40,000+ claims as zero, a larger sample may reveal some small positive value. The observed probability for each interval is computed by dividing the number of claims in each category by the total number of claims, as tabulated below.

Size of Claims (dollars)	Number of Claims	Observed Probability (%)	Cumulative Probability (%)
0–4,999	13	16.9	16.9
5,000–9,999	31	40.2	57.1
10,000–14,999	14	18.2	75.3
15,000–19,999	11	14.3	89.6
20,000–24,999	5	6.5	96.1
25,000–29,999	2	2.6	98.7
30,000–34,999	0	0	98.7
35,000–39,999	1	1.3	100.0
40,000+	0	0	100.0
Total	77	100.0	

Figure 2.1 depicts the observed probabilities in a histogram (shown in smaller intervals). Even more useful are the cumulative probabilities plotted in Fig. 2.2.

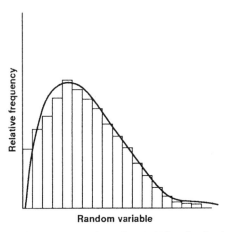

Figure 2.1   Construction of probability distribution.

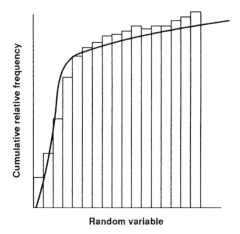

Figure 2.2 Construction of cumulative probability distribution function.

The cumulative distribution plot can be used for reading various risks at a glance. For instance, what are the chances (probability) that the next claim size will be below $22,000? The answer is obtained simply by projecting a vertical line from $22,000 until it intersects the plotted curve and reading the *y* value. The observed probability for this example is 93%.

Caution: Some statisticians recommend that the Bayes approach be used in estimating probability. This approach combines past experience and new data to generate a new distribution. Readers with adequate knowledge of statistics should take this into account.

For an entirely new product or for a system on which no past data are available, the failure modes are first determined by means of logic models, as in fault-tree analysis. The probabilities for various failures may be assigned by using expert opinions. The probability of an accident may be computed by using the rules explained in Chapter 5.

2.2.2 Computing Reliability

The reliability of a product is derived from the following relations:

Probability of product working + probability of product failure = 1

Therefore,

Probability of product working = 1 − probability of product failure

or

Reliability = 1 − probability of product failure

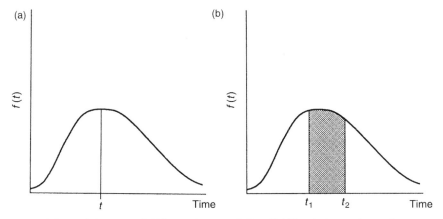

Figure 2.3   (a) Reliability at time t and (b) reliability during an interval.

Two types of reliability computations are generally required (Fig. 2.3):

1. Reliability up to time t is

$$R(t) = 1 - \text{probability of failure up to time } t$$

$$= 1 - \int_0^t f(t)\, dt = 1 - F(t)$$

where $F(t)$ is the cumulative probability up to time t, also called the *cumulative distribution function* (cdf).

2. Reliability during an interval $(t_2 - t_1)$ is defined as

$$R(t_2 - t_1) = 1 - \int_{t_1}^{t_2} f(t)\, dt$$

Sometimes the probability of failure during an interval can be defined in terms of reliability. For example,

$$P(t_1, t_2) = F(t_2) - F(t_1)$$
$$= [1 - R(t_2)] - [1 - R(t_1)]$$
$$= R(t_1) - R(t_2)$$

2.2.3   Failure Rate and Hazard Function

The interval failure rate during a time interval (t_2, t_1) is defined as the fraction (or percentage) of units failing per unit time (or distance or cycles). That is, in the interval, given that these units have not failed prior to time t_1

$$\text{Average interval failure rate} = \frac{N_1 - N_2}{N_1(t_1 - t_2)} \qquad (2.1)$$

where N_1 = number of survivors at time t_1, N_2 = number of survivors at time t_2, and $N_1 - N_2$ = number failed during the interval (t_2, t_1).

The *hazard function* is defined as the limit of the failure rate as $(t_2 - t_1)$ approaches zero. It is also called the *instantaneous failure rate*. That is, when the interval $(t_2 - t_1)$ is very small, the failure rate is denoted by the hazard rate, which can be written as [1] $h(t) = f(t)/[1 - f(t)]$ or $f(t)/R(t)$.

If one desires a probability of failure in a small interval, from t to $(t + \Delta t)$, the product $h(t) \, \Delta t$ is the probability. The hazard rate varies over time for most distributions. For an exponential distribution (Section 2.5), the hazard rate is constant.

2.3   NORMAL DISTRIBUTION

The normal distribution model typifies the concept of variation. It adequately describes many quality control measurements, test data, and wear-out failures for mechanical components. Some electronics characteristics also are adequately described by the normal distribution. Ideally, the products of a process will have exactly the same measurements, but this does not happen. The measurements will tend to crowd around a value. This value in the normal density function of Fig. 2.4 is represented by the *arithmetic average*, called $\overline{X}$. It is used as an estimate

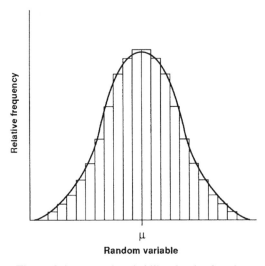

Figure 2.4   Normal probability density function.

of the population mean. The data are distributed symmetrically around the population average. In other statistical distributions, the data may not be spread symmetrically.

Consider a dimension on a part. All pieces do not measure exactly the same, and their departure from the average is called *deviation*.

This gives a measure of spread, which can be written as $(X_i - \overline{X})$, where X_i is the individual measurement and $\overline{X}$ is the average of all the measurements. The average deviation in this case cannot be computed arithmetically; the positive and negative deviations sum to zero. Thus, the average deviation is not a suitable measurement of spread. Statisticians, however, have a way to compute spread; they square each deviation, making it always a positive number. If a random sample of n units is drawn from a lot, the average of the squared deviation then is $(\sum(X_i - \overline{X})^2)/n$, where $\overline{X}$ is the arithmetic average of all the measurements in the sample. This positive value allows computation of the average deviation by taking the square root of this value. The average deviation computed this way is called *standard deviation*. For a sample, it is written as

$$s = \sqrt{\frac{\sum (X_i - \overline{X})^2}{n}} \tag{2.2}$$

For an unbiased estimate (see Ref. 2) of population when the sample size is less than 30, the formula is written as

$$s = \sqrt{\frac{\sum (X_i - \overline{X})^2}{n - 1}}$$

A property of the normal probability density function is that 67% of the population is within ± 1 standard deviation of μ, 95% is within ± 2 standard deviations of μ, and 99.7% is within ± 3 standard deviations (Fig. 2.5). The terms $\overline{X}$ and s usually refer to statistics computed from the sample data. Such values for population (called *parameters*) are represented by the Greek letters μ and σ.

The pdf for the normal distribution is

$$f(t) = \frac{1}{\sigma\sqrt{2\pi}} \exp\left(-\frac{(t - \mu)^2}{2\sigma}\right)$$

where t is the measurement under consideration, called the *random variable*.

Table A in the Appendix provides the numerical value of the cumulative area under the normal density curve for a given value of Z. It can be used to calculate the population percentage above or below a selected value of the variable. Typically, an engineer wants to estimate what percent of production is expected to be within or outside a desired specification. The only information needed is a score called Z, determined from the sample $\overline{X}$ and the number of standard deviations between $\overline{X}$

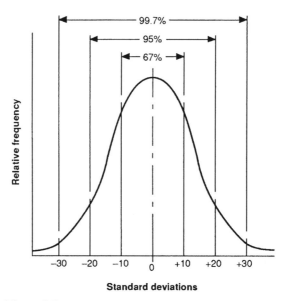

Figure 2.5   Percent area under the normal density curve.

and the specification:

$$Z = \frac{|\text{specification} - \overline{X}|}{\text{standard deviation}} \qquad (2.3)$$

With this score, one can look up the percentage area outside the specification in the table. The computation is also done for each side of the specification, if the specification is two-sided. The procedure is demonstrated in the following examples. The examples assume the data are normally distributed. The analyst should verify this assumption either analytically or graphically. See Ref. 4 for verification procedures.

Example 2.1   The data below were collected on a random sample from a production batch. The specification for the Rockwell hardness number is the range 54 to 58.

ROCKWELL HARDNESS NUMBERS

54	58	57	56
56	55	54	55
57	56	57	55
56	56	55	56
57	54	57	55

Compute:

1. The process capability defined as ± 3 standard deviations spread. (*Note:* A recent trend is to use ± 6 standard deviations.)

2. The percent production expected to be out of specification.
3. The ppm level (nonconforming parts per million).

Solution

X_i	$(X_i - \overline{X})^2$
64	3.24
66	0.04
67	1.44
66	0.04
57	1.44
68	4.84
66	0.64
66	0.04
66	0.04
64	3.24
67	1.44
64	3.24
66	0.64
67	1.44
66	0.04
66	0.64
65	0.64
66	0.04
66	0.64
67	1.44
1116	$\sum(X_i - \overline{X})^2 = 25.20$

$$\overline{X} = \frac{1116}{20} = 55.8$$

The estimate s of population standard deviation is

$$\sqrt{\frac{\sum(X_i - \overline{X})^2}{n-1}} = \sqrt{\frac{25.2}{20-1}} = \sqrt{1.33} = 1.15$$

Then

$$\begin{aligned}
\text{Process capability} &= 55.8 \pm 3 \text{ standard deviations} \\
&= 55.8 \pm 3 \times 1.15 \\
&= 55.8 \pm 3.45
\end{aligned}$$

This suggests that the process spread is expected to be in the range of 52.35 to 59.25 for 99.7% of the production (see Fig. 2.5). Only 3 out of 1000 are expected to be out of this range.

To estimate the percent production expected to be out of specification, compute the percentage outside each end of the specification. For the upper side of the pecification, from Eq. (2.3),

$$Z = \frac{|58 - 55.8|}{1.15} = 1.91$$

The area outside the specification for $Z = 1.91$, from Table A in the Appendix, is 0.0281 or 2.81%.

For the lower side of the specification:

$$Z = \frac{|54 - 55.8|}{1.15} = 1.57$$

The area outside the lower side from Table A for $Z = 1.57$ is 0.0582 or 5.82%. This process is therefore inherently expected to produce $2.81 + 5.82 = 8.63$ or approximately 9% units out of specification.

The out-of-specification parts in parts per million units of production are

$$\frac{8.63}{100} \times 1,000,000 = 86,300 \text{ ppm}$$

Example 2.2   Consider the pacemaker manufacturer that told doctors that the average battery life is 4 years and the doctors who told the patients that their batteries will work for 4 years. Assuming a normal distribution, estimate:

1. The number of patients whose batteries may fail in 4 years.
2. What proportion of batteries is expected to fail in the first 3 years if the standard deviation of battery life is 5.8 months.

Solution   For a normal distribution, 50% of the distribution lies on each side of the average (Fig. 2.6). Therefore, 50% of the patients are expected to see failures of the battery by 4 years. The doctors gave erroneous information to the patients that their batteries will last 4 years. Only half the patients are expected to find this statement true.

Note that the word *average* can be misleading. Many users think it means the minimum guaranteed value. When a light bulb bought from a store is said to have an average life of 2000 hours, many people expect every light bulb to last 2000 hours or more. They do not realize that half of them will fail before 2000 hours.

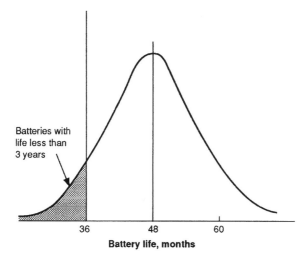

Figure 2.6 Failure distribution for Example 2.2.

The proportion of batteries expected to fail in the first 3 years is found from Eq. (2.3):

$$Z = \frac{|36 - 48|}{5.8} = 2.07$$

From Table A in the Appendix, the proportion of batteries failing in the first 3 years, for $Z = 2.07$, is 0.0192, or 2% rounded off.

Example 2.3 A specification requires that a device activate when a stress of 7 lb/in.^{2} or more is applied to it. The following data on activation stress in pounds per square inch were collected on 34 devices out of a lot of 300 devices. The data show that all devices were activated within the design limits. Is this design safe?

22.5	7.5	17.5	17.5	22.5
32.6	40.0	35.0	42.6	17.6
27.5	32.5	47.5	16.0	15.0
10.0	15.0	12.5	32.5	15.0
15.0	25.0	22.5	22.5	12.6
27.5	30.0	20.0	15.0	27.5
25.0	17.5	22.5	10.0	

Solution This device may not be safe. Even though all the values are within specification, the population of remaining untested devices $(300 - 34 = 266)$ is expected to have some defective devices. The estimates computed from the sample are $\bar{X} = 22.65$ and $s = 9.71$. Z for the 7.0 lb/in.^{2} specification is $(22.65 - 7.0/9.71)$ or 1.61. The fraction defective from Table A in the Appendix is 0.0537 or 5.37%. Therefore, $266 \times 0.0537 = 14$ devices can be unsafe.

2.4 LOG NORMAL DISTRIBUTION

In many applications, especially in reliability and maintainability, the data may not fit the normal distribution. However, the logarithms of the random variable may readily fit the normal distribution, hence the name *log normal*. If the log normal distribution applies, a plot of the data on log normal graph paper will be a straight line. The plotting procedure for this distribution and others is the same; it is discussed in Section 2.7. The analytical procedure consists of computing the average and the standard deviation of the logarithm values and taking the antilogarithms of the final results. This will become clear with the following example.

Example 2.4 The data below were collected for type 7075.T6 clad aluminum tested at 12,000 lb/in.^{2} The aluminum was tested in the laboratory and in the field.

	Field Results			Laboratory Results	
i	Probability (%)	Life in Cycles	i	Probability (%)	Life in Cycles
1	2.08	277,000	1	2.0	212,500
2	6.25	303,600	2	6.0	234,800
3	10.42	311,800	3	10.0	249,500
4	14.68	321,500	4	14.0	253,300
5	tt.75	322,800	5	18.0	256,900
6	22.92	332,200	6	22.0	259,400
7	27.08	337,500	7	26.0	260,400
8	31.25	339,400	8	30.0	260,900
9	35.42	355,000	9	34.0	265,500
10	39.58	364,000	10	38.0	265,500
11	43.76	372,200	11	42.0	266,400
12	47.92	384,600	12	46.0	268,200
13	52.08	396,300	13	50.0	269,800
14	56.25	405,800	14	54.0	272,600
15	60.42	430,900	15	58.0	279,700
16	64.58	431,000	16	62.0	279,900
17	68.75	439,600	17	66.0	286,300
18	72.92	505,500	18	70.0	292,000
19	77.08	518,700	19	74.0	299,500
20	81.25	519,900	20	78.0	319,900
21	85.42	547,200	21	82.0	324,600
22	89.58	579,200	22	86.0	327,800
23	93.75	643,100	23	90.0	329,100
24	97.92	887,400	24	94.0	364,800
			25	98.0	367,300

1. Calculate the number of cycles at which the chance of failure is less than 6.0%. Assume a log normal distribution. (This information will be used in planning preventive maintenance.)
2. Plot the data on log normal graph paper to make sure the model is valid. Calculate the life multiplication factor if the laboratory results are to be used to predict field results. The failure mode in the laboratory is identical to that in the field. Is the log normal model valid? Is the prediction valid?

Solution   Converting the field life-cycle data to logarithms (base 10) gives

$$\overline{X} = 5.4477$$
$$s = 0.0566$$

For the chance of failure to be less than 5.0%, the value of Z from Table A in the Appendix is 1.65 (for area 0.0495). The cycles to failure at 5% probability (the specification in this case) can be calculated from the value of Z:

$$Z = 1.65 = \frac{\overline{X} - \text{cycles to failure at 5\% probability}}{\text{standard deviation}}$$

Therefore,

$$\text{Cycles to failure at 5\% probability} = \overline{X} - (1.65)(\text{standard deviation})$$
$$= 5.4477 - (1.65)(0.0566)$$
$$= 5.3543 \quad \text{(in terms of log scale)}$$

The value is converted to true cycles to failure at 5% probability by taking the antilogarithm:

$$\text{antilog } 5.3543 = 226{,}100 \text{ cycles}$$

The cumulative probabilities for field and lab are drawn by plotting the cumulative probability of failure on the vertical axis and cycles to failure on the horizontal axis, as shown in Fig. 2.7. The procedure explained in Section 2.7, may also be used. If the two lines are parallel (or coincident), the correlation between the two tests is considered good. In this case, they are not quite parallel, indicating that the field environment is loosely related to the lab test. But since the failure modes are identical, the correlation can be considered acceptable (see Chapter 3 for a discussion of this point). Correlation can be quantified, if desired, by using the least squares method covered in standard books on statistics.

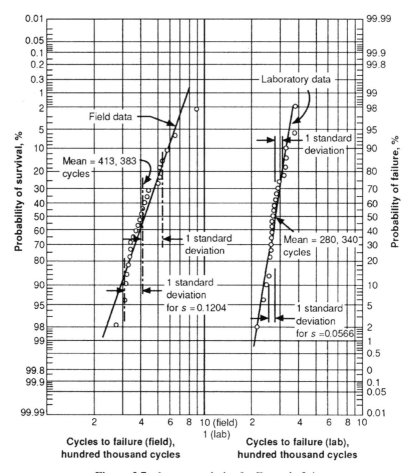

Figure 2.7   Log normal plot for Example 2.4.

To establish the multiplication factor between the two plots, one can compare the means:

$$\text{Multiplication factor} = \frac{\text{mean of field plot}}{\text{mean of laboratory plot}}$$

$$= \frac{413,383 \text{ cycles}}{280,340 \text{ cycles}}$$

$$= 1.475$$

The log normal model seems valid since most of the points are on a straight line.

Note: If the two plots are not parallel, the validity of the multiplier becomes doubtful. See the discussion of accelerated testing in Chapter 3.

2.5 EXPONENTIAL DISTRIBUTION

The exponential distribution is the most widely used (and misused) distribution for reliability analysis. It applies when the failure rate is constant. Figure 2.8 shows some properties of this distribution. The failure rate is the fraction of failure per unit of time: say, percent per 1000 hours, per million hours, or per billion device hours. Other units, such as miles or cycles, can be used in place of hours. When the failure rate is defined as number of failures per billion device hours, it is expressed as FITS (failure units); 15 FITS means 15 failures per billion hours. Stewart Peck originated the FITS concept while at AT&T. That company now uses the concept widely, as do Matsumura [5] and others, because the failure rates on many of their components are very low.

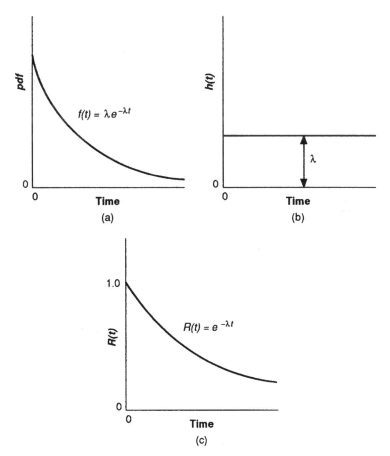

Figure 2.8 Some properties of the exponential distribution: (a) the probability density function, (b) the failure rate, and (c) the reliability function.

The failure rate λ is a parameter of the exponential distribution. The pdf for this distribution is given by the equation

$$f(t) = \lambda e^{-\lambda t}$$

where t is the time to failure. Failure rate for exponential distributions is often estimated as

$$\hat{\lambda} = \frac{\text{number of devices that failed}}{\text{total operating hours}}$$

Since, the probability of failure up to time t is

$$F(t) = \int_0^t \lambda e^{-\lambda t} = 1 - e^{-\lambda t}$$

the reliability up to time t is

$$R(t) = 1 - \int_0^t \lambda e^{-\lambda t} = e^{-\lambda t}$$

In this equation, λt is the average number expected to fail before time t. The only parameter of the exponential distribution is the failure rate λ.

Since λ is constant, most contracts use the reciprocal of failure rate, the mean time between failures (MTBF), as a measure of reliability. MTBF is estimated as

$$\text{MTBF} = \hat{\theta} = \frac{\text{total operating hours}}{\text{number of devices failed}} = \frac{1}{\hat{\lambda}} \tag{2.4}$$

The equation for reliability can be written in terms of MTBF as

$$R(t) = e^{-t/\text{MTBF}} \tag{2.5}$$

Note: The term MTBF is used for repairable systems. For others, mean or minimum life may be appropriate.

Example 2.5   Ten devices were tested. The test was stopped at the occurrence of the fourth failure. The hours to failures were 16, 40, 180, and 300. Six devices ran 300 hours without failure.

1. Calculate the MTBF.
2. If the sample MTBF is 584 hours, should the manufacturer guarantee repairs for failures up to 584 hours in the field?

Solution The MTBF is estimated from Eq. (2.4):

$$\text{MTBF} = \frac{16 + 40 + 180 + 300 + (6 \times 300)}{4} = \frac{2336}{4} = 584 \text{ hours}$$

(*Note*: This estimate is appropriate only for an exponential distribution.)

Suppose that the repairs are to be paid for by the manufacturer for the 584-hour duration. The estimated reliability is

$$R(t) = e^{-t/\text{MTBF}} = e^{-584/584} = e^{-1} = 0.37$$

This shows that only an estimated 37% of the devices will survive 584 hours; 63% may fail during the warranty. Since most manufacturers will not be able to afford that many repairs, the manufacturer should not guarantee them for 584 hours.

Example 2.6 Suppose that the mean cycles between failures (equivalent to MTBF) of a product is 100,000 cycles. What should be the warranty period if the manufacturer wishes to pay for repair of no more than 10% of the devices?

Solution Since $R(t) = e^{-t/\text{MTBF}}$

$$t = \text{MTBF} \times \ln\left(\frac{1}{R}\right)$$

where ln is the natural logarithm. In this example, $R = 1 - 0.10 = 0.90$. Therefore,

$$t = 100{,}000 \ln\left(\frac{1}{0.90}\right)$$
$$= 100{,}000 \times 0.1054$$
$$= 10{,}540 \text{ cycles}$$

Example 2.7 If a customer requires a guarantee during 100,000 cycles, what should the MTBF goal be for the supplier. The supplier is willing to pay for repairs of 5% of its product.

Solution The reliability goal R is $1 - 0.05 = 0.95$. Since $\hat{R} = e^{-t/\text{MTBF}}$

$$\hat{\theta} = \frac{t}{\ln(1/R)}$$
$$= \frac{100{,}000}{0.0513} = 1{,}949{,}317 \text{ cycles}$$

Statistical Estimation of the Exponential Parameter   The above calculations yield point estimates. If the limits are desired, then the following calculations can be used.

For *failure-truncated testing*, in which the test is terminated at a predetermined number of failures:

1. Two-sided limits for MTBF are

$$\frac{2T}{\chi^2_{\alpha/2,2r}} \le \theta \le \frac{2T}{\chi^2_{(1-\alpha/2),2r}} \tag{2.6}$$

where θ = true population MTBF; α = probability that the true value of MTBF is outside the confidence limits ($\alpha = 1 -$ confidence); $T =$ total operating hours, cycles, or miles; $r =$ number of failures; and χ^2 = chi-square statistic for $2r$ degrees of freedom obtained from Table C in the Appendix (see Ref. 2 for more details).

2. The one-sided lower limit for MTBF is

$$\frac{2T}{\chi^2_{\alpha,\,2r}} \le \hat{\theta} \tag{2.7}$$

For *time-censored testing*, in which the test is terminated at a predetermined time [note the $2(r+1)$ degrees of freedom in the lower bound]:

1. Two-sided limits for MTBF are

$$\frac{2T}{\chi^2_{\alpha/2,2(r+1)}} \le \hat{\theta} \le \frac{2T}{\chi^2_{(1-\alpha/2),2r}} \tag{2.8}$$

2. The one-sided lower limit for MTBF is

$$\frac{2T}{\chi^2_{\alpha,2(r+1)}} \le \hat{\theta} \tag{2.9}$$

Example 2.8   In Example 2.5, the test was failure-truncated; it was terminated at the occurrence of the fourth failure with a MTBF estimate of 584 hours. What are the 90% confidence limits for the true MTBF?

Solution   Equation (2.6) applies. Then the lower bound for $\hat{\theta}$ is

$$\frac{2T}{\chi^2_{.05,8}} = \frac{2 \times 2336}{15.507} = 301.28 \text{ hours}$$

and the upper bound on $\hat{\theta}$ is

$$\frac{2T}{\chi^2_{95.8}} = \frac{2 \times 2336}{2.733} = 1709.48 \text{ hours}$$

Example 2.9 Fifty taxi drivers used a new component on automobiles to assess its MTBF. Together they accumulated 2,686,442 miles. At least 10 drivers drove more than 80,000 miles. There were no failures. The test was stopped at 93,000 miles. What is the minimum expected MTBF with 95% confidence?

Solution The test is time- (miles-) censored, and Eq. (2.9) applies. Therefore, the one-sided lower limit for MTBF is

$$\frac{2 \times 2,686,442}{\chi^2_{0.05,2(0+1)}} = \frac{5,372,884}{5.991} = 896,825 \text{ miles}$$

Note: The information about 10 drivers driving 80,000 miles was not used in the computation. This information is important for those who want to have an estimate of the component life. *MTBF* and *life* are not the same. See Chapter 3 for estimating life.

Example 2.10 What is the 100,000-mile reliability for the component in Example 2.9, if the customer is satisfied with a 70% confidence limit?

Solution The $\hat{\theta}$ limits for 70% confidence is

$$\frac{2 \times 2,686,442}{\chi^2_{0.30,2(0+1)}} = \frac{2 \times 2,686,442}{2.408} = 2,231,564 \text{ miles}$$

At the lower 70 percent confidence limit, the reliability is, from Eq. (2.5),

$$R(100,000) = e^{-100,000/2,231,264} = e^{-0.0448} = 0.956$$

2.6 WEIBULL DISTRIBUTION

There is a new trend toward use of the Weibull distribution. The distribution is not new; the theory was advanced in 1951, but applications have been mostly in the aerospace and automotive industries. The reason this distribution is getting wider acceptance is that it adequately fits many data. In addition, the exponential distribution is a special case of the Weibull distribution. Figure 2.9 shows some curves for the Weibull model. The commonly used Weibull model

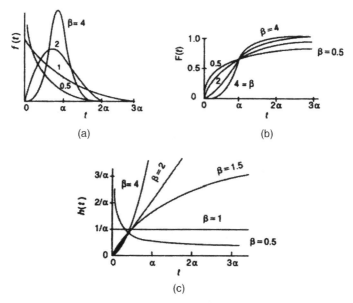

Figure 2.9   Weibull family of curves: (a) probability densities, (b) cumulative distributions, and (c) hazard functions. (From Wayne Nelson, *How to Analyze Reliability Data*, American Society for Quality Control, Milwaukee, WI, 1983.)

has two parameters:

β, *the shape parameter.* The value of β determines the shape of a Weibull curve.

α, *the scale parameter*, also called the *characteristic life.* It is the life at the 63.2 percentile of the distribution. Curves of all shapes intersect at a common value of α when they are plotted as a cumulative distribution, as shown in Fig. 2.9b.

Noteworthy properties of the Weibull distribution are:

1. When the value of the shape parameter β is greater than 1, the rate of failure increases with time.
2. When β is less than 1, the failure rate decreases with time.
3. When β equals 1, the failure rate is constant; that is, the distribution is exponential.

Most Weibull distributions fitted to data have β values between 0.5 and 4. Values above 4 indicate a fast wear-out, or the data may be corrupted.

The probability density function is

$$f(t) = \frac{\beta t^{\beta-1}}{\alpha^\beta} \exp\left[-\left(\frac{t}{\alpha}\right)^\beta\right] \quad \text{for } t > 0 \tag{2.10}$$

$$= 0 \quad \text{otherwise}$$

where t is the time to failure.

Equation (2.10) is elaborate, but fortunately a simpler equation, that for the cumulative distribution function, is the one that is usually employed to calculate reliability. The cdf is

$$F(t) = 1 - \exp\left[\left(-\frac{t}{\alpha}\right)^{\beta}\right]$$

Then reliability is

$$R(t) = \exp\left[-\left(\frac{t}{\alpha}\right)^{\beta}\right] \tag{2.11}$$

This equation is revealing. Earlier it was said that the exponential distribution is a special case of the Weibull distribution. Substituting $\beta = 1$ in Eq. (2.11) reduces it to the traditional exponential equation

$$R(t) = \exp\left[-\left(\frac{t}{\alpha}\right)\right]$$

Note that α is used instead of the usual θ.

Since the hazard rate is the ratio of $f(t)$ and $R(t)$, the Weibull hazard rate from Eqs. (2.10) and (2.11) can be written as

$$h(t) = \frac{\beta}{\alpha^{\beta}}(t)^{\beta-1} \tag{2.12}$$

Example 2.11   Failure data for a component show $\beta = 1.5$ and characteristic life $\alpha = 2000$ hours (a two-parameter Weibull distribution). Find:

1. The reliability at $t = 2000$ hours.
2. The hazard rate at 500 hours.
3. The hazard rate at 2000 hours.

Solution   The reliability, from Eq. (2.11), is

$$R = e^{-(2000/2000)1.5} = e^{-1} = 337$$

The 500-hour hazard rate, from Eq. (2.12), is

$$h(500) = \frac{1.5}{2000^{1.5}}(500)^{1.5-1.0} = \frac{1.5}{89{,}442.72} \times 22.361$$

$$= 0.000375 \text{ failure/hour}$$

The 2000-hour hazard rate is

$$h(2000) = \frac{1.5}{2000^{1.5}}(2000)^{1.5-1.0}$$

$$= \frac{1.5}{89{,}442.72} \times 44.721 = 0.00075 \text{ failure/hour}$$

Example 2.12   What values would be calculated in Example 2.11 if the failure density distribution was exponential?

Solution   For an exponential distribution, $\beta = 1$. Then the reliability is

$$R = e^{-(2000/2000)^{1.0}} = e^{-1} = 0.37$$

Similarly,

$$h(500) = \frac{1.0}{2000^{1.0}} (500)^{1.0-1.0}$$

$$= \frac{1}{2000} \times 1 = 0.0005 \text{ failure/hour}$$

$$h(2000) = \frac{1}{2000^{1.0}} (2000)^0 = \frac{1}{2000} = 0.0005 \text{ failure/hour}$$

The values of $h(500)$ and $h(2000)$ are the same, since hazard rate is constant in an exponential distribution.

2.7   DATA ANALYSIS WITH WEIBULL DISTRIBUTION

Mathematical estimation of the Weibull parameters α and β is cumbersome. Fortunately, a simple graphical method has been developed and is described here. Readers interested in mathematical estimation and other graphical methods, such as hazard plotting, can find more information in Refs. 3 and 6.

   Weibull analysis applies chiefly to life test data from fatigue tests, although other models such as the log normal and gamma distributions may also apply [6–9]. The Weibull distribution can be used for many kinds of reliability data when a single failure mode is being analyzed. There have been some misapplications in industry, and cautions will be indicated later.

   Graphical analysis uses a special Weibull probability chart based on the Weibull mathematical model [3, 6]. The chart in Fig. 2.10 is from an automotive company, but similar graph papers may be purchased.[1] The vertical axis is the cumulative probability plotting position $F(i)$. There are several models for $F(i)$ The most frequently used is the median rank model $F(i) = (i - 0.3)/(n + 0.4)$, where i is the order number when the failure times are arranged (ordered) from the

---

[1]A source is Technical and Engineering Aids for Management (TEAM), P.O. Box 25, Tamworth, New Hampshire 03886. Sample charts are also found in J. R. King, *Probability Charts for Decision Making*, Industrial Press, New York, 1971.

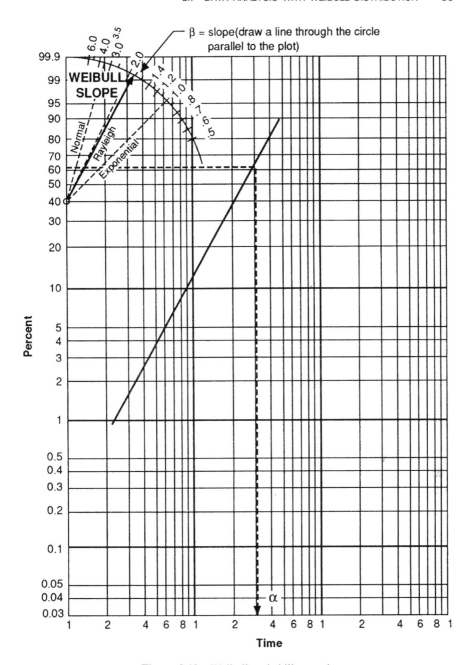

Figure 2.10 Weibull probability graph.

lowest time to failure to the highest and n is the sample size [7]. Table D in the Appendix contains computed values of the median rank for various sample sizes.

The x axis is a log scale for plotting time to failure (or cycles to failure or miles to failure). Each cdf shown in Fig. 2.9 plots as a straight line on the Weibull probability paper with a slope equal to its value of the Weibull parameter β. The life corresponding to 63.2% cumulative probability of failure is the value of α.

In graphically assessing reliability for a component or for a failure mode, one does not require the values of α and β. The $F(t)$ at any time t can be obtained directly from the graph. The reliability is then found by a straightforward calculation: $R = 1 - F(t)$.

One of the advantages of Weibull analysis is that many times it does not require a large sample size. If a straight line is obtained with a sample size of 10, the chances are that the plot for 20 points will be very close to it. Of course, the plot with higher sample size is always more accurate. However, this luxury is not always available with engineering prototypes such as the experimental components of a large turbine for a power station. Sample sizes less than 10 have been used on prototype parts.

Plotting the Data   The plotting procedure below is used when a complete sample has failed. (For a partially failed sample, the procedure for censored data should be used, as described in Ref. 3.) An example will follow.

1. Put the times to failure in order, from lowest to highest. Assign order number to each failure sequentially.
2. Look up the median rank value from Table D in the Appendix for each order number. The median rank is a plotting position on the y axis for $F(i)$. (Other models may be used such as those in Refs. 3 and 8.)
3. Plot all the points according to median rank value and time to failure.
4. Draw the line of best fit through the plotted points by using a transparent straight edge so that a roughly equal number of points is on each side of the line. The least squares method may be used for more accuracy.
5. For a given time t on the x axis, find the value of $F(t)$ on the y axis.
6. Compute $R(t) = 1 - F(t)$.

Example 2.13   A specification requires that no part should fail for 26 hours. Plot the hours-to-failure data for the part, tabulated below, on Weibull plotting paper. Using the graph, estimate:

1. The fraction expected to fail in 25 hours.
2. The reliability if the customer wants failure-free operation for 25 hours.

Specimen Number	Hours to Failure
1	90
2	128
3	180
4	112
5	254
6	35
7	168
8	65
9	220
10	108

Solution   Order the data and assign the median rank from Table D in the Appendix for sample size 10 as follows:

Hours to Failure	Order Number	Median Rank
35	1	6.7
65	2	16.2
90	3	25.9
108	4	35.5
112	5	45.2
128	6	54.8
168	7	64.5
180	8	74.1
220	9	83.8
254	10	93.3

Plot the graph using hours to failure on the x axis and median rank on the y axis. Figure 2.11 shows the resulting Weibull plot.

The fraction expected to fail in 25 hours, from the graph, is 2.3% (point A on the y axis). The reliability for 25 failure-free hours $= 1 - 0.023 = 0.977$.

Caution: Weibull analysis is best suited for fatigue tests. Therefore, it is best suited for component testing. At system level, the failure may be not from fatigue but from infant mortality problems of soldered and welded joints (Chapter 3). The Weibull model might be expected to suit the system failure data, but the analyst should judge cautiously.

Moreover, at system level the nature of failures differs from time to time. The model may therefore be an acceptable indicator of past performance, but its validity as a predictor of later life is questionable.

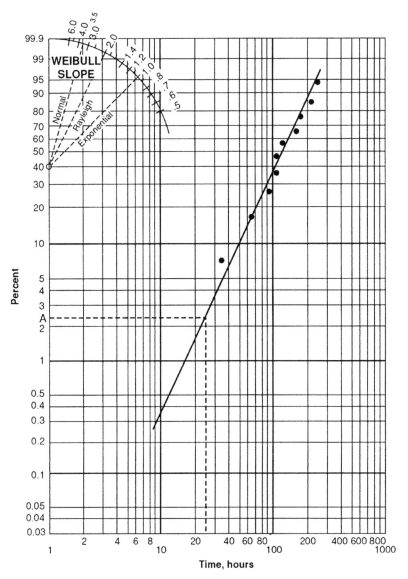

Figure 2.11 Weibull plot for Example 2.13.

2.8 DISCRETE DISTRIBUTIONS

Most reliability work is based on continuous distributions. Sometimes, however, the variables have certain discrete values but cannot have the values in between. For example, the number of failures of a system cannot be a fraction. Some qualitative features also take certain discrete values. For example, a product may have only two reliability levels—acceptable and unacceptable. These two states can be assigned values of 1 and 0, but it is not meaningful to assign values between these two numbers. This section covers two such discrete distributions—the binomial and Poisson distributions—because they are widely used. References 1 and 3 cover these and other distributions in more detail.

2.8.1 Binomial Distribution

The binomial distribution applies to situations where there are only two mutually exclusive outcomes. For example, a product is reliable or not reliable; a product meets the specification or does not meet the specification; blood samples are positive or negative; a malfunction is caused by high pressure or low pressure. The term *mutually exclusive* means the two outcomes cannot occur simultaneously.

The binomial distribution can be applied to modeling of the reliability (or the probability of failure) of fault-tolerant systems. These systems usually contain hardware or software redundancy. If there are five computers on a space shuttle and if at least three must work during a mission, the binomial distribution can be used to compute the system reliability. The general expression for reliability of a system with exactly x components out of N working is

$$R(s) = \sum \binom{N}{x} R^x (1 - R)^{N-x} \tag{2.13}$$

where R is the component reliability and $\binom{N}{x}$ represents the expression

$$\frac{N!}{[(x!)\,(N - x)!]}$$

Example 2.14 A computer has five processors. At least three must be in working condition during a mission. If each processor has an estimated reliability of 0.9, what are the chances of the system's success?

Solution The system will work if three processors survive or four processors survive or all five survive. Thus, from Eq. (2.13),

$$R(s) = \binom{5}{3}(0.9)^3(0.1)^{5-3} + \binom{5}{4}(0.9)^4(0.1)^{5-4} + \binom{5}{5}(0.9)(0.1)^{5-5}$$

$$= \frac{5}{3!2!}(0.9)^3(0.1)^2 + \frac{5}{4!1!}(0.9)^4(0.1)^1 + \frac{5!}{5!0!}(0.9)^5(0.1)^0$$

$$= (10)(0.729)(0.01) + (5)(0.6561)(0.1) + (1)(0.5905)(1)$$

$$= 0.9915$$

Example 2.15 The space shuttle has three surviving processors. If the output of at least two processors must match, what is the system reliability if each processor is 0.99 reliable?

Solution $N = 3$, since only three processors are available. But since the outputs of only two processors are used, we consider only $x = 3$ and $x = 2$ in calculating system reliability from Eq. (2.13):

$$R(s) = \binom{3}{2}(0.99)^2(0.01)^{3-2} + \binom{3}{3}(0.99)^3(0.01)^{3-3}$$

$$= \frac{3!}{2!1!}(0.99)^2(0.01) + \frac{3!}{3!0!}(0.99)^3(0.01)^0$$

$$= (3)(0.9801)(0.01) = (1)(0.9703)(1)$$

$$= 0.9997$$

2.8.2 Poisson Distribution

The Poisson distribution is used for estimating the number of random failures or occurrences in a given time when their likelihood is small. It should not be used for predicting fatigue-related failures or premature failures from manufacturing faults. It is applied when the sample size is at least 20 and the probability of occurrence is less than 0.05. The population size can be very large—as much as infinity. Under certain conditions, it approximates the binomial distribution.

The Poisson distribution assumes that the probability of occurrence is exponential, but while the exponential distribution is continuous, the Poisson is a discrete distribution. The general expression for x occurrences in a given time is

$$f(x) = \frac{a^x e^{-a}}{x!} \tag{2.14}$$

where a is the expected number of occurrences based on historical data and x is the number of occurrences of a failure.

Example 2.16 Historical records show the following information on unscheduled outages of a power plant:

Number of Outages in a Week	Frequency
0	2
1	9
2	4
3	2
4	1

1. What is the chance of no outage in a week?
2. What is the probability of two or less outages?

Solution   First find a, the expected number of outages per week:

$$\hat{a} = \frac{\text{Total number of outages}}{\text{Total number of weeks}}$$

$$\hat{a} = \frac{2 \times 0 + 9 \times 1 + 4 \times 2 + 2 \times 3 + 1 \times 4}{2 + 9 + 4 + 2 + 1} = 1.5$$

The probability of two or less occurrences is then calculated from Eq. (2.14) as the sum of the probabilities of zero, one, and two failures:

$$f(2 \text{ or less}) = \frac{(1.5)^0 e^{-1.5}}{0!} + \frac{(1.5)^1 e^{-1.5}}{1!} + \frac{(1.5)^2 e^{-1.5}}{2!}$$

$$= e^{-1.5}(1 + 1.5 + 1.125)$$

$$= 0.223(3.625) = 0.8084$$

The chance of no outage in a week is the first term of the sum above:

$$f(0) = \frac{(1.5)^0 e^{-1.5}}{0!} = 0.2231$$

2.9   TOPICS FOR STUDENT PROJECTS AND THESES

1. Explore why textbooks emphasize the exponential distribution in reliability models. Is it a correct approach? (Read Chapter 3 for more information.)
2. Describe at least four statistical models not covered in this chapter.
3. All quality control work is based on the assumption that a variable is distributed normally (or skewed). Why are the exponential, log normal, and Weibull distributions not considered? (See Chapter 6.)
4. Develop applications of the Poisson distribution for early design work.

REFERENCES

1. G. J. Hahn and S.S. Shapiro, *Statistical Models in Engineering*, John Wiley & Sons, Hoboken, NJ, 1967.
2. H. M. Wadsworth, *Handbook of Statistical Methods for Scientists and Engineers*, McGraw-Hill, New York, 1990.
3. W. Nelson, *Applied Life Data Analysis*, John Wiley & Sons, Hoboken, NJ, 1982.

4. S. S. Shapiro, *How to Test Normality and Other Distributional Assumptions*, American Society for Quality Control, Milwaukee, 1980.

5. H. Matsumura, N. Maki, and I. Matsumoto, Strategic Performance of Reliability Evaluation for New Sophisticated VLSI Product. In: *Proceedings of International Conference on Quality Control*, Union of Japanese Scientists and Engineers, Tokyo, 1987, pp. 649–650.

6. W. Weibull, A Statistical Distribution Function of Wide Applicability, *Journal of Applied Mechanics*, September 1951, pp. 293–297.

7. C. Lipson and N. Sheth, *Statistical Design and Analysis of Engineering Experiments*, McGraw-Hill, New York, 1973.

8. L. G. Johnson, *The Statistical Treatment of Fatigue Experiment*, Elsevier, New York, 1964.

9. M. O. Locks, *Reliability, Maintainability, and Availability Assessment*, Hayden, Rochelle Park, NJ, 1973.

FURTHER READING

Dixon, W. J., and F. J. Massey, Jr., *Introduction to Statistical Analysis*, 4th ed., McGraw-Hill, New York, 1983.

Gross, A. J., and V. A. Clark, *Survival Distributions: Reliability Applications in the Biomedical Sciences*, John Wiley & Sons, Hoboken, NJ, 1976.

Kapur, K. C., and L. R. Lamberson, *Reliability in Engineering Design*, John Wiley & Sons, Hoboken, NJ, 1977.

Lewis, E. E., *Introduction to Reliability Engineering*, John Wiley & Sons, Hoboken, NJ, 1987.

Nelson, W., *Accelerated Life Testing*, John Wiley & Sons, Hoboken, NJ, 1990.

O'Connor, P. D. T., *Practical Reliability Engineering*, 2nd ed., John Wiley & Sons, Hoboken, NJ, 1985.

CHAPTER 3

RELIABILITY ENGINEERING AND SAFETY-RELATED APPLICATIONS

3.1 RELIABILITY PRINCIPLES

Lord Kelvin once said, "I often say that when you can measure what you are talking about and express it in numbers, you know something about it." This is certainly true of product design. An engineer may make a product work in the laboratory, but the real criterion is how many of the units produced work in the field. If less than half the units work, as in the case of the Adam computer,[1] the engineering department can hardly expect accolades.

A typical flaw in thinking is that we design for averages. When an automotive engineer talks of 100,000 miles life, it implies the average life is 100,000 miles. What is not consciously understood is that about 50% may fail before the product has reached 100,000 miles. A customer's interpretation is quite the contrary. The customer thinks the automobile will not fail for 100,000 miles. A good designer will be clear in defining life as minimum life or the time to first failure. Therefore, a product has to be designed for much higher than the advertised life.

Even though some prototypes may work well in the laboratory, others may have seemingly minor problems. Such problems are often ignored and attributed to human error. Similar errors and new ones can occur on the manufacturing floor. They compound, resulting in very low reliability and eventual loss of market share; they can reduce reliability to a very low number.

---

[1]Computer introduced in the United States by Coleco Industries in 1986.

Assurance Technologies Principles and Practices: A Product, Process, and System Safety Perspective, Second Edition, by Dev G. Raheja and Michael Allocco
Copyright © 2006 John Wiley & Sons, Inc.

Therefore, the reliability measurements on products should be documented in design, in manufacturing, and in the field. The documentation reveals which life-cycle phase is responsible for the downfall.

Reliability engineering in the literature is perceived mostly as a statistical tool. The truth is that statistics is used mostly for measuring, analyzing, and estimating various parameters. Most of the true reliability work is done during the development of a design, long before test data are available for statistical manipulation. True reliability is that built into the design and is called inherent reliability. It is the highest reliability a product will have. Neither responsive manufacturing nor preventive maintenance can raise it above this level.

As the product passes through the production stage, its inherent reliability tends to be degraded, although good screening and process control programs can reduce or prevent degradation. Reliability is further degraded in the field because of unanticipated stresses or environments, inadequate procedures, and poor maintenance. If the product's inherent reliability is extremely high, it can sometimes allow for field degradation and still meet the customer's expectation.

Reliability is defined as the probability that a product will perform its intended function for a given time period under a given set of conditions. The definition assumes the product is in working condition at the outset. Time in this definition refers to such durations as mission time, warranty period, a predefined number of cycles, and even a complete lifetime. A product can also have several missions. In that case, a separate set of requirements may be developed for each mission.

To understand reliability, it is necessary to understand the concept of the so-called bathtub curve. This is a plot of hazard rate versus life (Fig. 3.1). It consists

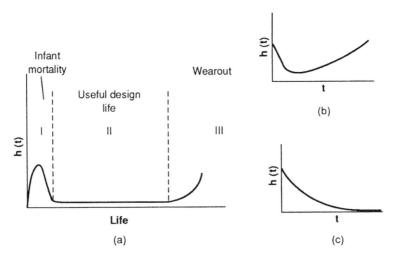

Figure 3.1 Typical failure rates versus life—bathtub profiles: (a) electronic equipment, (b) mechanical equipment, and (c) software.

of three regions, each having different hazard rates. In combination, the distributions form a single smooth curve that resembles a bathtub profile. Reliability is a function of hazard rate in each region.

The failures in region I (Fig. 3.1a) are premature and are likely to be present in the field, unless due care is taken in manufacturing. They are usually a result of shoddy manufacturing, poor workmanship, and excessive variation in the material. They can be mitigated by design changes and screening programs. Failures in this region are called infant mortality failures.

Region II, called the useful design life period, represents the real failure rate the customer can expect. There is a general misunderstanding in industry about this region. For a good design, this region has a very low failure rate and the curve is approximately flat (Fig. 3.1a), indicating a constant failure rate. The misunderstanding is that many practitioners assume that the region always has a constant failure rate even though many times it does not (Fig. 3.1b,c). To make matters worse, some practitioners assume a constant failure rate even in region I, where such a condition is a rarity. No wonder that failure predictions made in industry rarely correlate with field results. To make meaningful predictions, a reliability analyst must consider the impact of infant mortality failures (usually from production causes). Because production problems are different at various times, the distribution of infant mortality failures is unstable. If region I failures are allowed to go to the customers, reliability prediction is optimistic.

To simplify computations, the industry uses mean time between failures (MTBF), the reciprocal of failure rate, as a measure of reliability. Many misunderstand this term (including many writers of government documents) as an equivalent of life. Actually, MTBF has absolutely no relation to life. MTBF is only a function of the gap between the x axis and the bottom of the bathtub. This gap is independent of the life or the length of the bathtub! This gap (the failure rate) is usually a function of design.

Region III is the wear-out phase. Its shape is a function of design. At system level, the onset of wear-out can be pushed to the right if the weak components are replaced before they fail. In general, preventive maintenance can extend the useful life of a system.

There are many design improvement techniques for reliability. The widely used techniques are given below.

Reliability Improvement Techniques

- Zero failure design, where causes of critical failures are entirely eliminated by design. (The product can fail from unanticipated causes.) A similar concept in the aircraft industry is called safe life design, where the components are qualified for twice the expected life. This allows a failure to occur but not during the useful life.
- Fault tolerance, where redundant elements are used to switch over to a backup or alternative mode.
- Derating, where a component is used much below its capability rating.

- Durability, where a component is designed to have longer useful life or is designed for damage tolerance (such as aircraft structures).
- Design with product safety margins for all applicable worst-case stresses and environments.

Major analytical tools needed to achieve these objectives are:

- Design reviews
- Reliability allocation, prediction, and modeling
- Failure-mode, effects, and criticality analysis
- Fault-tree analysis (see Chapter 5)
- Sneak circuit analysis
- Worst-case analysis
- Statistical analysis of failure distributions
- Quality function deployment (see Chapter 6)
- Robust design methodology using design of experiments (see Chapter 6)

Most of these techniques are covered in this chapter; others are covered in the chapters indicated in the list above. References and Further Reading at the end of the chapter give sources for more information.

3.2   RELIABILITY IN THE DESIGN PHASE

Our experience shows that as much as 60% of failures can be prevented by making changes in the design. In all design improvements, the designer can choose a strategy for preventing failure according to the following order of precedence:

1. Change design to remove the failure mode.
2. Design for fault tolerance.
3. Design for fail-safety.
4. Design early warnings of failure to the user.

If these strategies are not viable, the designer may choose reliability-centered preventive maintenance (see Chapter 7) and issue special maintenance instructions to the user. These instructions should be presented in such a way that users will be sure to read, understand, and follow them.

The activities in the design stage start with specifying the reliability requirements. These specifications are implemented in design through reliability analysis and are verified by testing.

3.2.1 Writing Reliability Specifications

Reliability requirements should be an integral part of engineering specifications. As the definition of reliability implies, at least four features must be identified:

1. Probability of successful performance of the product
2. Functions (missions) to be performed
3. Intended time (such as mission time)
4. Operating conditions

These requirements give a skeleton structure for a reliability specification. But a design engineer needs ideas to meet the specification requirements.

Consider a reliability specification for a robot for welding automobile chassis. The probability of successful performance can be specified in several ways. It may be specified as availability, the probability that the robot is in working condition during the scheduled hours only. It may be specified as several different probabilities for various missions; some functions may be less important than others. Limits on applications may be specified; for example, the robot may not be able to weld certain steels.

The intended time can be specified as hours of actual operation, number of cyclical loads, or calendar time. The operating environment should be specified so that qualification tests can be developed. In this case, the environment would include dust, chemicals, sparks, humidity, vibration, shock, and perhaps an unskilled machine operator. The specification should also include software reliability. Chapter 9 contains a section on this aspect.

Example 3.1 An automotive cranking system consists of engine, starter motor, and battery. The reliability requirement for the system in an engineering performance standard reads: "There shall be a 90% probability [probability of success] that the cranking speed is more than 85 r/min after 10 seconds of cranking [mission] at $-20\,°F$ [environment] for a period of 10 years or 100,000 miles [time]. The reliability shall be demonstrated at 95% confidence."

3.2.2 Conducting Design Reviews

Design reviews are the most important part of product engineering. They include discussions of the results from various analyses, informal brainstorming, and review of checklists and project controls.

Design reviews should be a team effort headed by someone who can provide unbiased leadership. With proper leadership, design productivity will increase significantly. The purpose of design reviews is to challenge the design from different viewpoints. This can happen only when designers and nondesigners are encouraged to think creatively and to advance fresh views. The biggest benefit of design reviews is that engineering changes can be made early, when they will have the greatest effect in reducing life-cycle costs. Darch [1] presents several noteworthy viewpoints on design reviews:

- Design reviews provide a forum for discussion of interface problems and their resolution.
- Design reviews provide a common base for future work.
- As design proceeds toward completion, design reviews continue to reduce life-cycle costs.
- The nature of the design review process is such that, after a stage is completed, the results from that stage are accepted as final and the next stage proceeds from there.

Decisions resulting from design reviews at any stage should be given serious consideration at the top management level. A design review is not merely a review of product performance, but a risk assessment process with major effects on future costs. The decisions will determine the inventory of spares, skill levels of operating and maintenance personnel, test equipment required in the field, and many support requirements. Therefore, before design review decisions are approved, the life-cycle costs they will incur should be estimated.

3.2.2.1 *Preliminary Design Review* A preliminary design review should be held to select strategies for developing design improvements. The following strategies are available:

1. Modular design
2. Fault-tolerant design
3. Fault isolation
4. Fault masking
5. Built-in testing
6. Self-checking
7. Self-monitoring
8. Derating
9. Self-healing
10. Design for ease of maintenance
11. Design for ease of inspection
12. Design for throwaway instead of repair (e.g., light bulbs)
13. Testable design
14. Maintenance-free design
15. Zero-failure design

16. Built-in retry (software)
17. Design for damage detection
18. Top–down structure (software)

3.2.2.2 Lessons Learned and Checklists It is always a good idea to start a design review with two important items: (1) the lessons learned since the previous review and (2) a checklist subject to ongoing development.

The lessons-learned discussion includes examinations of failure histories, competitive analyses, and interviews with dealers, distributors, and users of similar products. It is important to eliminate from the discussion, as far as possible, failure modes common to the designs of all competitors. At the same time, the strengths of the competitors should be discussed. It may be possible to introduce new strengths into a product, such as maintenance-free, fault-tolerant, or modular design.

Checklists are powerful tools in getting a review done faster, but they do not cover all design concerns. Additional brainstorming is required. The following generic checklist may be useful as a start.

1. Are reliability, maintainability, availability, and safety features included in the specification?
2. Is failure-mode, effects, and criticality analysis (FMECA) planned in early design?
3. Can fault-tree analysis be used to improve the design?
4. Are component interchangeabilities analyzed?
5. Are safety margins adequate?
6. Is software reliability specified?
7. Is testability analysis required?
8. Is fault-isolation capability needed?
9. Do electronic circuits have adequate clearances between them?
10. Are software logic concerns independently reviewed?
11. Has software coding been thoroughly reviewed?
12. Are qualification test plans done well in advance?
13. Are production tests planned and reviewed?
14. Are redundancies considered for software?
15. Are redundancies considered for hardware?
16. Has the design been critiqued for human mistakes?
17. Are designers familiar with the human engineering guidelines contained in Mil-Hdbk-1472?
18. Is reliability-centered maintenance considered in design?
19. Is competitive analysis for reliability done?

20. Are there long-term problems with design such as dendritic growth on integrated circuits?
21. Is the product analyzed for corrosion failures?
22. Is there a statistical model for analyzing test data?
23. Are the switches for backup devices reliable? Do they need maintenance?
24. Are protective devices such as fuses, sprinklers, and circuit breakers reliable? Do they need reliability-centered maintenance?
25. Is the product performance evaluated against radiation and accidental damages?
26. Does the product need to withstand earthquakes and unusual loads?
27. Are component tolerances optimized to give the desired reliability?
28. Has the product been tested for environments of the countries to which it will be exported?
29. Can manufacturing personnel introduce any defects? Can they be prevented by design?
30. Can maintenance personnel introduce defects? Can they be prevented by design?
31. Can the operator introduce wrong inputs? If so, can the product be designed to switch to a fail-safe mode?
32. Can a single component cause the failure of a critical function?
33. Is crack growth and damage tolerance analysis required?
34. Do components need corrosion protection?
35. Can applied loads be larger in the future (such as heavier trucks on a bridge)?
36. Are there unusual environments not already considered?
37. Are inspection provisions made for detecting cracks, damage, and flaws (for aircraft)?
38. Does design require crack-retardation provisions?

3.2.3   Reliability Allocation

The process of allocation establishes a hierarchy of design requirements. The purpose is to distribute operational reliability goals to subsystems, subassemblies, and all the way down to components. Companies that do not allocate reliability rarely achieve the desired overall reliability. For example, a company making electronic warfare equipment struggled to get a MTBF of 360 hours but rarely achieved more than 40 hours MTBF. It always complained about suppliers' poor performance. Yet, on investigation, it was found that suppliers were never told their MTBF goals.

Allocations start with the system goal. For example, as shown in Fig. 3.2, designers of an engine with 70% reliability over 10 years will have to allocate higher reliabilities among subsystems, since the product of subsystem reliabilities is equal to the system reliability. Similarly, each subsystem reliability should be allocated among its subassemblies (only one breakdown is shown in the figure).

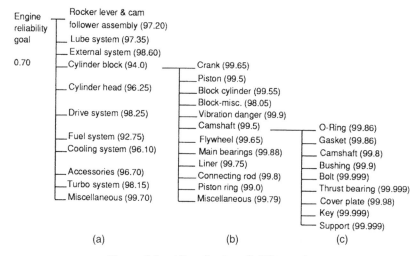

Figure 3.2   Allocation by reliability goals.

The allocation continues until all the components are accounted for. Note that reliabilities as high as 99.999% are required to get only 70% reliability at engine level.

Where subsystems and components have constant failure rates, the allocation can be in terms of failure rate. This is more convenient than using reliabilities because failure rates for subsystems can be added instead of multiplied. For a serial system, the MTBF is first converted to a failure rate, which is allocated among the subsystems and components. At the completion of the analysis, the failure rates are converted back to MTBF. Figure 3.3 shows such an allocation for a serial system.

3.2.4 Reliability Modeling

Reliability modeling is done in early design to assess alternative product reliability approaches. There are two common models: (1) the basic reliability model, popularly called the series model, and (2) the fault-tolerance model, popularly called the parallel model, in which some or all portions may be redundant.

3.2.4.1 Series Model   If a product will fail when any of its components fails, it is called a series model. For example, if any component of a bicycle fails, then the entire bicycle fails. Such a model can be represented by a chain. If any link in the chain fails, the entire chain has failed. Figure 3.4 shows series as well as several parallel models.

In the series model, since all the components must work, the system reliability is a function of all the reliabilities. The system reliability is represented by the equation

$$R(s) = R_a \cdot R_b \cdot R_c \cdots R_n \tag{3.1}$$

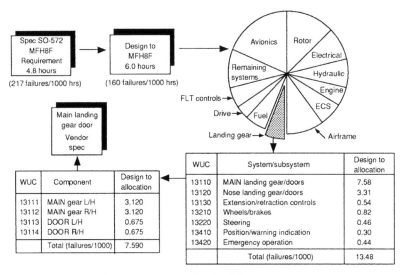

WUC	Component	Design to allocation
13111	MAIN gear L/H	3.120
13112	MAIN gear R/H	3.120
13113	DOOR L/H	0.675
13114	DOOR R/H	0.675
	Total (failures/1000)	7.590

WUC	System/subsystem	Design to allocation
13110	MAIN landing gear/doors	7.58
13120	Nose landing gear/doors	3.31
13130	Extension/retraction controls	0.54
13210	Wheels/brakes	0.82
13220	Steering	0.46
13410	Position/warning indication	0.30
13420	Emergency operation	0.44
	Total (failures/1000)	13.48

Figure 3.3   Allocation by failure rates [2].

If there are four components with reliabilities 0.90, 0.80, 0.90, 0.70, respectively, the resultant reliability is

$$R(s) = 0.90 \times 0.80 \times 0.90 \times 0.70 = 0.4536$$

This computation shows that the more components there are of a given type, the worse the system reliability will be. Therefore, a design goal is to minimize the number of components for a series system. For a parallel system this may not be the case.

3.2.4.2 *Parallel Model*   In a series system, the reliability degrades rapidly with increase in number of components; each fraction multiplied by another fraction results in a smaller fraction. To increase reliability, one can back up a component of a series (Fig. 3.4b), or back up the entire chain with another chain (Fig. 3.4c), or individually back up each component (Fig. 3.4d). The reliability of a parallel

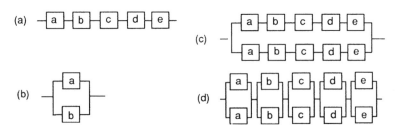

Figure 3.4   (a) Series system and (b, c, d) parallel systems.

system is computed from the relation

Probability of system working + probability of not working = 1

For Fig. 3.4b, the only time the system will not work is when both components have failed. The probability of the entire system not working is $(1 - R_a)(1 - R_b)$. From the above relation, the probability of the system working is

$$R(s) = 1 - (1 - R_a)(1 - R_b)$$

In general,

$$R(s) = 1 - (1 - R_a)(1 - R_b)(1 - R_c) \cdots (1 - R_n) \tag{3.2}$$

From Eq. (3.2), the calculations for parallel models are:

$$R(s) = 1 - (1 - 0.9)(1 - 0.9) = 0.99 \quad \text{(for Fig. 3.4b)}$$

$$R(s) = 1 - (1 - R_a \cdot R_b \cdot R_c \cdot R_d)(1 - R_a \cdot R_b \cdot R_c \cdot R_d)$$
$$= 1 - (1 - 0.9 \times 0.8 \times 0.9 \times 0.7)(1 - 0.9 \times 0.8 \times 0.9 \times 0.7)$$
$$= 0.7014 \quad \text{(for Fig. 3.4c)}$$

and

$$R(s) = [1 - (1 - R_a)(1 - R_a)][1 - (1 - R_b)(1 - R_b)]$$
$$\times [1 - (1 - R_c)(1 - R_c)][1 - (1 - R_d)(1 - R_d)]$$

$$R(s) = [1 - (1 - 0.9)(1 - 0.9)][1 - (1 - 0.8)(1 - 0.8)]$$
$$\times [1 - (1 - 0.9)(1 - 0.9)][1 - (1 - 0.7)(1 - 0.7)]$$
$$= 0.99 \times 0.96 \times 0.99 \times 0.91$$
$$= 0.8562 \quad \text{(for Fig. 3.4d)}$$

3.2.5 Reliability Prediction

Reliability modeling establishes how a system is configured. Reliability predictions estimate the reliability of the configuration to assure it will satisfy the allocated operational goals. Most practitioners of electronic equipment design use Mil-Hdbk-217 [3], or a similar database, to predict reliability. They assume the system is a series system and usually overlook redundant components. It is amazing that some experienced engineers, not just beginners, do this. Perhaps no harm is done because this practice gives the worst reliability estimate.

One reason this practice is so well accepted, even when the series assumption is wrong, is that it makes the reliability calculation so simple. The reliabilities can simply be multiplied to estimate system reliability, as in Eq. (3.1). If the components are also assumed to have an exponential failure distribution, the model becomes dramatically simple:

$$R(s) = R_a \cdot R_b \cdot R_c \cdots R_n = e^{-\lambda_a t} \cdot e^{-\lambda_b t} \cdot e^{-\lambda_c t} \cdots e^{-\lambda_n t}$$
$$= e^{-(\lambda_a + \lambda_b + \lambda_c + \cdots + \lambda_n)t}$$

This model shows that the failure rates can simply be added to give a total failure rate. The final model is

$$R(s) = e^{-(\Sigma \lambda)t}$$

The truth is that a good product should have a very low constant (or decreasing) failure rate, and therefore this model should be the goal. If (1) the infant mortality failures are completely screened and (2) the wear-out process does not accelerate until after useful life is experienced, this model will apply, even for mechanical components. Most practitioners do not verify the validity of this model. They assume that, if there are hardly any failures during the useful life, the failure rate can be regarded as a very small constant.

Caution: Estimating reliability by the procedure above is only for the hardware. Field reliability is also degraded by manufacturing errors, maintenance errors, improper use, and software-induced errors.

In estimating failure rates from Mil-Hdbk-217, reasonable care should be exercised. The failure rate data can be 8–10 years old, the data are collected under varied conditions, and factors such as transient line voltages and electromagnetic interference are not considered. The best way to estimate failure rates is to build an in-house data bank using models similar to those in Mil-Hdbk-217. Also, before the estimates are made, one should consider the fact that the hazard rates for some electronic components are not constant. A military specification on capacitors [4] claims the failure rate for such devices has never been constant. It has always been a decreasing function of time. These considerations suggest that the predictions from Mil-Hdbk-217 are not reliable. The reader also should recognize that field reliability is influenced by human error, software, preventive maintenance, and the operating procedures. These features are so qualitative that it is impractical to predict field reliability.

One big advantage of failure rate models is that they make designers aware of some factors influencing the failures and their impact on the product. Figure 3.5 gives failure rate models for an electronic and a mechanical component. The electronic component model is from Mil-Hdbk-217. The model for the mechanical component is from a draft copy of a handbook for mechanical components being developed under contract to the U.S. Navy and U.S. Army [5].

Example 3.2   A system for the U.S. Air Force consisted of 10 circuit boards to monitor 10 facilities independently (Fig. 3.6a). Each circuit board had about 160

Part operating failure rate model (λ_p):

$$\lambda_p = \pi_Q (C_1\pi_T\pi_V + C_2\pi_E) \, \pi_L \text{ failures/10}^6 \text{hours}$$

where

λ_p is the device failure rate in F/10^6 hours

π_Q is the quality factor

π_T is the temperature acceleration factor, based on technology

π_V is the voltage stress derating factor

π_E is the application environment factor

C_1 is the circuit complexity factor based on transistor count as follows:

for 1 to 100 transistors, $C_1 = 0.01$

for > 100 to 300 transistors, $C_1 = 0.02$

for > 300 to 1000 transistors, $C_1 = 0.04$

C_2 is the package complexity failure rate

π_L is the device learning factor

(a)

$$\lambda_{BE} - \lambda_{BE,B} \left(\frac{L_A}{L_S}\right)^\gamma \left(\frac{A_E}{.006}\right)^{2.36} \left(\frac{v_o}{v_L}\right)^{0.54} \left(\frac{C_{AL}}{60}\right)^{2/3} \left(\frac{M_O}{M_F}\right) \cdot C_{CW}$$

where λ_{BE} – predicted failure rate of bearing, using actual conditions, per million hours of operation

γ – 3.33 for roller bearing; 3.0 for ball bearings

L_A – actual radial load, lb

L_S – specification radial load, lb

A_E – alignment error, rad

v_o – specification lubricant viscosity, lb-min/in.^{2}

v_L – operating lubricant viscosity, lb-min/in.^{2}

M_O – material factor, base material, lb/in.^{2}, yield strength

M_F – material factor, operating material, lb/in.^{2}, yield strength

C_{AL} – actual contamination level, micrograms/meter^3

C_{CW} – water contamination factor

(b)

Figure 3.5   Failure rate models: (a) electronic component and (b) mechanical component.

components in the power supply region, about 100 components in the input–output region, and 70 components in the communication portion.

Since each facility had a separate circuit board assigned to it, troubleshooting was easy. The maintenance people knew right away which circuit board had failed. The reliability model for this system is a series model. Prediction from Mil-Hdbk-217 showed the circuit board reliability over a 6-year period to be 0.83. This gave a system reliability $R = 0.155$. It was later decided that the cause of the low system reliability (the chance of failure was 15.5%) was the presence of too many components in series. A design engineer proposed that, since the power supply requirement is the same for all the circuit boards, why not have one common external power supply. But if the single power supply failed, all the circuit boards would become nonfunctional. To overcome this problem, a redundant power supply was added. The new configuration is shown in Fig. 3.6b. The reliability is now 0.68 versus 0.166 before. This is a good example for those who do not believe that reliability

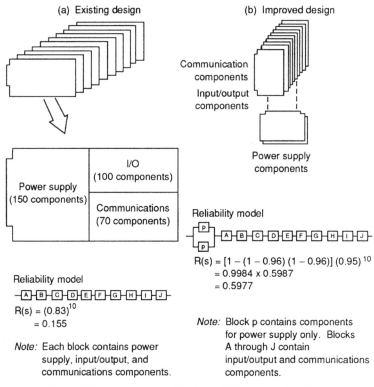

Figure 3.6   An example of higher reliability at reduced cost.

reduces cost: in the improved version there are 2000 components versus 3200 in the original design. In addition, there is a 339% gain in system reliability. (*Note:* The maintenance cost may go up slightly to assure the redundant power supply is always in working condition.)

3.2.6   Failure-Mode, Effects, and Criticality Analysis

When all obvious design concerns are seen to, it is time to run a detailed analysis (Fig. 3.7). Failure-mode, effects, and criticality analysis (FMECA) is a technique for macro and micro level analysis of the design. It starts the system level or the component level to determine what could go wrong with each function and its effects on the system. Unfortunately, many companies spend a substantial amount of money on FMECA but do not know what to do with the information (a gross misuse of this technique). An engineer showed one of the authors a FMECA of a circuit board having over 4000 failure modes. When asked what he did with the information, the engineer replied: "Nobody asked me to do anything with it." It was another nice-looking report that, after a 3-month effort, never saw daylight.

System_____
Indenture level_____
Reference drawing_____
Mission_____

Criticality Analysis

Date_____
Sheet_____ of_____
Compiled by_____
Approved by_____

Identification number	Item/functional identification (nomenclature)	Function	Failure modes and causes	Mission phase/ operational mode	Severity class-ification	Failure probability / Failure rate data source	Failure effect probability (β)	Failure mode ratio (α)	Failure rate (λ_p)	Operating time (t)	Failure mode criticality no. $C_m = \beta\alpha\lambda_p t$	Item criticality no. $C_? = \Sigma(C_m)$	Remarks

Figure 3.7   One of the FMECA formats in Mil-Std-1629.

Such a response is fairly common. Some use FMECA to predict reliability only. The right way to use FMECA is to improve the product design with this knowledge. Prediction should be treated as a by-product.

The original FMECA technique, suggested in Mil-Std-1629 [6], is to systematically analyze each component by using a format such as that in Fig. 3.8. The number and headings of the columns will vary with the application.

① Item part no.	② Block diag. ref.	③ Functions	④ Failure modes	⑤ Failure mechanism (Cause)	⑥ Effect on			⑦ Interface effects	⑧ Criticality			⑨ Corrective action		⑩ Item close sign off or rationale for other action
					Local	Higher levels	End item		Severity	Freq.	Detect-ability	Time avail. for corrective action	Recommendation	

Figure 3.8   FMECA Matrix.

Further improvements in FMECA methodology are proposed here. For example, Mil-Std-1629 suggests identifying failure modes related to hardware failures only, while these authors recommend treating human factors and workmanship problems, such as poor solder joints and welded joints, as a part of FMECA. This can be accomplished by dividing the analysis into two units: one for the product design, called Design FMECA, and the other for manufacturing, called Process FMECA. This chapter covers Design FMECA. Process FMECA is covered in Chapter 6.

The documentation in forms like that in Fig. 3.8 is essential because it can be used for several other analyses such as safety, maintainability, quality, and logistic support. The procedure for filling in the form follows.

Column 1: Item part number. Write the system element or the component identification number. Make sure all components are listed, including bolts, nuts, washers, wires, and screws; the purpose of the form is to help engineers analyze systematically. A part such as a nut may seem trivial, but improper clearance with a bolt might result in a catastrophic accident. (Several engines in U.S. Air Force jets fell off because of improper clearance. In fact, in 1987, 60% of the Air Force's engine inventory was found suspect of such problems). Most defense contractors perform analysis only on components they think are important at a functional level, but they risk overlooking some critical components. This latter approach is also acceptable in Mil-Std-1629, especially for systems with components in the thousands.

Column 2: Block diagram reference. This column is for identifying where the component exists in the block diagram, logic diagram, or assembly drawing. This column is optional.

Column 3: Functions. The function is described to give the analyst clues to failure modes. Any defect or event that prevents the function from being performed is a failure mode. For example, if the function of a transistor is to act as a switch, then insufficient current, in addition to opens and shorts, constitutes a failure mode.

Column 4: Failure modes. A part will have several failure modes that can impede its function. The simplest example is a bolt whose function is to hold a cover plate in a vibration environment, in combination with a nut. Failure modes such as loose joint from a wrong pitch diameter, too much clearance between it and the nut, and improper plating can impede its function. Robust design considers not only how the part can fail or what can go wrong with it, but also the effects of human operators such as an assembler who forgets to assemble the bolt altogether.

Describing failure modes is often difficult and complex. They can be different in various mission phases. For example, a spacecraft can have one failure mode on the launching pad, another in geosynchronous orbit, and yet another in the reentry phase. Such was the case with a very tough and lightweight Kapton-insulated wire developed for Spacelab. The wire had excellent endurance properties, but it burst like fireworks in the zero gravity of space in the presence of moisture or any liquid, according to a NASA internal memo.

Failure-mode prediction can be difficult even in a single-mission-phase product. Figure 3.9 shows an unpressurized joint on solid booster rocket. It shows two seals. The secondary seal is supposed to protect the system if the primary seal fails.

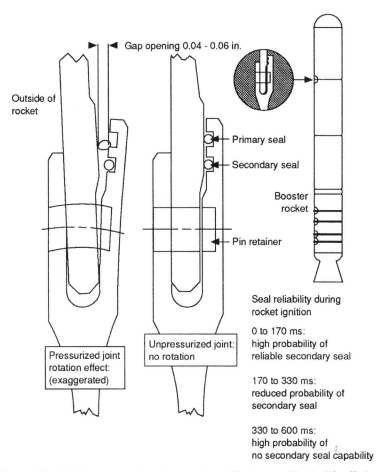

Figure 3.9 Solid booster rocket showing loss of backup seal in the lift-off phase.

Unpressurized, the secondary seal does what it is supposed to do. But pressurized, the joint becomes subject to rotational forces, which open up the secondary seal. The joint no longer has seal redundancy. The figure also shows that there is a high probability of secondary seal failure after the first 330 milliseconds of rocket firing. A design fix was to keep the primary seal warm with a heating element.

Column 5: Failure mechanism (cause). The information on causes has two purposes. First, identifying the root cause helps engineers eliminate it by redesigning the product. Second, identifying the stress (e.g., thermal, shear, tensile, compressive) that causes the failure makes it possible to validate a test program that will assure robustness of the component for this stress. In some cases, identifying the stress may not be possible.

Column 6: Effects of failure. The effects of the functional failure on the product should be considered all the way to the user level. To do that, the effect has to be traced to higher-level assemblies. For example, a failure of the connecting rod in an automobile engine will have some effect on the piston performance, which in turn will result in automobile failure. The purpose of this column in the FMECA is to get some idea about the risk.

Column 7: Interface effects. This information is not required in the normal FMECA but is important added information to the FMECA analyst. It shows how the end product interfaces with software or with other products. If the effect of the failure under consideration causes the software or other product to malfunction, then the risk must be evaluated. A possible redesign may be considered.

Column 8: Criticality. The automotive industry and many commercial organizations follow the risk priority number system. In this method, each failure is rated in three categories and the ratings are multiplied together to arrive at an index called the risk priority number (RPN). The three categories are severity, frequency, and detectability. The rating scale in each is from 1 to 10, where 10 is the worst rating. The detectability category has a special meaning. If an incipient failure is detectable before it becomes a failure, then the detectability rating is low. The worst rating of 10 implies absolutely no advance warning of a catastrophic failure. For example, an automobile that goes into sudden acceleration without any warning would be rated 10. This feature makes designers aware of the need for a safety or warning device.

In this system, RPN can be anywhere between 1 and 1000. Note that one has to be cautious about a potential catastrophic failure being masked by low ratings on frequency and detectability. For example, a failure with a rating of $10 \times 2 \times 2$ has a rating of only 40 and may appear to be unimportant. One must make sure that every failure with a severity rating of 9 or 10 is assessed for safety no matter what the overall rating.

Another popular system of criticality ranking is by the magnitude of failure rates only. This does not necessarily reflect safety risks.

The U.S. government uses a system slightly different from the RPN system. The criticality is related to risk analysis. The two primary components of the risk are severity and frequency of the failure. Whether a team takes an action or not depends on this information. There are several approaches to how the information is used. The Department of Defense assigns qualitative ratings to severity and frequency. The severity ratings are:

I—Catastrophic failure or loss of system

II—Critical failure or major loss to system

III—Minor failure or minor injury

IV—Incidental failure

Similarly, the frequency ratings are from A through E:

A—Very likely to fail

E—Almost impossible to fail

The inherent weakness in this system is that if there are 800 category I failure modes, it is difficult to rank them according to priority.

Column 9: Corrective action. This column is the one that is often misused. The purpose is to eliminate category I and II failures by design. Many analysts instead recommend inspection and testing (which may be acceptable for category III and IV failure modes). Many times a fault-tree analysis will help in coming up with the appropriate design change. If the failure is not eliminated by design, then this column should indicate the rationale for handling this failure any other way. To assure that at least a sincere attempt is made to improve design, some companies provide reevaluation of severity, frequency, and detectability as in Fig. 3.10.

Column 10: Item close sign-off. Often the analysis is done but no action is taken. This column is introduced as an optional column to assure the important failure modes are taken care of. In some countries, criminal action can be brought against engineers and managers for ignoring safety issues.

FMECAs have been modified from time to time. They can be used for different purposes as indicated earlier. Actually, the FMECA should be updated as the product develops.

Cautions on FMECA: One of the features that creates confusion about criticality is fault tolerance. A system can tolerate a fault if there is a redundant backup. Ordinarily, a reliability engineer will regard the failure of a component in redundancy as not critical. A safety engineer, on the other hand, may call it critical because, once the primary component has failed, the system no longer has a backup. The system at NASA overcomes this problem by adding R for redundancy in the criticality rating— for example, IR, IIR—where a redundant element is present. This is an indication to the designer to consider providing a fault alarm for the failure of the primary component. The alarm will call for restoring the redundancy.

Part name/ part number	Potential failure modes	Causes (failure mechanism)	Effects	Risk priority rating				Recommended improvement	Improved rating			
				Sev*	Freq	Det	RPN		Sev	Freq	Det	RPN
Pipe	Leakage in pipe	1. Corrosion	Loss of Freon	4	3	8	96	Use stainless steel pipe.	4	1	8	32
		2. Temperature cycling	Loss of Freon	4	3	8	96	Monitor temperature with thermocouples.	4	2	2	16
	Leakage at the joint	1. Cumulative fatigue	Loss of Freon	4	4	8	128	Monitor vibration with acceleration	4	2	2	16
		2. Poor soldering	Loss of Freon	4	4	5	80	Update process FMEA	TBD	TBD	TBD	TBD
Valve	Sticky, intermittent	1. Dirt or foreign objects	Loss of control of temperature	7	3	8	168	Electronic redundant valve action and fault identification	7	1	1	7
	Stuck open	1. Component wearout	Loss of control of temperature	7	2	8	112	Electronic redundant valve action and fault identification	7	1	1	7
	Stuck closed	1. Component failed	Loss of control of temperature	9	2	8	144	Electronic redundant valve action and fault identification	7	1	1	7
		2. Component expansion and contraction	Loss of control of temperature	9	2	8	144	Electronic redundant valve action and fault identification	7	1	1	7

* Severity ratings 8 to 10 request special effort in design improvement regardless of RPN rating.

Figure 3.10   FMECA with reevaluation of risks.

A bigger problem is that a vast majority of companies violate the first principle of FMECA—that an analysis be done by a designer and a nondesigner together [7]. This gives a more thorough critique of the design. Usually, a reliability engineer does the analysis alone. Except in rare cases, this is a waste of valuable resources because a reliability engineer alone is not qualified to judge the details of design. As a team, the design engineer and the reliability engineer can truly develop a highly reliable product.

The future role of FMECA will be bigger than it is today. Signs of increasing acceptance abound. The mass transit system in Atlanta uses FMECA in writing maintenance procedures. At a Polaroid plant, FMECA is used for training repair personnel. At a candy mass production plant, maintenance technicians use the tool for solving engineering problems. Its role will expand to maintainability analysis, process control, testability analysis, and software reliability analysis. Some commercial companies are already performing process FMECA. However, most aerospace and defense contractors are still unaware of the benefits of applying FMECA to manufacturing processes.

FMECA is a versatile tool. For software, it can identify missing requirements in the specifications and point out sneak paths in the code. For maintainability analysis, FMECA can identify critical microprocessor paths that must be tested (but which are often untestable).

The authors have performed FMECA at various levels and in different forms. The hierarchy in Fig. 3.11 has been found highly productive for high-technology products.

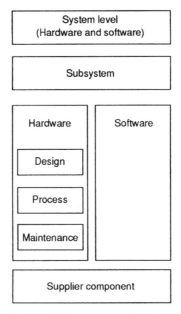

Figure 3.11   FMECA hierarchy.

3.2.7 Worst-Case Analysis

Worst-case analysis can be done in many ways, although the objective always is to determine whether a product will still work under the worst possible conditions. For electronic circuits, effects on the entire circuit are evaluated for component values at the lowest and highest ends of the tolerance. This also allows prediction of drifts in the engineering parameters. For mechanical components, similar evaluations are made to determine the effects on interchangeability and on the overall assembly performance. In all cases, the worst effect of all the possible environments can also be evaluated.

Worst-case analysis can also be used for quickly determining the effects of engineering changes. For example, an engineer who is not sure whether a new tolerance on a component is adequate can test some components at the low end of the tolerance and some at the high end. If both ends give satisfactory performance, there is perhaps no need to test components in the middle of the distribution. This form of testing will give much higher confidence with fewer components tested. Figure 3.12 shows such an approach. Worst-case analysis should go beyond tolerance analysis, however. Failures can be caused by electrical noise, race conditions, storage in a warehouse, electrostatic discharge, electromagnetic interference, and human errors. The analysis should be done for static as well as dynamic conditions.

3.2.8 Other Analysis Techniques

Fault-tree analysis and sneak circuit analysis are covered in Chapter 5. The robust design approach and the quality function deployment technique are covered in Chapter 6. Statistical analysis of failure distributions is covered in Chapter 2.

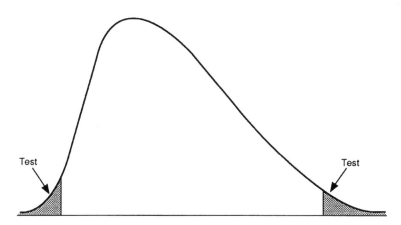

Figure 3.12 Worst-case testing.

3.2.9 Design Improvement Approaches

Derating and fault-tolerant design are the most frequently and gainfully used design improvement techniques. Zero-failure design is covered in Chapter 5 and modular design is covered in Chapter 4.

3.2.9.1 Derating This technique became popular in electronic design, but it is used to provide design safety margins in other fields as well. The concept can be explained from an *S-N* diagram, which is a graph of stress (*y* axis) versus number of cycles to failure (*x* axis). For a given stress, the diagram gives the cycles to failure. Today, the *x* axis may be plotted in miles or hours to failure, as in Fig. 3.13a, or in the traditional units of number of cycles. On log-log paper, the diagram is usually a straight line plot with a knee at some low value of stress, as in the figure. This diagram shows that if the stress on the component is lower, the component will last longer. If the stress is below the knee level, the component can last almost forever (100 million cycles is considered an infinite life for practical purposes). The reduction of stress is expressed as a derating factor and can be roughly equated with reliability level:

Derating Factor	Reliability Level
50% of maximum component rating	High reliability
30% of maximum component rating	Very high reliability

The 30% rating usually is equivalent to a stress level below the knee. At this stress the component in Fig. 3.13a, a high-voltage cable, shows a life of more than 30 years.

Caution: There are some exceptions to the above rule. For example, an aluminum electrolytic capacitor will have lower reliability if the stress falls below a certain threshold. Frozen batteries may not work at all.

3.2.9.2 Fault Tolerance Fault tolerance usually implies two devices (one a backup to assure functional performance). However, for measurement or computation in such applications as aircraft instruments or software control, an odd number of devices may be used, and a decision is made by majority vote. This type of odd redundancy is often called voting redundancy. Whether two or three or even more are required will depend on the savings assessed from the life-cycle cost projection for each option.

Redundancy may be active or standby in nature. In active redundancy, the primary and the backup devices both operate at the same time. They may share the load. In standby mode the secondary device is switched on only when the primary device fails. The advantages and disadvantages are compared below.

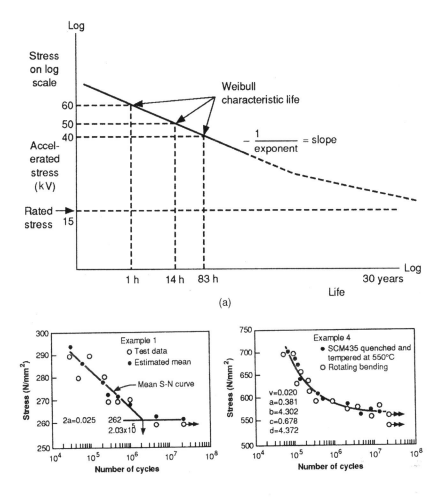

The proposed *S-N* curve may be expressed as

$$y = a\left(\sqrt{(x-d)^2 + c^2} - x\right) + b$$

where $y = \log S$, $x = \log N$, and a, b, c, and d are constants to be optimized. When $c = 0$, this equation yields the simplest bilinear relation composed of inclined and horizontal straight lines; the slope is then given by $-2a$, the knee position by d, and the fatigue limit by $b - ad$.

(b)

Figure 3.13 (a) *S-N* diagram for a high-voltage component based on tests conducted by the authors. (b) *S-N* diagrams reported by Nishijima. (From T. Tanaka, S. Nishijima, and M. Ichikawa. *Statistical Research on Fatigue and Fracture*, Elsevier Applied Science, New York, 1987.)

I. Active Redundancy

A. *Advantages*
1. The system is automatically derated, since each device is sharing the load. This increases the life of the devices.
2. Availability is increased because the failed device stops sharing the load and gets immediate attention by repair personnel.
3. The reliability of an individual device may be inherently higher when it is in the operating condition at low stress than when it is in storage.

B. *Disadvantages*
1. The cost of two devices is higher than that of one. This disadvantage is not significant if the whole equipment does not need to be duplicated. Many times a subassembly or just a few components are required to be redundant. For example, a company provided a redundant component on a circuit board instead of the whole board. By the same token, one could include a redundant circuit board instead of duplicating the whole black box. (One medical company provided a redundant fan to keep circuitry cooler instead of providing a redundant black box.)
2. More maintenance may be required because of additional devices.

II. Standby Redundancy

A. *Advantages*
1. The backup can take the full load when the primary has failed.
2. Backup maintenance can be performed without disrupting the system.

B. *Disadvantages*
1. Backup equipment may fail to turn on. Switching reliability is important.
2. Backup may fail without maintenance personnel knowing it. Therefore, standby systems require frequent monitoring or automated diagnostic checks.
3. Storage and the nonoperational standby condition may introduce additional failure modes.

As noted, using only one backup device may not be satisfactory where numerical data are involved. For example, if one computer gives a wrong output because of a malfunction, and its backup gives the right answer, the user may not know which one is right. If there were three computers, the two right ones could vote by giving a matching output. This is called two-of-three voting redundancy. On the NASA space shuttle that exploded in 1986, there were five computers to allow for sufficient voting in case one or two of them failed totally. This approach is often used for checking errors caused by noise or line-voltage transients. In general, if k out of n devices must work, the model is called k-of-n redundancy. For computation of reliabilities, refer to Chapter 2.

Caution: Redundancy is not always the answer if the redundant devices have a serial element that limits reliability, such as a common power supply. The reliability of the power supply then becomes critical. The system may require an independent power supply for each device. Analysis of such problems is called common-cause analysis. Usually, fault-tree analysis and minimum cut set analysis are required. This information can easily be obtained by a glance at the block diagrams for system reliability. This approach is covered in Chapter 5.

3.3 RELIABILITY IN THE MANUFACTURING PHASE

The high early failure rate of the bathtub curve is mostly from manufacturing discrepancies, although some manufacturing discrepancies can increase failure rates in the useful life and wear-out portions of the curve as well. Figure 3.14 shows the effect of manufacturing screening on field failure rates.

The objective of designing for reliability in manufacturing is to control the manufacturing-related causes of failure. To identify these causes, a FMECA for the manufacturing process should be done before the process is installed. This is covered in Chapter 6. Process FMECA shows the causes of process discrepancies during the design of the process. Many additional causes of failures will be discovered during a pilot run. It is at this time that process control is needed most, before any defective product goes out the door. As an example, consider an operator who takes too long to solder a joint, thereby producing a poor connection. One solution to the problem would be to study how long (and short) a time the solder gun should be kept on to produce a reliable joint. This would establish the tolerances for the solder time. If adhering to soldering tolerances still does not produce reliable

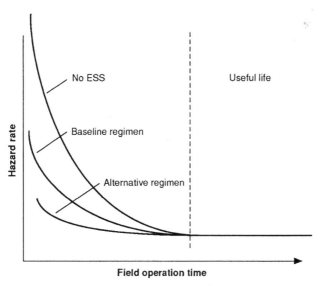

Figure 3.14   Effect of screening on field failures. (From Ref. 2.)

joints all the time, then it may be necessary to change the soldering process itself, or to use a different process altogether.

The action required to ensure reliability is not necessarily the same action required for traditional quality control. Not all discrepancies or defects lead to low reliability. For example, the following defects may degrade quality but not reliability:

The wrong shade of a color

A light dent on the surface of a casting

A scratch on the paint

A poor surface finish

The wrong plating on screws

However, the following defects or flaws usually reduce reliability:

A poor weld

A cold-soldered joint

Leaving out a lockwasher

Using the wrong flux

Not cleaning the surfaces to be joined

A large dent on a spring

An improper crimp on a wire joint

Every failure in a test and every nonconformance to specifications should be evaluated for effects on reliability. This area has received little attention, but it has the highest influence on product costs. If manufacturing defects are controlled, then burn-in and screening test costs can drastically be reduced. One-hundred percent screening of parts may not be required. Only an audit may be required to make sure no new errors are introduced.

3.4   RELIABILITY IN THE TEST PHASE

The test phase of reliability starts in the product development stage and continues in the production phase at less rigorous levels. The following reliability tests are done as a minimum:

During Design

1. Reliability growth tests
2. Durability tests
3. Reliability qualification tests (same as qualification tests during manufacturing)

During Manufacturing

1. Burn-in and screening
2. Failure-rate tests, also called MTBF tests

3.4.1 Reliability Growth Testing

In complex equipment, good designers make few mistakes while hasty ones leave many problems unsolved. To catch such problems, the reliability growth concept was developed. The concept calls for identifying problems and solving them as the design progresses. To accomplish this, a closed-loop corrective action procedure called TAAF—for test, analyze, and fix—is used. The idea is that, if environmental tests are done in early design, the product should mature faster and achieve higher reliability, in what James T. Duane called a learning curve approach. In 1964, Duane published a simple model that showed cumulative failure rate linearly related to cumulative operating hours on log-log paper [8]. Other models have appeared since then (see Mil-Hdbk-189 [9]), but because of its simplicity the Duane model has prevailed.

Duane Model   The Duane model is characterized by the equation

$$\text{Cumulative } \lambda_c = kt^{-b}$$

This can also be written as

$$\text{Cumulative MTBF} = \frac{1}{k}t^{b}$$

where b = slope, k = starting failure rate, and t = cumulative test time.
    When logarithms of both sides of the equation are taken, the equation becomes linear:

$$\log \lambda_c = \log k = b \log t$$

Data on three groups of vehicles showed the following relations:

$$\text{Group A:}\quad \lambda_c = 5.47t^{-0.6632}$$

$$\text{Group B:}\quad \lambda_c = 0.719t^{-0.3595}$$

$$\text{Group C:}\quad \lambda_c = 0.562t^{-0.3189}$$

Today, most companies track MTBF rather than λ since it is easier to communicate whole numbers instead of fractions (assuming constant failure rate). The Duane model shows that the design fixes are effective and implemented before the next test.

If many engineering changes have been made, credit for them should be taken by computing instantaneous MTBF, which can be derived from the cumulative failure rate equation as follows. λ_c can be expressed as the number of failures $N(t)$ within time t ($t > 0$) divided by t, or

$$\lambda_c = \frac{N(t)}{t} \quad \text{and} \quad N(t) = \lambda_c t = kt^{(1-b)}$$

The change per unit time of $N(t)$ is the instantaneous failure rate λ_i, which when $t \to 0$ is

$$\lambda_i = \frac{d[N(t)]}{dt} = k(1-b)t^{-b} = (1-b)\lambda_c$$

Since instantaneous MTBF is the reciprocal of the instantaneous failure rate, the instantaneous MTBF (M_i) is

$$M_i = \frac{1}{\lambda_i} = \frac{1}{(1-b)\lambda_c} = \frac{M_c}{(1-b)}$$

where M_c is the cumulative MTBF. Figure 3.15 shows cumulative and instantaneous MTBF curves.

AMSAA Model   The Duane model assumes constant reliability growth. The growth actually can accelerate or decelerate, depending on the competency of the design team and the number of engineering changes made after each test. Concern over variable reliability growth led the U.S. Army to a new path. The Army Material Systems Analysis Activity (AMSAA) model assumes that the reliability growth is a step function from one test phase to another, as shown in Fig. 3.15b. At a workshop of the Institute of Environmental Sciences in March

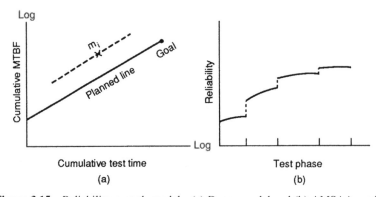

Figure 3.15   Reliability growth models: (a) Duane model and (b) AMSAA model.

1988, the AMSAA model was recommended by the Army and the Air Force Rome Air Development Center. The AMSAA model is broader in application, and the Air Force step model [2] is not greatly different. This approach reflects better fit. For details of the AMSAA model the reader is referred to Mil-Hdbk-189 [9].

Regardless of the reliability growth model, two fundamental related questions must be answered in a growth program: What should be the starting MTBF? What should be the growth rate?

Traditionally, projects have used a starting MTBF value between 10% and 30% of the goal. Proponents say that this is realistic, according to experience. An Air Force manager, however, favors going to a higher starting MTBF, since starting at a low percentage indicates a need for much more design work. He prefers to enter the growth test with a well-engineered design. The authors agree with him. Reliability is a mature science now. Commercial manufacturers would be out of business if their reliability grew so slowly. Most engineers in aggressive commercial sectors try to achieve more than the 100% MTBF goal at the outset. Some succeed, some do not. One manufacturer of Winchester disk drives had a MTBF of 4000 hours when a competitor came up with a MTBF of 5000 hours in a drive at half the price. To compete head-on, the manufacturer went into a reliability growth program with a starting MTBF of 6000 hours and a goal of 10,000 hours over 3 years. But the competitor introduced a product with 20,000-hour MTBF in 1 year—at a lower price, too!

The answer to how fast the reliability should grow depends on many factors. Some of the factors are discussed in the following sections.

TAAF Versus ATAF   The truth is that reliability growth in a complex product can be faster if most of the design changes are made because of proactive analysis instead of reactive testing. The intent of reliability growth programs is to encourage a thorough analysis before testing, but usually the sequence is reversed. Microanalysis takes place after testing instead of macroanalysis before. FMECA and fault-tree analysis are often used too late, if used properly at all. Good design principles call for detailed analysis first so that most of the design improvements are made while the design is still on the drawing board. It is much cheaper to make changes at this stage than after the hardware and tooling have been fabricated. What is needed is ATAF (analyze first, then test and fix) instead of TAAF. In other words, always perform FMECA before implementing reliability growth tests.

Reliability Growth Through Accelerated Testing   The commercial sector has fostered reliability growth for many years without calling it by that name. When one of the authors worked in the commercial sector, he had to accept (out of necessity, of course) the challenge of improving the MTBF of a product line tenfold in 1 year. The only way to succeed was to resort to highly accelerated but valid tests, described later in this chapter.

Without support from design engineering, this product line would have vanished from that company. The company did achieve a major breakthrough that enabled it to regain all of its lost market share. It even guaranteed the product for 15 years at no charge to the customer. As a result of the improvements, the return on investment for that business increased from 17% to 71%!

Companies like Cray Research, Apple Computer, IBM, Polaroid, Hewlett-Packard, Motorola, and Intel have stunned the industry with reliability growth many times. Many agree to the need for higher acceleration in testing, but not enough is published about it.

Controlling Warfare Over Failure Causes Most projects are plagued by disagreements over the causes of failure. Software engineers tend to ignore hardware failures. Bob Cruickshank of IBM Federal Systems makes this point clearly, with a chuckle: "How many programmers are needed to change a light bulb? None! Because this is a hardware problem!" Hardware engineers often do the same; they work independently of the software engineers. The systems then inherit interface problems that sneak up at a much later stage. Reliability predictions are then shaky, to say the least. As long as major engineering errors and single-point failures remain in the system, reliability growth is hard to control. In addition, the system cannot be designed to fail gracefully.

Reliability growth testing is necessary, otherwise design and management may overlook reliability problems. The growth management clause in contracts gets attention upfront during R&D. Then more models will come, such as software growth models, maintainability growth models, and availability growth models. As reliability lecturer Allan S. Golant says of the Duane growth model: "Its inherent simplicity makes it the preferred method if communication with your management is the prime requisite."

3.4.2 Tests for Durability

Durable life can be defined as the life to the point at which the wear-out starts to accelerate, where the third region of the bathtub curve starts. To establish this point, the data from fatigue testing only are considered. (The presence of infant mortality failures will corrupt the data.) Depending on the severity of failure, this point is selected so that the cumulative failure probability is less than 1%, 5%, or 10%. The lives are then commonly referred to as B_{01} life, B_{05} life, or B_{10} life.

Many think that MTBF is a measure of durable life, but there is absolutely no relation between the two. As noted earlier, MTBF is strictly a function of failure rate, while durable life is a function of length of the bathtub. Two bathtubs can have the same failure rate (therefore the same MTBF) but different lengths.

Some consider durable life as the mean life. This concept has little value in the reliability literature, because by the time the mean life is achieved, roughly 37–68% of the products would be expected to fail, depending on the probability distribution. Who wants a product with a 63% chance of failure? It is amazing how much business is conducted on the basis of advertisements of mean life without manufacturers or customers realizing what it means. A pacemaker manufacturer told doctors that the battery in one product had an average life of 4 years. The doctors thought each battery would last 4 years, and so their patients were told. No wonder that many patients sued the doctors and the doctors sued the manufacturer. As might be expected, a high percentage of the patients found that their batteries failed before 4 years, some in less than 6 months. Since the batteries did not give advance warning of failure, some patients died. Such cases abound. Think of a light bulb. When you buy a lamp with an average life of 2000 hours, do you expect each lamp to last 2000 hours? How many people know that only one out of every two such lamps may last more than 2000 hours? The same is true for products ranging from cadmium batteries to satellites; it should not have been a surprise that a weather satellite with a mean life of 5 years failed in 2 years.

Durability is assessed in many ways. Usually some empirical model is used. This chapter covers several models that are often used. But first, discussion of accelerated life tests is in order. Accelerated tests are necessary because testing life under normal conditions would take months or years.

Accelerated Life Testing   The purpose of accelerated testing is to duplicate the field failures by providing a harsher—but nonetheless representative—environment. The product is expected to fail in the lab just as it would have failed in the field—but in much less time. Then an empirical relation can be used to predict field life from lab life. Before any prediction of life, it is necessary to make sure certain requirements are met. The test should not create failure modes that the customers will never see. The four basic requirements, based on the authors' long experience, are the following:

1. The predominant failure mode under field stress and under accelerated stress should be the same. This is an indication of the validity of the accelerated test profile. The stress (such as voltage or temperature) that creates such a failure is called the dominating stress.
2. The engineering properties of a material under accelerated stress should be the same before and after the test, except at the point of failure. Properties such as lattice structure or grain structure, for example, can usually be defined by a metallurgist. This means that the range of stress application is within the elastic limits of the material.
3. The shape of the failure probability density function at rated and higher stress levels should be the same. This is demonstrated by the curves having the same slope on a Weibull graph. See Chapter 2.
4. The Weibull characteristic life should be repeatable among production lots, within 5%.

As long as the above four conditions are satisfied, one can select higher stresses to reduce test time. The most popular approach is to construct an *S-N* diagram as a test model. Figure 3.13a shows an *S-N* diagram for a high-voltage connector made from the same material as high-voltage cables. The customer uses the connector at 15 kV, but it cannot be practically tested at that voltage (it would take 15–30 years to fail).

The voltage therefore was increased to 40 kV, and, since one unit cannot be representative of production, a random sample was tested at that voltage. Usually Weibull characteristic life at 63.2% failure gives a good representative point for the S-N diagram. Similarly, Weibull characteristic lives were plotted at 50 and 60 kV. These three Weibull characteristic lives fell on a straight line on log-log paper. The characteristic life at 70 kV did not fit on this line, indicating that that stress was too high for the material. At this level, the failure mode also changed, so it was clear that one of four principles cited has been violated. Once the S-N diagram is constructed, the graph can be projected to give a life estimate at 15 kV.

Accelerated Test Models for Durability Prediction The two most popular models in use are the inverse power law model and the Arrhenius model. The inverse power law model is a versatile model for many different mechanical and electrical stresses. It states that life is inversely proportional to the stress raised to the power N, where N is the acceleration factor derived from the slope of the S-N diagram. It is computed from the equation

$$\text{Slope} = -\frac{1}{N}$$

(The slope in the graph is negative.)
    The model can be written as

$$\frac{\text{Life at rated stress}}{\text{Life at accelerated stress}} = \left(\frac{\text{Accelerated stress}}{\text{Normal stress}}\right)^{N}$$

For the product in Fig. 3.13a, the value of N was found to be 8.2 from the slope of the S-N diagram.
    How long should the test last? This too can be found from the inverse power law. For the product in Fig. 3.13a,

$$\frac{30 \text{ years} \times 365 \text{ days} \times 24 \text{ hours}}{\text{Test hours at accelerated stress}} = \left(\frac{60}{15}\right)^{8.2}$$

The test time at the accelerated stress is therefore 4.01 hours.
    Caution: Make sure all four accelerated test principles are valid at each accelerated stress.
    The Arrhenius model has been widely misused. The intent of the model is to predict life at normal temperatures from accelerated test data. However, it applies only to failures from chemical reactions, or metal diffusions, which are accelerated by temperature. It does not apply to failures generated by burn-in and screening for infant mortality.

The Arrhenius model can be applied to insulating fluids, lubricants, battery cells, plastics, and some failure mechanisms of electronic components. But care should be exercised when applying this model to electronics because most so-called electronic failures are really mechanical failures. The Arrhenius model is given by

$$\text{Life} = Ae^{E/(kT)}$$

where A = empirical constant, T = temperature in kelvins (for electronics, the junction temperature; for fluids and other products, the hot spot temperature), k = Boltzmann's constant = 8.62×10^{-5} eV/K, and E = activation energy.

Arrhenius graph paper (Fig. 3.16) has a special grid for plotting the Arrhenius equation as life (time to failure) versus temperature in degrees Celsius.

The value of E is obtained from the relation

$$E = -k \times \text{slope}$$

The value of A is obtained by selecting any point on the graph and using its x and y coordinates (life, temperature) and the slope to compute A from Eq. (3.3).

Once these values are known, the model can be used for testing at high temperature to predict life at lower temperatures. Extrapolation can be done toward lower temperatures, but not below the knee temperatures and certainly not below freezing temperatures.

Comparison of life at two temperatures from the graph will lead to the following equation, which is used by some organizations:

$$\frac{L_1}{L_2} = e^{[(E/k)(1/T_1 - 1/T_2)]}$$

Many in the electronics industry assume that the life and failure rates are inversely proportional. (The authors disagree. See the argument in Section 3.1 on life versus MTBF.) If the assumption is valid,

$$\frac{\lambda_2}{\lambda_1} = e^{[(E/k)(1/T_2 - 1/T_1)]}$$

Caution: (1) The Arrehenius model is an empirical model. It should be applied to failure rates with proper understanding. Chemical reaction rate and mechanical failure rate on electronic devices are not related. The failures may result from nonchemical causes such as a transient voltage. Certainly, infant mortality failures do not belong in this model. (2) Some practitioners determine the value of E only once and apply it across the board to other components. This practice is not recommended. Failure rates vary from one component to another, even though they may have similar designs. Units with the same part number manufactured in two geographical locations can have two different activation energies.

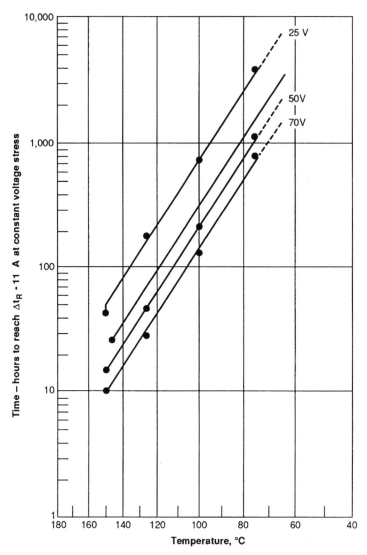

Figure 3.16   An Arrhenius plot for a discrete semiconductor device. (From Erwin A. Herr, Reliability Evaluation and Prediction for Discrete Semiconductors, *IEEE Transactions on Reliability*, August 1980.)

Models for Specific Applications   A component may go through cyclical or varying loads, and the models just described may not apply. However, the literature contains several models for such components. Among them are Goodman diagrams [10] and Miner's rule [11,12]. Chapter 5 covers discussions on safety margins. An illustration of Miner's Rule follows below.

Miner was a design engineer at Westinghouse. He postulated that a component subjected to different cyclical loads accumulates damage and that failure can be expected when the summation of incremental damage from each load equals unity. Failure is likely when

$$\frac{C_1}{N_1} + \frac{C_2}{N_2} + \frac{C_3}{N_3} + \cdots + \frac{C_i}{N_i} = 1$$

where C_f = number of cycles applied at stress S and N_i = capability of the material in cycles at S, determined from an S-N diagram.

This equation shows that tests to establish an S-N diagram are required for this approach.

Example 3.3   A component is expected to see the following stresses during its lifetime:

150,000 cycles at 2600 lb/in.^{2}
100,000 cycles at 2900 lb/in.^{2}
25,000 cycles at 3200 lb/in.^{2}

The capability as determined from the S-N diagram at these stress levels is:

400,000 cycles at 2600 lb/in.^{2}
250,000 cycles at 2900 lb/in.^{2}
190,000 cycles at 3200 lb/in.^{2}

From Miner's Rule, the cumulative damage is

$$\frac{150,000}{400,000} + \frac{100,000}{250,000} + \frac{25,000}{190,000} = 0.907$$

Since the cumulative damage is less than 1, failure is not expected.

Caution: Studies have shown that failure takes place in the range of 0.7–2.3 cumulative damage value.

3.4.3   Testing for Low Failure Rates

For simplicity, MTBF has been used as a measure of failure rate even though it is based on the assumption that failure rate is constant. In practice, failure rate is rarely constant. Infant mortality failures may be present in the field for any product, and for mechanical products the failure rate is almost always an increasing function of time. But industry marches ahead with the assumption of constant failure rate. Industry should accept failure rate for what it is: it can be increasing, constant, or decreasing with time. In such situations, Weibull analysis may be the appropriate choice for failure tests. MTBF should be used only when the failure rate is truly constant. Failure rate tests are done both for reliability qualification of new products and for reliability demonstration during production.

If the failure rate is increasing or decreasing with time, the times at which the failure rate should be reported must be specified. The Weibull failure rate model from Chapter 2 can be used. The goal should be to reduce failure rate to a small constant value during the useful life.

Reliability Qualification Tests   Reliability qualification is done on production devices but should also be done on prototype models. It should be an accelerated test if the failure rate for the entire expected life must be verified. Many companies are not quite sure how to accelerate the tests (see Section 3.4.2 for guidance). Others are afraid to overstress beyond the published ratings. Most companies qualify only for a portion of the bathtub curve.

Qualification tests consist of stressing the product for all the expected failure mechanisms. The test can be stopped if there are no failures during the expected life. The test conditions for qualification are generally severe. Some may not be performed for production demonstration tests, such as thermal shock, impact, and constant acceleration. Some special stresses may have to be applied to duplicate field conditions. For example, a bias voltage may be needed to create the worst electron migration in integrated circuits. Similarly, tensile and compressive forces may be needed to stress a spring for a mechanical apparatus. An impact load may be required for an aircraft structure.

Example 3.4   A high-voltage fuse was qualified for 30-year life under these accelerated test conditions:

1. Environmental reliability
   a. Lightning surges and continuous dielectric stress
   b. Environmental such as moisture, oil, dust, and fungi
   c. Ultraviolet rays from sunlight
   d. Current in the customer's system, which could be continuous or transient during load application or fault interruption
2. Mechanical reliability
   a. Vibrations
   b. Mechanical switching
   c. Magnetic forces
3. Thermal reliability
   a. Operating temperature
   b. Temperature cycling
   c. Rate of change of temperature
4. Operational reliability
   a. Physical operation of components

   b. Installation practice
5. Interface reliability
   a. Interface with customer's system
   b. Interfaces between components, such as soldered, welded, and mechanically connected joints and connectors for electronics
   c. Interface between subassemblies
   d. Interface between subassemblies and final assembly
   e. Interface with software
6. Use reliability
   a. Field instructions
   b. Repair practices
   c. Maintenance practices
   d. Logistics support practices
   e. Handling and storage
   f. Tools

Reliability demonstration tests, on the other hand, usually require stressing during only a portion of the useful life. They do not duplicate all the tests done during product qualification. Since reliability demonstration tests are not very severe, qualification tests should be done at preset intervals. For electrical relays, for example, Mil-Std-454 [13] calls for qualifying every 2 years.

Caution: The interval for qualification tests should be chosen carefully. For example, one of the authors had to investigate a relay failure that took at least 11 lives. The relay was qualified for leakage of salt water every 2 years. The leak tests in production were not very severe, and defective relays went out to the field. If the qualification tests had been done more frequently on a few units, most of the defective units would have been detected earlier.

Qualification test data should also be analyzed to determine whether the product meets the qualification criteria. The basic criteria are:

1. No infant mortality failures should go to the customer.
2. Failure rate should be below a target percentage over the useful life (such as less than one failure in 10 years or, for an automobile, 100,000 miles).

Example 3.5 The data in the table below were recorded for an accelerated life test of Winchester disk drives for mainframe computers. The sample of 16 drives was drawn from a finished-goods warehouse. Each hour of accelerated testing represents approximately 0.4 month in the field. The quality criteria were that customers should not experience early failure and that the failure rate during the expected 7-year life should be less than 0.5 failure per 100 disk drives. Assuming the test environment is a very close duplicate of field environment, is this product qualified?

Failure Rank Number	Failure Mode	Hours to Failure
1	Servo	7
2	Read/write	12
3	Power supply	49
4	Read/write	140
5	Head crash	235
6	Head crash	260
7	Servo	320
8	Head crash	320
9	Head crash	380
10	Head crash	388
11	Personality	437
12	Microprode	472
13	Head crash	493
14	Read/write	524
15	Head crash	529
16	Head crash	592

Solution   Since the Weibull distribution fits over 90% of life test data, it will be appropriate to plot the tabulated data on a Weibull graph as in Fig. 3.17. The graph shows that part of the data fits on one slope and the remaining data on another slope. This is typical of a product infested with manufacturing-induced "weak" failures. The first slope reflects infant mortality failures, and this should be verified with failure analysis tools. The weak failures are often a result of excess variability in quality; the failure mechanisms are usually obvious: cold-soldered joints, loose lockwashers, voids, and nonhomogeneous material. The second slope shows that the product goes straight into wear-out phenomena. There are hardly any random failures in the laboratory for mechanical products. In electronic products, of course, three slopes are possible, one for each region of the bathtub curve.

The plot shows that this product is not qualified for market. Since these units were drawn from the finished-goods warehouse, the infant mortality failures should not have been present. Apparently, the screening was insufficient. The graph shows the change in slope at 200 hours. The failures during the initial 200 hours are likely to be from production problems, and this should be verified. The knee therefore signifies that the product can be shipped after every unit goes through an additional screening of at least 200 hours. The alternative is to solve manufacturing problems to eliminate causes of infant mortality.

The second criterion, that failure rate be below a certain target value, cannot be determined from the data. In the data, the "lemons" (infant mortality failures) and the "pineapples" (inherent design failures) are mixed. When two such populations are mixed, prediction is risky. In fact, since the infant mortality failures change their personality from day to day, prediction is almost impossible.

Assuming the causes of infant mortality failures are eliminated (or an expensive screening program is in place), the data would have been without the first four failures. The plot of data without infant mortality failures (sample size 12 instead of 16 as in Fig. 3.17b) shows that the expected failures in 7 years (84 months = 210 accelerated hours) is 5%, or 5 out of every 100 disk drives. This is not acceptable, since the goal is to have only 0.5 failure per 100. The product therefore fails to qualify.

Caution: Many users apply Weibull analysis at the system level, where there are mixed failure modes. Weibull analysis is intended for fatigue experiments, which

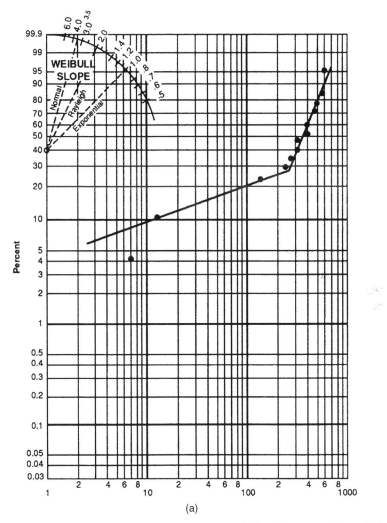

(a)

Figure 3.17 Weibull plot for disk drive data. (a) Plot including infant mortality and (b) plot without infant mortality.

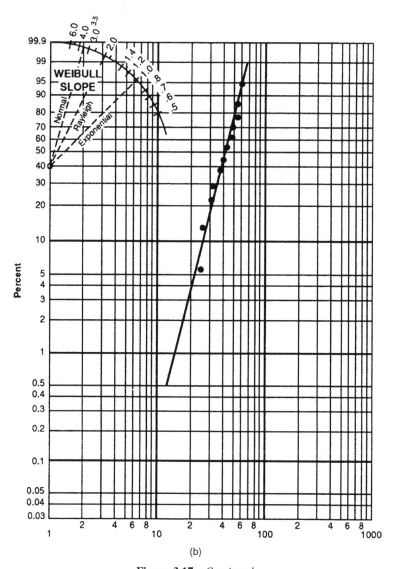

(b)

Figure 3.17 *Continued.*

can be applied only at component level, or for an assembly where the same failure mode appears repeatedly, such as failures in solder joints, welded joints, and a simple assembly of two mating parts.

It is easy to get a straight line on a Weibull graph at system level. But the slope of the line will change from production lot to production lot (unless production is so highly mechanized that the same defects occur all the time). The question then is: How does one predict field reliability at system level? The answer is: One doesn't. At system level one can only postulate, because many failure mechanisms are neither part of the reliability model nor the same every time. Some examples of these variables are bad wire connections, miswiring, electric power transients, electrostatic discharge, electromagnetic interference, poor maintenance, failure to follow instructions, human errors, and software errors. Weibull software programs should also be used with great caution. Most of them fit a line through two or more distributions of the bathtub. Statistically, a model can be fitted to mixed populations (some fit a lognormal model) but may violate basic engineering principles for solving problems. The critical failure mechanisms may no longer be visible. A single model for mixed populations may be a good indicator of the past but is an uncertain indicator of the future.

Reliability Demonstration Tests   Under less severe conditions, the production lots are accepted on the basis of failure rate tests. Most tests are operated under the assumption that failure rate is constant. Therefore, MTBF is the common measurement. Department of Defense Mil-Hdbk-H108 [14] contains three types of sampling plans:

- Feature truncated: Test is terminated at a predetermined number of failures.
- Time censored: Test is terminated at a predetermined time.
- Sequential testing: Covered in Mil-Std-781 [15].

In these plans the customer and the supplier decide on four factors: the producer's risk (chance of rejecting a good product), the consumer's risk (chance of accepting a rejectable product), the minimum MTBF, and the average MTBF.

In a failure-truncated plan, the MTBF is computed at the occurrence of a predetermined number of failures. If this MTBF (operating hours divided by number of failures) is higher than the value suggested by the handbook, the lot is accepted.

In time-censored plans, the handbook suggests the number of units to be tested for a predetermined number of hours. It also gives the number of failures permitted.

A sequential test plan requires little testing at one time. If a product is not acceptable or rejectable, then more units are tested. This plan is the most popular. The main advantage of this plan is that a decision can be reached very quickly if the product is very good or very bad.

Before a sequential test, the customer specifies producer's risk, con-Burner's risk, and discrimination ratio (ratio of average MTBF and minimum acceptable MTBF). From these values, a chart such as that in Fig. 3.18 is selected. A customer may also specify minimum test duration and minimum number of pieces to be tested. If such conditions are not specified, it is up to the supplier to test any number of pieces (as few as one). At each failure, the cumulative failures are plotted against cumulative test hours (in multiples of minimum acceptable MTBF). If the point of their intersection crosses the accept lines, the entire lot is accepted. If this point crosses the reject lines, the entire lot is rejected. If the point is in the continue test area, the tests are continued.

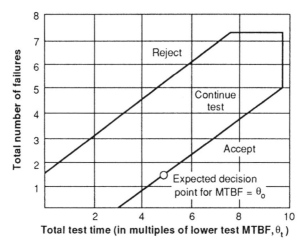

Figure 3.18 A reliability demonstrations plan for Mil-Std-781.

3.4.4 Burn-in and Screening

Burn-in and screening tests are 100% production tests designed to catch infant mortality failures. If a component is manufactured properly, it should not require these tests, and the aim should be to eliminate them by removing the causes of the problems. But, since new causes can crop up, audits should always be conducted. In burn-in tests, devices are subjected to a constant high temperature. Screening tests use a profile of operating environments such as temperature cycling, vibration, and humidity. Screening may require special tests other than environmental tests. For example, leakage current was monitored in the tests illustrated in Fig. 3.19 to identify weak components that might not have been noticed otherwise.

Many companies are forced into doing these tests because the supplier does not. These tests are also done at circuit board level and assembly level to reveal poor workmanship (such as defective solder joints) or loose connections. These tests also show whether the design can withstand vibrations and temperature cycling.

The type of stress and the length of the test are generally suggested by the appropriate military standard, which is based on experience. Unfortunately, many practitioners take these suggestions as the final word; as a result, screening may not be long enough and weak products may get out to the customer. The manufacturer should first find the stresses that will shake out all weak components and second make sure the test duration extends to the point where failure rate is constant.

To reduce test time, one can use accelerated test conditions to flush out weak failures faster. The only requirement is that the failure mode on accelerated and unaccelerated tests should be the same. Some use combined environmental reliability tests (CERTs) for screening. These tests combine temperature cycles, vibration, humidity, dust, and other environments with the equipment powered. Weak failures have been known to appear faster. Automotive products, for example, see these environments together in use.

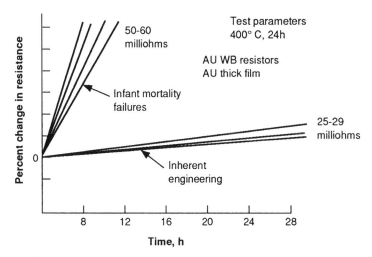

Figure 3.19   Weak components identified by functional screening.

Testing Philosophy   The nature of screening tests, which are often called environmental screening, depends on the use profile of the system. A test may include vibration, temperature cycling, burn-in, run-in, humidity, and special stresses unique to the system use. Most electronics manufacturers screen 100% of the product. A wise strategy, before embarking on a test program, is to first determine which tests are necessary, according to the use profile, and which components must be tested. If a good manufacturing process is in place, sample testing instead of 100% testing will be adequate insurance.

There should also be a rational analysis of whether the test should be done at component level, board level, or system level. The following cost information, derived from electronics industry rules of thumb, can aid in making the right choice:

Test Level	Approximate Cost of Finding and Replacing a Defective Component
Component	$5.00
Board	$50.00
Black box	$500.00
System	$2500.00 and up

These numbers show that, if system reliability is low, testing should be done at the black-box level. If the black-box reliability is low, testing should be done at the circuit board level. Not all the boards need to be 100% tested—only the questionable ones.

Similarly, only the questionable components need to be tested 100%. Conversely, if reliability at system level is very high, system-level testing may be unnecessary. The same test philosophy applies at black-box level and below.

Testing at board level is generally desirable, even though a few manufacturers trust their suppliers enough to ignore board-level screening. But, if there are 100 components on a board, each having 99% reliability, the board reliability before soldering (assuming the components are mounted properly) is 36.6%. This means that only one out of three boards can be expected to work satisfactorily, and this is not an uncommon situation. The reliability is further reduced by such elements as insertion of wrong components, reversed polarities, and poor soldering. If one out of four boards contains some assembly defect, the board yield is shaved by another 25%. For such situations, screening at board level is costeffective.

Role of Automated Testing   If the chance of finding a reliable board is roughly 1 in 3 before soldering, it pays to do screening at the component level because, as the preceding table shows, it is 10 times cheaper ($5 versus $50) to replace a bad component at that level. Functional testing at the component level is rarely justified, however, unless it is done on automated test equipment (ATE). Whether testing is automated or not, the prudent course is to first determine whether the supplier is already doing the screening. Some components are sensitive to environments such as electrostatic discharges and fail in spite of the screening done by the supplier. If these "dead on arrival" (DOA) components are numerous, rescreening may be desirable.

The investment in intelligent, programmable automated test equipment is justified when:

- The volume of components is large.
- There are numerous tests for components and boards. The parallel test capabilities of the equipment can drastically cut test time.
- Reliability can be improved by diagnostic testing.
- The component location and the cause of its failure must be identified quickly.

Automated testing permits effective screening at the component level because the test time is short enough to fit into a production schedule. If screening is not done at this level, much more time must be spent at board level. Also, at board level, the repair process can cause failure of other components.

Whether testing is done at board level or component level, the total cost of inspection, troubleshooting, test, and repair should be kept as low as possible. If a board contains about 200 components and is produced at a rate of about 100,000 a year, our experience shows that automated test equipment can pay for itself within 2 years. The savings come primarily from a reduction in test and repair time.

ATE offers many other advantages. It can test a component alone or on a circuit board. When a complete board with 200 or so components is tested, the ATE can isolate defective components. A major benefit to designers and quality control engineers is that ATE summarizes vast amounts of data on design parameters such as leakage currents and various voltages—valuable information for determining statistical distributions for vendors' products.

Dynamic Burn-in   Dynamic burn-in requires functional testing during burn-in. Software built into automated test equipment can control the functional tests. Dynamic burn-in greatly improves the field reliability of the boards because they have, in effect, gone through a system test. Boards tested this way tend to be less than 2–10% defective as shipped. A simple system of five printed circuit boards showed the following reliabilities after standard and dynamic burn-in:

Test Reliability	Outgoing Board Reliability (%)	System Reliability (%)
Standard burn-in	91.41	63.82
Dynamic burn-in	98.55	92.96

Economies from Choosing the Right Tests   Most manufacturers do not use the correct tests; they are happy when they do not find failures. Absence of failure is not an indication of a good circuit board. Burn-in at low temperature may help weed out very weak failure modes, but the product leaving the door may still be unreliable. Performing a good test on a statistically selected sample is a good strategy. A good test forces the unreliable components to fail. It may require 100% testing initially and may reduce the manufacturing yield at first, but it improves the yield for the customer. Later, manufacturing yield can be improved and 100% testing can be eliminated with a good failure analysis and management program. The savings accomplished in this way can be enormous, and customer satisfaction can be expected to soar.

Caution: The correct test will be an accelerated test based on use profile. Effective screening can take months if the test is not accelerated.

Example 3.6   A computer display oscillated frequently in the humid summer environment only. This is an indication that humidity testing is one of the right tests.

Economies from Trade-offs   Some manufacturers do not perform environmental screening; they test circuit boards only functionally. Weak components pass the quality control tests but fail later, in the field. These manufacturers do not understand that a trade-off can be made between the cost of screening and the increase in reliability. An increase in reliability means lower warranty costs for the manufacturer and lower repair and maintenance costs for the customer.

One method of evaluating a trade-off is to compute life-cycle costs (LCCs) for alternatives. The following is a simplified example of such a model:

$$LCC = \text{initial cost} + \text{screening cost} + \text{warranty failures cost}$$
$$+ \text{repair and maintenance costs to customer}$$
$$+ \text{potential loss of business because of poor reliability}$$

For a medium-size business, the following costs were calculated for a printed circuit board having a service life of 5 years:

Market price: $2500
Screening cost: $15
Warranty cost without screening: $45.95
Warranty cost with screening: $27.57
Repair and maintenance cost (charged to customer): $95.40
Repair and maintenance cost (with screening): $66.78
Potential loss of business (because of low reliability): $56.25
Potential loss of business if board is screened but still not competitive in design: $33.75

LCC without screening $= \$2500 + \$0 + \$45.95 + \$95.40 + 56.25 = \$2697.60$
    LCC with screening $= \$2500 + \$15.00 + \$27.57 + \$66.78 + \$33.75 = \$2643.10$

The savings per board ($2697.60 − $2643.10 = $54.50) may not appear significant, but for 200,000 boards, the total savings are $54.50 × 200,000 = $10,900,000! Of the savings, 47.5% went to the manufacturer and the remainder went to the customers. Such a situation, in which both the supplier and customers save, is a major benefit of assuring quality.

3.5 RELIABILITY IN THE USE PHASE

The usual tests are unable to catch all failure modes, many of which are discovered in the field. Too many of these failures indicate shortcomings in the reliability effort in the earlier stages. A closed-loop failure management system should not only solve the problem permanently but should also improve the design review process and the design of the tests.

The biggest challenge facing quality producers is competition on longer warranties. When warranties are involved, low reliability can easily destroy a business. At the time of this writing, the following warranties were in effect:

Automotive: 3-year warranty on all parts; 7 years or 70,000 miles on power train and corrosion
Water meters: 25 years warranty on any malfunction or failure
Power transformers: 15 years
Radio-frequency tags: 100 years; 150 years under consideration

These warranties are standard features of the products. The manufacturers do not charge their customers extra for them, in the hope of getting more business. The reliability effort in the use phase, therefore, must be aimed at detecting trends as early as possible and making design changes expediently. A good tool for establishing trends early is Weibull analysis. A Japanese company gets information on worldwide failures immediately through communication satellites and performs Weibull analysis weekly. With Weibull analysis, warranty costs can be predicted easily.

Trends like this are spreading globally. The best practice is to track the distribution of field failures. If the distribution starts during the warranty period, the failure modes should be thoroughly analyzed and removed permanently.

3.6 RELIABILITY AND SAFETY COMMONALITIES

Complex systems have evolved into sophisticated automated multifaceted systems with many interactions and interfaces. These systems can be comprised of vast subsystems of hardware, firmware, software, electronics, avionics, hydraulic, pneumatic, biomechanic, ergonomic, and the human. Some conventional concepts of reliability and safety may be inappropriate when considering these complex technical risks. Reliability and safety disciplines have evolved—and it is apparent that past notions have changed.[2]

3.6.1 Common System Objective

At first glance, to the layperson, there apparently is very little difference between the disciples of reliability and safety, or any other system engineering practices like maintainability, quality, logistics, human factors, software performance, and system effectiveness. However, from the system engineering specialist viewpoint, there are many different specialty objectives to consider and these objectives must be in concert with the overall system objective—designing an effective reliable complex system with acceptable risks.

A shared relationship between reliability and safety equates to the potential unreliability of the system and associated adverse events. Adverse events can result in potential system accidents. A system that functions satisfactorily should be an indication that the system should fail-safe, and failures will not present hazards, and adverse events such as accidents should not occur. The safety objective addresses "the optimum degree of safety," and since nothing is perfectly safe the objective is to eliminate or control known risk to an acceptable level.

3.6.2 Unreliability and Hazards

Hazards are unsafe acts and unsafe conditions with the potential for harm. Unsafe acts are human errors that can occur at any time throughout the system life cycle.

---

[2]This topic was originally addressed in the following paper: M. Allocco, Appropriate Applications Within System Reliability Which Are in Concert with System Safety: The Consideration of Complex Reliability and Safety-Related Risks Within Risk Assessment, In: *Proceedings of the 17th International System Safety Conference*, System Safety Society, August 1999.

Human reliability addresses human error or human failure. Unsafe conditions can be failures, malfunctions, faults, and anomalies that are initiating or contributory hazards. An unreliable system is not automatically hazardous—systems can be designed to fail-safe, or their failure may not have a safety consequence. Procedures and administrative controls can be developed to accommodate human error or unreliable humans. The ideal choice is to design out the susceptibility to human errors as discussed in Chapter 8.

3.6.3 Complex Risks

Consider a system as a composite, at any level of complexity. The elements of this composite entity are used together in an intended environment to perform a specific objective. There can be risks associated with any system and complex technical systems are everywhere within today's modern industrial society. They are part of everyday life, in transportation, medical science, utility, general industry, the military, and the aerospace industry. These systems may have extensive human interaction, complicated machines, and environmental exposures. Humans have to monitor systems, pilot aircraft, operate medical devices, and conduct design, maintenance, assembly, and installation efforts. The automation can be comprised of extensive hardware, software, and firmware. There are monitors, instruments, and controls. Environmental considerations can be extreme, from harsh climates, to outer space, to ambient radiation. If automation is not appropriately designed from a reliability and safety perspective considering potential risks, system accidents can result. The overall system objective should be to design a complex system with acceptable risks. Since reliability is the probability that a system will perform its intended function satisfactorily, the safety-related risks should also be addressed, which directly equate to failures or the unreliability of the system. This consideration includes hardware, firmware, software, humans, and environmental conditions.

3.6.4 Potential System Accidents

For discussion, consider the potential loss[3] of a single-engine aircraft due to engine failure. Simple linear logic would indicate that a failure of the aircraft's engine during flight would result in possible uncontrolled flight into terrain. Further multi-event logic, which can define a potential system accident, would indicate additional complexities such as loss of aircraft control due to inappropriate human reaction, deviation from emergency landing procedures, less than adequate altitude, and/or less than adequate glide ratio. The reliability-related engineering controls in this situation would be just as appropriate to system safety. Consider the overall reliability of the engine, fuel subsystems, and the reliable aerodynamics of the

---

[3]Dr. Perrow discusses the concept of a normal accident, which is the result of many failures throughout a system. See the following reference for more information: C. Perrow, *Normal Accidents—Living With High-Risk Technologies*, Basic Books (A Division of HarperCollins Publishers), New york, 1984, p. 70.

aircraft. The system safety-related controls would further consider other contributory hazards: inappropriate human reaction and deviation from emergency procedures. The additional controls are administrative in nature—the design of emergency procedures, training, human response, communication procedures, and recovery procedures.

In this example, the controls above would decrease the likelihood of the event and possibly the severity. The severity would decrease as a result of a successful emergency landing procedure, where the pilot walks away and there is minimal damage to the aircraft.

This has been a review of a somewhat complex potential system accident in which the hardware, the human, and the environment were evaluated. There would be additional complexity if software were included in the example. The aircraft could have been equipped with a fly-by-wire flight control system, or an automated fuel system.

3.6.5 Software Reliability and Safety

Software does not fail but the faults in it can disable the important functions or create mishaps by performing functions at an undesired time. Hardware and firmware can fail. Humans can make software-related errors. Design requirements can be inappropriate. Humans can make errors in coding. The complexity or extensive software design could add to the error potential. There could be other design anomalies, sneak paths, and inappropriate do-loops. The sources of software error can be extensive [16, p. 269]: "Studies show that about 60 percent of software errors are logic and design errors; the remainder are coding- and service-related errors." There are specific software analysis and control methods that can be applied successfully to contributory hazards, which are related to software.

3.6.6 Reliability and Safety Trade-offs

What appears to be a design enhancement from a reliability standpoint will not inherently improve safety in all cases. In some instances, risk can increase. In situations where such assumptions are made, it can be concluded that safety will be improved by application of a reliability control, for example, redundancy has been added within a design. After all, it is a redundant system so it must be safe. Be cautious of such assumptions. The following discussions present the argument that an apparent enhancement from a reliability view will not necessarily improve safety. Risk controls in the form of design and administrative enhancements are discussed along with associated trade-offs, in support of this assumption.

3.6.7 Reliability and Safety Misconceptions

A common misconception that has been known in the system safety community for many years was discussed by Hammer [17, p. 21]. It is that by eliminating failures, a product will not be automatically safe. A product may have high reliability but it may be affected by a hazardous characteristic.

Another misconception is that conformance to codes, standards, and requirements are assurance of acceptable risk. As indicated, appropriate hazard analysis is needed to identify hazards, and that associated risk should be eliminated or controlled to an acceptable level. Codes, standards, and requirements are usually insufficient or they may be inadequate for the particular design. Therefore, risk control may be inadequate. Good engineering practice is required in all design fields. Certain basic practices can be utilized, but a careful analysis must be conducted to ensure that the design is suitable for its intended use.

Consider other inappropriate assumptions: the system is redundant and monitored so it must be safe. Unfortunately, this may not be true. Proving that each redundant subsystem, or string or lag is truly redundant may not be totally possible. Proving that the system will work, as intended, is also a concern.

Defining what is acceptable risk is dependent on the specific entity under analysis—the project, process, procedure, subsystem, or system. Judgment has to be made to determine what can be tolerated should a loss occur. What is an acceptable catastrophic event likelihood? Is a single fatality acceptable, if the event can occur in a one in one million chance? This risk assessment activity can be conducted during a system safety working group effort within a safety review process. The point to be made here is that a simplistic assumption, which is based on a single hazard or risk control (redundancy and monitoring), may be oversimplistic.

3.6.7.1 *Redundancy*   Proving true redundancy is not cut and dry in complex systems. It may be possible to design a hardware subsystem and show redundancy, that is, redundant flight control cables, redundant hydraulic lines, and redundant piping. When there are complex load paths, complex microprocessors, and software, true independence can be questioned. The load paths, microprocessors, and software must also be independent. Ideally, different independent designs should be developed for each redundant leg.

The concepts of redundancy management should be appropriately applied.[4] Separate microprocessors and software should be independently developed. Boeing 777 flight controls have triple redundancy in hardware as well as three different versions of software. Single-point failures should be eliminated if there are common connections between redundant legs. The switch-over control to accommodate redundancy transfer should also be redundant. System safety should be concerned with the potential loss of transfer capability due to a single common event. Common events can eliminate redundancy. The use of similar hardware and software presents additional risks, which can result in loss of redundancy. A less than adequate process, material selection, common error in assembly, material

---

[4]Redundancy management requirements were developed for initial space station designs.

degradation, quality control, inappropriate stress testing, calculation assumption—all these can present latent risks that can result in common events. A general rule in system safety states that the system is not redundant unless the state of the backup lag is known and the transfer is truly independent, available when needed, and reliable.

Physical location is another important element when evaluating independence and redundancy. Appropriate techniques of separation, protection, and isolation are important. In conducting common cause analysis (a technique described in the *System Safety Analysis Handbook*),[5] not only is the failure state evaluated but possible common events are also part of the equation. The analyst identifies the accident sequence in which common contributory events are possible due to physical relationships.

Other analysis techniques also address location relationships, for example, vicinity analysis and zonal analysis. The analyst determines what the possible outcome is is should a common event occur and how this can affect all lags of redundancy simultaneously, for example, a major fire within a particular fire division, an earthquake causing common damage, fuel leakage in an equipment bay of an aircraft, or an aircraft crash into a hazardous location.

The designers of the *Titanic* considered compartmentalization for watertight construction. However, they failed to consider latent common design flaws, such as defects in the steel plating, the state of knowledge of the steel manufacturing process, or the effects of cold water on steel.

3.6.7.2 Monitoring

Monitoring devices can be incorporated into the design to make sure that conditions do not reach dangerous levels. Monitors can be used to indicate [16, p. 262]:

- Whether or not a specific condition exists. If the indication is erroneous contributory hazards can result.
- Whether the system is ready for operation or is operating satisfactorily as programmed. An inappropriate "ready indication" or inappropriate "satisfactory indication" can be a problem from a safety view.
- If a required input is provided. An erroneous input indication can cause errors and contributory hazards.
- If the output is being generated

3.6.7.3 Concepts of Probability

A probability is the expectancy that an event will occur a certain number of times in a specific number of trials. Probabilities provide the foundations for numerous disciplines, scientific

---

[5]*System Safety Analysis Handbook*, 2nd edition, System Safety Society, 1997, pp. 3-37 and 3-38.

methodologies, and risk evaluations. Probability is appropriate in reliability, statistical analysis, maintainability, quality, and system effectiveness.

Over time, the need for numerical evaluations of safety has generated an increase in the use of probabilities. In 1972, Hammer [16, pp. 91–92] expressed concerns and objections about the use of quantitative analysis to determine probability of an accident. These concerns and objections are based on many reasons:

- A probability, such as reliability, is not a 100% guarantee. Actually, a probability indicates that a failure, error, or accident is possible, even though it may occur rarely over a period of time or during a considerable number of operations. Unfortunately, a probability cannot indicate exactly when, or where, during which operation, or to which person an accident will occur.
- It is ethically unjustifiable to permit a hazard to exist unless maximum effort is applied to eliminate it, control it, or limit any damage that it could possibly produce, no matter how high the probabilistic safety level. The loss of the *Titanic*, considered the safest ship in the world at the time, can be blamed on the fact that hazards were ignored because the ship was considered safe.
- Probabilities are projections determined from statistics obtained from past experience and future predictions. Although equipment to be used in automated operations may be exactly the same as that with which the statistics were obtained, the circumstances under which it will be operated will probably be different. In addition, variations in production, maintenance, handling, and similar processes generally preclude two or more pieces of equipment being exactly alike. There are numerous instances in which minor changes in methods to produce a component with the same or improved design characteristics as previous items have instead caused failures and accidents. If an accident has occurred, correction of the cause by change in the design, material, code, procedures, or production process may immediately invalidate certain data.
- Reliability is the probability of successful accomplishment of a mission within prescribed parameters over a specific period of time. It may become necessary to operate the system outside those prescribed parameters and time limits, as in an emergency. A catastrophic event can inadvertently eliminate redundancy, or monitors, or alarms—all considerations for reliability calculations.
- Human error can have damaging effects even when equipment or system reliability has not been lessened. A common example is a weapon system. It is highly reliable, but people have been killed or wounded when maintaining and testing weapons.
- Probabilities are predicated on an infinite or large number of trials. Probabilities, such as reliabilities, for complex systems may be based on very small samples that result in low confidence levels. What distribution do we trust our lives with?

3.6.7.4 Familiarization to Automation

Fortunately, humans usually try to familiarize themselves with automation prior to its use. Depending on the complexity of the system, familiarity will take resources, time, experience, training, and knowledge. Automation has become so complex that knowledge of system design and operation has turned out to be only possible by integration conducted by a system committee.[6] Specialists are needed in operations, systems engineering, human factors, system design, training, maintainability, reliability, quality, logistics, automation, electronics, software, network communication, avionics, and hardware. Detailed instruction manuals, usually with cautions and warnings, in appropriate language, are required. Simulation training may also be required.

The interaction of the human and machine if inappropriate can also introduce additional risks. The human can become overloaded and stressed due to inappropriately displayed data, an inappropriate control input, or similar erroneous interface. The operator may not fully understand the automation, due to its complexity. It may not be possible to understand a particular system state. The human may not be able to determine if the system is operating properly, or if malfunctions have occurred.

Envision relying on an automated system and, due to malfunction or inappropriate function, artificial indications are displayed and the system is inappropriately communicating. In this case, the human may react to a false situation. The condition can be compounded during an emergency and the end result can be catastrophic. Consider an automated reality providing an artificial world and the human reacts to such an environment. Should we trust what the machines tell us in all cases?

The concept of undertrust or overtrust of a system is the subject of human factors research [18, pp. 148–149]. There is a need for balance between having enough trust in an automated system to make effective use of the advice provided and having enough distrust to stay alert and know when to ignore the advice.

There are parts of this equation concerning human and contingency, which can accommodate system risks. Once the system has become unbalanced due to a failure, malfunction, or human error, the human and system must respond quickly and appropriately. An operator may not have time to make a phone call to the system design committee to figure out what is going on in an artificial environment. These risks can also be designed out or controlled to an acceptable level. Contingency design, planning, simulation, and training must be part of the efforts—appropriate loss, risk, and damage control.

3.6.7.5 *Reliable Software and Safety Considerations* Software reliability is the probability that software will perform its assigned function under specified conditions[7] for a given period of time [16, p. 262]. To develop reliable or safer software, many concerns and observations should be addressed:

- Software does not degrade over time.
- Since software is the written word, it does not fail in the form but can fail functionally.

[6]The complications posed by automation in regard to system safety were discussed in the following: M. Allocco, Automation, System Risks and System Accidents. In: *Proceedings of the 17th International System Safety Conference*, System Safety Society, August 1999.
[7]The list indicated was first discussed in the following: M. Allocco, Computer and Software Safety Considerations in Support of System Hazard Analysis. In: *Proceedings of the 21st International System Safety Conference*, System Safety Society, August 2003.

- The testing of software is not an all-inclusive answer to solve all potential software-related risks. Software applied within an air bag system will have millions of combinations and therefore will be almost impossible to test completely.
- Software will not get better over time. Usually it gets worse over time because of inadequately implemented engineering modifications.
- Software can be overly complex.
- Systems can be overly complex.
- There should be concern with how the human designs such vast complex systems.
- Humans are the weak link in complex systems since they can make many unpredictable errors.
- Humans can be and often are unreliable.
- If designers do not consider system risks, accidents will occur.
- Cookbook and generic approaches do not work when there are system accidents and system risks to consider.
- It is not possible to totally segregate software, hardware, humans, and the environment, in the system.
- It may not be possible to determine what went wrong, what failed, or what broke.
- The system does not have to break for a system accident to occur.
- Planned functions can be hazards under certain circumstances.
- Software functions can be inadequate or inappropriate.
- It is unlikely that you can change only part of the software and not affect system risk.
- If you change the application just a little, you will change the risk.
- Software is not universal in all circumstances.
- The system can be spoofed or jammed.
- One single error can propagate throughout the complex system.
- Any software, no matter how apparently inconsequential, can cause hazards. Consider a process tool, automated calculations, automated design tools, and safety systems.
- It is very hard to appropriately segregate safety-critical software in open, loosely coupled systems.
- Combinations of contributory events (hazards) can have catastrophic results.

It appears at times that specialty engineers have different objectives (for a particular discipline or field of practice). Designers should not lose sight of what should be the main goal—to apply science for the benefit of humankind. The objective should be accomplished without harm or needless risk.

3.6.7.6 Reliability Analyses and Safety Applications When evaluating complex systems of hardware, automation, and software, it is important to have an understanding of how failures may propagate from a latent design error that could result in failures, which are considered initiating hazards. Failure-mode, effects and criticality analysis (FMECA) is a powerful and universally used reliability technique. It depicts error and failure propagation. These methods are inductive in nature and are used in system and subsystem failure analysis. FMECA attempts to predict possible sequences of events that lead to subsystem and system failure, to determine their consequences, and to devise methods to minimize their occurrence or reoccurrence.

General FMECA Approach The FMECA procedure consists of a sequence of steps starting with the analysis at one level or a combination of levels of defined subsystem or system abstractions, such as functions, operations, or architectural depictions. The analysis may be conducted at system, subsystem, component, element, or device levels.

The abstraction is described in a manner that allows the analyses to be performed logically. In other words, the analysis should be conducted at the same level as the supportive description. Functional, operational, or architectural block diagrams can represent the descriptions that describe the system.

FMECA can also be conducted from constructed reliability block diagrams (RBDs), which are used to model the effect of item failures on system performance. They often correspond to the physical arrangement of items in the system. Reliability block diagrams show system configurations: series, parallel, complex series/parallel, and standby redundant systems. The analyst systematically follows logical flow through the diagrams to identify the failure modes and causes. The manner of failure of the function, subsystem, component, or part is identified.

Integration Considerations Between FMECA and Hazard Analysis (HA)
System failures that present system hazards should be identified within the hazard analysis—all unsafe conditions, initiators, or contributors. In some cases, the system safety efforts are conducted simultaneously with reliability and both analyses will identify these hazards. A FMECA cross-check analysis should be conducted to ensure consistency between reliability and safety efforts and to validate hazard identification and control.

Consider an analysis of a particular subsystem, for example, an automated brake system for an automobile or train. It is important to determine the hardware-related failures that may prevent operation of the brakes when required. It is apparent that this failure is a hazard, so consequently a FMECA on this subsystem is needed. The hazard analysis will further expand considerations addressing human error, software malfunction, logic error, decision errors, and design errors.

Remember that the safety engineer is usually concerned with hazards and the reliability engineer addresses failures. Failures do not automatically result in unsafe conditions. The system could be designed to fail-safe rather than present accidents. Within the hazard analysis, reference should be made to related analyses such as FMECA. There also should be consistency between hazard scenario severity and reliability criticality. The elements of a potential catastrophic scenario should equate to catastrophic criticality. Disconnects between analyses will present problems.

Critical Items List   The safety-critical failures developed in the FMECA can be presented within a critical items list (CIL) (see Fig. 3.20). When critical items fail, there are safety-related consequences. Certain components, parts, and assemblies are critical from a system safety view and must be given special care and attention. Criteria for critical items should consider:

- Single-point failure items whose loss or malfunction could contribute to an accident
- Hazard controls, safety devices, guards, interlocks, and automated safety controls
- Materials and components that are considered dangerous, such as radioactive materials, toxic materials, and explosives
- Items that are considered hazardous, special materials, and critical parts

Criticality Ranking   Criticality ranking may be used to determine the following:

- Which items should be given more intensive study or analysis for elimination of a hazard that could cause the harm
- Which items require special attention throughout the life cycle, require stringent quality control, and need protective handling

Critical Items List					
System: Date: Analyst:					
Item	Description of Item	Mode of Failure	Failure Probability	Effect on System	Criticality Ranking Within Subsystem (0 to 10)
Item name	Describe the item; provide part number, detailed drawing number	List possible failure modes	List probability for each failure mode	Describe effects should failure occur, (i.e., abort, degradation of performance, or damage)	Indicate ranking

Figure 3.20   Example of CIL.

- Special requirements to be included in specification for suppliers concerning design, subsystem hazard analysis, performance, reliability, safety, or quality
- Acceptance standards to be established for components received at a plant from subcontractors and for parameters that should be tested most intensively
- Where special procedures, tasks, safeguards, protective equipment, monitoring devices, or warning systems should be provided
- Special handling, transportation, packaging, marking, or labeling
- Where safety efforts and resources could be applied most effectively

Integration Considerations Between CIL and Hazard Analysis   Safety-critical items can be identified during hazard analysis. Hazard scenarios are to be developed to address single-point failures that are initiators or contributors. The analysis will identify hazard controls, safety devices, guards, interlocks, and automated safety controls that will be considered safety-critical. Furthermore, consider items that are hazardous, special materials, critical parts, materials and components that are extremely dangerous, radioactive materials, toxic materials, and explosives.

If the CIL has been developed from a FMEA/FMECA, cross-checking can be conducted between hazard scenarios identified and safety-critical items. All risks associated with safety-critical items must be identified.

Energy Analysis (EA)   One concept of system safety is based on the idea that unwanted energy flow (or uncontrolled energy) presents unsafe conditions, which can be initiators and contributors within system accidents. At such times, energy is transferred or lost in uncontrolled, undesirable manners. System safety can therefore be evaluated and improved by analyzing the following:

- Sources of available energy existing in a system or subsystem or in its environment

		Energy Trace and Barrier Analysis		Block Diagram	
System: Subsystem: System State: Date: Analyst:					
Energy Source, Quantity, and Location	Initiator and Contributors (Barrier breakdown, Less than adequate barriers)	Hazard Scenario (Energy contacts target and results in harm)	Initial Risk	Controls Precautions and Recommendations	Residual Risk

Figure 3.21   Example of Energy Trace and Barrier Analysis Worksheet.

- Means of reducing and controlling the level of energy
- Means of controlling the flow of energy
- Methods of absorbing free energy to prevent or minimize damage should loss of control occur

Flow Analysis Approach   The approach for analyzing unwanted energy transfer or liquid flow associated with a particular system may be accomplished as follows:

- Review all fluid and energy within the system under consideration for initiators or contributors.
- Develop hazard scenario themes associated with potential adverse fluid dynamics and chemical reactions.
- Determine whether the adverse flow could affect the surroundings or other equipment with which it may come in contact.
- Determine if any incompatibility would occur.
- Review the proximity and relationship between lines, containers, cables, and equipment containing incompatible fluids/energy.
- Establish the level of leakage or seepage that could constitute a problem and the allowable level.
- Provide engineering and administrative controls to eliminate or minimize accident risk due to leakage, incompatibility, flammability, explosion, and material degradation.
- Refine hazard controls into program safety requirements.
- Validate and verify hazard controls.

Bathtub Curve   Energy will affect systems and they will constantly degrade until they fail. In evaluating the life of a system, the hazard rate function is plotted against time. (Do not confuse hazard rate function with hazard; not all failures are considered hazards and not all hazards equate to failures.) Because of its shape, the plot is commonly referred to as the bathtub curve. Systems having this hazard rate function experience decreasing failure rates early in their life cycle (infant mortality), followed by a nearly constant failure rate (useful life), followed by an increasing failure rate (wear-out), (See Fig 3.1, page 42).

Energy-Induced Failure Mechanisms and Initiating Hazards   Specific examples of uncontrolled energy-induced failure mechanisms[8] are:

---

[8]For additional information concerning the physics of failure, consult the following: C. E. Ebrling, *An Introduction to Reliability and Maintainability Engineering*, McGraw-Hill, New York, 1997, pp. 124–141.

- Mechanical stress. Excessive or continued vibration may cause cracking of integrated circuits, metal supports, and composite materials.
- Substandard or defective parts. Parts may have been physically damaged due to poor handling, poor design, and inappropriate material selection.
- Fatigue. Physical changes in material may result in fracture of metals or composites.
- Friction. Movement between materials may cause excessive heat and deformation.
- Contamination. Foreign materials may be introduced within a safety-critical structure, subsystem, system, component, or part.
- Evaporation. Filaments age because of filament molecules evaporating and flammables outgassing and vaporizing.
- Aging and wear-out. Material degradation results from prolonged exposures or use.
- Temperature cycling. Repeated expansion and contraction will weaken materials.
- Operator- or maintenance-induced error. The human adversely causes uncontrolled energy degradation, fracture cracking, poor processing, welding, temperature damage, physical damage, and inconsistent applications.
- Corrosion. Chemical changes result in weakened material.
- Abnormal stress. Stress is excessively applied externally or environmentally such as a power surge or water hammer effect.

Material Properties and Hazards Uncontrolled energy can have adverse effects on materials, which can degrade and result in hazards. Knowledge of material properties and the external stresses to which the system will be exposed is important.

- Fatigue life is the number of cycles until failure for a part subjected to repeated stresses over an extended period of time. The fatigue strength is generally less than would be observed under a static load.
- Tensile strength is the ability to withstand a tensile or compressive load. Material will deform first elastically and then plastically. The deformation is elastic if after the load has been removed, the material returns to its original shape. When a material is stressed beyond its elastic limit, a permanent, or plastic, deformation occurs.
- Impact value is a measure of the toughness of the material under sudden impact. Toughness refers to the amount of energy absorbed before fracture.
- Hardness is the resistance of material to the penetration of an indenture. Hardness measurements are useful in analyzing service wear of material.

- Creep is the progressive deformation of material under a constant stress. Creep is a design consideration when the component will be operating at moderate or high temperatures.

Integration Considerations Between Energy Analysis and Hazard Analysis

Failure mechanisms cause failures, which then result in unsafe conditions—hazards, initiators, or contributors. If hazard scenarios have been developed within the energy trace and barrier analysis, they should be integrated into the system-level hazard analysis. However, if single hazards have been identified, they have to be integrated into specific hazard scenarios.

3.7 TOPICS FOR STUDENT PROJECTS AND THESES

1. Explore why industry and universities use an exponential model for reliability predominantly when practically no system has demonstrated exponential probability. Many distributions are used to describe failure probability. (*Note*: This represents opposing views of the authors' opinions drawn from many years of experience. It is recognized that there are complementary views; readers should draw their own conclusions.)

2. Most popular standard sampling plans for reliability are based on exponential distribution (e.g., Mil-Std-781, Mil-Hdbk-H108). If real distributions in the field are not exponential (see Ref. 19), should industry still use the standard plans? Support your argument.

3. Does the bathtub profile seem to apply to every product, process, or system? Are there products, processes, or systems where it does not apply? What types of distributions would you expect in the first region—infant mortality?

4. Can you combine the three regions of the bathtub profile into one model? If so, what model would you propose?

5. Each distribution on a Weibull graph has a unique slope. Will the same line represent two normal distributions with identical means but different standard deviations? If so, how can you differentiate the two distributions on the same graph?

6. Why does industry use MTBF for mechanical products when it is known that the failure rate is not constant? Can you use MTBF if the failure rate is not constant?

7. Explain the differences and similarities between reliability engineering and safety engineering.

8. Select a sample complex system and develop a system description addressing the system functions, operations, and architecture. Develop a reliability block diagram. Use the material to conduct a system-level FMECA. Describe how the FMECA would be used in support of hazard analysis.

9. Describe when it is appropriate and when it is not appropriate to apply probability calculations in support of systems analysis.

10. Select a sample complex system and develop a system description addressing the system functions, operations, and architecture. Conduct a subsystem FMECA and make a CIL. Discuss how the items in the CIL would be used as input to conduct hazard analysis.

REFERENCES

1. M. B. Darch, Chapter 3, in *Reliability and Maintainability of Electronic Systems*, J. E. Arsenault, and J. A. Roberts (Eds.), Computer Science Press, Potomac, MD, 1980.

2. *USAF R&M 2000 Process, Handbook SAF/AQ*, U.S. Air Force, Washington, DC, 1987.

3. *Reliability Prediction of Electronic Equipment*, Mil-Hdbk-217.

4. *Specification for Capacitor, fixed, Tentalum*, Mil-C-39003.

5. *Handbook of Reliability Prediction Procedures for Mechanical Equipment*, Eagle Technology for Logistics Division, David Taylor Research Center, Carderock, MD, and Product Assurance and Engineering Directorate, Belvoir Research Development and Engineering Center, Fort Belvoir, VA, 1995.

6. *Procedures for Performing a Failure Mode, Effects, and Criticality Analysis*, Mil-Std-1629.

7. D. Raheja, Failure Mode and Effects Analysis—Uses and Misuses, *ASQC Quality Congress Transactions*, 1981.

8. J. T. Duane, Learning Curve Approach to Reliability Monitoring. *IEEE Transactions on Aerospace*, April 1964.

9. *Reliability Growth Management*, Mil-Hdbk-189, 1988.

10. C. R. Lipson, and N. Sheth, *Statistical Design and Analysis of Engineering Experiments*, McGraw-Hill, New York, 1979.

11. P. D. T. O'Connor, *Practical Reliability Engineering*, John Wiley & Sons, Hoboken, NJ, 1986.

12. *Low Cycle Fatigue*, Special Technical Publication 942, ASTM, Philadelphia, 1988.

13. *Standard General Requirements for Electronics Equipment*, Mil-Std-454, Naval Publications and Forms Center, Philadelphia, 1984.

14. *Sampling Procedures and Tables for Life and Reliability Testing*, Mil-Hdbk-H108, Naval Publications and Forms Center, Philadelphia, 1960.

15. *Reliability Testing for Engineering Development, Qualification, and Production*, Mil-Std-781D, 1996.

16. Dev G. Raheja, *Assurance Technologies—Principles and Practices*, McGraw-Hill, New York, 1991.

17. Willie Hammer, *Handbook of System and Product Safety*, Prentice-Hall, Englewood Cliffs, NJ, 1972.

18. T. B. Sheridan, *Humans and Automation: System Design and Research Issues*, John Wiley & Sons, Hoboken, NJ, 2002.

19. K. Wong, The Roller-Coaster Curve Is In, *Quality and Reliability International*, January–March 1989.

FURTHER READING

Arsenault, J. E., and J. A. Roberts, *Reliability and Maintainability of Electronic Systems*, Computer Science Press, Potomac, MD, 1980.

Automated Electronics Reliability Handbook, Publication AE-9, SAE, Warrandale, PA, 1987.

Bazovsky, I., *Reliability Theory and Practice*, Prentice-Hall, Englewood Cliffs, NJ, 1961.

Carter, A. D. S., *Mechanical Reliability*, Macmillan, New York, 1986.

Definition of Terms for Reliability and Maintainability, Mil-Std-721; *Reliability Design Qualification and Production Acceptance Tests*, Mil-Std-781; *Reliability Program for Systems and Equipment Development and Production*, Mil-Std-785.

Fuqua, N. B., *Reliability Engineering for Electronic Design*, Marcel Dekker, New York, 1987.

Ireson, G. W., and C. F. Coombs, *Handbook of Reliability Engineering and Management*, McGraw-Hill, New York, 1988.

Jensen, Finn, and N. E. Petersen, *Burn-In*, John Wiley & Sons, Hoboken, NJ, 1982.

Johnson, L. G., *Theory and Technique of Variation Research*, Elsevier, Amsterdam, 1964.

Johnson, L. G., *The Statistical Treatment of Fatigue Experiments*, Elsevier, Amsterdam, 1964.

Kapur, K. C., and L. R. Lamberson, *Reliability in Engineering Design*, John Wiley & Sons, Hoboken, NJ, 1977.

Lewis, E. E., *Introduction to Reliability Engineering*, John Wiley & Sons, Hoboken, NJ, 1987.

Lipson, C., and N. J. Sheth, *Statistical Design and Analysis of Engineering Experiments*, McGraw-Hill, New York, 1979.

Murphy, R. W., *Endurance Testing of Heavy Duty Vehicles*, Reliability Assurance Program, Mil-Std-790, Publication SP-506, SAE, Warrandale, PA, 1982.

Nelson, W., *Applied Life Data Analysis*, Wiley-Interscience, New York, 1982.

Shooman, M., *Probabilistic Reliability: An Engineering Approach*, McGraw-Hill, New York, 1968.

Singpurwalla, N. D., Inference from Accelerated Life Tests and When Observations Are Obtained from Censored Samples, *Technometrics*, February 1971, pp. 161–170.

CHAPTER 4

MAINTAINABILITY ENGINEERING AND SAFETY-RELATED APPLICATIONS

4.1 MAINTAINABILITY ENGINEERING PRINCIPLES

Reliability deals with reducing the frequency of breakdowns. *Maintainability* deals with the duration of breakdowns. Many practitioners as well as academicians think maintainability refers to easy repair and maintenance. But the aim of maintainability is to prevent the occurrence of downtime and prevent the need for maintenance altogether. Only when this is not feasible (because of limited-life components) does the goal become to minimize downtime. Maintainability includes prognostics design in which sensors and software are used for predicting remaining life in a component and warning the user to check the condition prior to failure.

A report released by the U.S. Government Accounting Office in 1989 revealed dismal statistics on maintenance costs. The government paid large sums for maintenance, which was supposed to be covered by warranties. During a 12-month warranty period, the Navy reported 251 failures on antimissile systems, but the contractor paid none of the repair costs. The warranty stated that the company was responsible for repairs only if the Navy experienced more than 5238 failures. In 1987, the Army reported that it paid $9.9 million in warranties for engines on its M-1 tanks but was reimbursed for only $10,453 worth of claims. The contractor refused to pay many claims because they were not filed within the required 90 days. Such examples abound not only in government but also in private industry. One of the big reasons is lack of organization for tracking warranty agreements. These examples show that the customer often pays twice for corrective maintenance,

Assurance Technologies Principles and Practices: A Product, Process, and System Safety Perspective, Second Edition, by Dev G. Raheja and Michael Allocco
Copyright © 2006 John Wiley & Sons, Inc.

at least during the warranty period. The cost of the associated downtime is huge and unfortunately does not even get quantified. Sooner or later, these costs become a big burden. The solution here is to increase investment in maintainability, not in maintenance.

Maintainability is mainly a design function; most of the effort should be expended in the design phase. It should be part of every design review. One cannot emphasize enough that maintainability has tremendous influence on maintenance and other support costs, which are usually the largest segment of life-cycle costs.

Maintainability, according to Mil-Hdbk-472 [1], is the ability of an item to be retained in or restored to specific conditions when maintenance action is performed by personnel having specified skill levels and using prescribed procedures and resources at each prescribed level of maintenance and repair.

This definition does not emphasize design enough. As a result, many design engineers have diverted their attention from minimizing maintenance requirements, often to the point where maintenance becomes extremely difficult. In a large manufacturing machine, for example, it was necessary to inspect the condition of a gear periodically. The preventive maintenance procedure took 8–16 hours because the technician had to open about 30 bolts, then call an electrician (as required by union rules) to remove electrical connections, then remove several parts before obtaining a good view of the gear. With maintainability analysis, the design was improved to eliminate all such extra work. The new equipment had a camera installed inside, and the gear was visible much more clearly on a television screen. Downtime due to routine inspection: zero hours!

With this example in mind, it is time to redefine maintainability. The definition these authors have used for years is: *maintainability is the science of minimizing the need for maintenance and minimizing the downtime if maintenance action is necessary*. This definition emphasizes the aggressive role of the designer in reducing the operating and support costs. It is up to the reader to decide which definition to choose. In either case, the maintainability objectives can be met.

Maintainability itself is not quantified, since it depends on qualitative parameters "using prescribed procedures and resources." The argument is that rarely are the prescribed procedures adequate, and the resources such as spares and trained personnel are not available all the time. Three basic indexes of maintainability are:

- Equipment repair time
- Inherent availability
- Mean downtime for maintenance

Equipment repair time (ERT) is the time an average technician will take to repair an item. It is the time from preparing for repair to testing to make sure the equipment is working right. Mil-Hdbk-472 contains the detailed listing of tasks. Equipment repair time does not include waiting time and logistic delays.

ERT is probabilistic in nature. Some technicians will make repairs quickly while others will be slow. If the majority of technicians are highly trained or not trained enough, then the distribution is skewed. One can use the median value or the following rules of thumb [2]: if the time-to-repair distribution is symmetrical and bell-shaped, the typical ERT is equal to the arithmetic average:

$$ERT = \text{mean time to repair} = MTTR$$

If the time-to-repair distribution is exponential,

$$ERT = 0.69 \, MTTR$$

If the time to repair has a log normal distribution, as is usually the case,

$$ERT = 0.45 \, MTTR$$

The second measure of maintainability, *inherent availability*, actually is a measure of maintainability and reliability together. It is the uptime ratio designed into the system and can be defined as

$$A_i = \frac{\text{uptime}}{\text{total scheduled time}} = \frac{\text{uptime}}{\text{uptime} + \text{active downtime}}$$

Most defense contracts assume that uptime is proportional to MTBF and active downtime is proportional to MTTR. The inherent availability then becomes

$$A_i = \frac{MTBF}{MTBF + MTTR}$$

This definition, even though used industry-wide, has several flaws. MTBF, the mean time from one failure to another, may include logistic delays as well as downtime. Generally, the records for computing MTBF are not adequate. Chapter 3 shows some problems in using the MTBF.

The definition of inherent availability is the ideal case. It assumes the spares are available all the time and repair is done immediately. It does not include waiting or logistic delay time because the design engineer has no way of predicting them, unless some complex mathematical modeling is done. The design engineer can predict MTTR on the assumption that the spares and the trained personnel are available. There is, however, another side to the coin. The customer is concerned about actual availability, called *operational availability*, in which waiting time and delays are a rule rather than the exception. A design engineer must also try to maximize operational availability if possible. Operational availability can be defined as

$$A_o = \frac{\text{uptime}}{\text{uptime} + \text{active repair time} + \text{mean logistic downtime}}$$
$$= \frac{\text{MTBF}}{\text{MTBF} + \text{MTTR} + \text{MLDT}}$$

Quantifying the terms in this equation requires a planned data collection effort that very few are willing to mount. If one wants to know the actual availability, then accurate logs of uptime or downtime must be kept. Some companies have running clock-hour logs built into their equipment. They can then compute operational availability, which may be redefined as

$$A_o = \frac{\text{uptime}}{\text{total scheduled time}} = \frac{\text{scheduled time} - \text{downtime}}{\text{scheduled time}}$$

The third measure of maintainability, *mean downtime for maintenance* (MDT), is the mean downtime spent on corrective as well as preventive maintenance:

$$\text{MDT} = \frac{F_c M_{ct} + F_p M_{pt}}{F_e + F_p}$$

where F_c = frequency of corrective maintenance, F_p = frequency of preventive maintenance, M_{ct} = average time for corrective maintenance, and M_{pt} = average time for preventive maintenance.

4.2 MAINTAINABILITY DURING THE DESIGN PHASE

4.2.1 Developing Maintainability Specifications

Typical maintainability requirements such as equipment repair time, inherent availability, and mean downtime are not sufficient for specifying maintainability for complex equipment, which today contains complex software and microprocessors. There are many false fault alarms—as much as 50% false in some cases. No longer can a technician isolate a fault without very expensive automated test equipment. Therefore, fault diagnosis and fault isolation capability must be designed into the product. This will permit maintenance to be performed quickly.

If the equipment is designed with backup features, maintenance can be performed during the off shifts, without inconvenience to users. Such equipment forgives failures from many causes, but requires more maintenance on backup components.

Mil-Hdbk-338 [2] suggests MTTR computed from the log normal distribution (if applicable) as the primary measurement for maintainability. In addition, the following indexes may be used:

- Maximum corrective maintenance time at a specified percentile, such as the 90th or 95th percentile.
- Percent faults that can be isolated to a single replaceable item.
- Mean maintenance worker-hours per repair.
- MTBM—mean time between maintenance.
- MTBM(C)—mean time between corrective maintenance (corrective maintenance may not be related to a component failure).
- MTBM(P)—mean time between preventive maintenance.
- MTBA—mean time between adjustments.
- %NFF—percent no faults found (e.g., applies to a circuit board that is found to be functional but did not work in a system because of a temporary fault; it may not respond at system level because of vibration, shock, etc.). Sometimes these faults are as much as 50% of total faults.
- %CND—percent faults that cannot be duplicated at the repair facility.
- %BCS—percent faults bench check serviceable. These are faults checkable at the depot.
- Testability index—(1) percent digital "stuck-at" 0 or 1 faults testable for microprocessors; (2) a quality rating of the ability to make testing easier for any equipment. A comprehensive checklist for a testability rating system is given in Section 4.2.9.
- Mean downtime for software errors.
- %DTS—percent downtime due to software errors.
- %DTH—percent downtime due to hardware failures.
- %DTS—percent downtime from system failures (both hardware and software faults count as failures).
- Safe life for safety-critical component replacement.
- Safe life for calibration of safety devices.
- Safe life for damage tolerance (crack growth).

It is advisable that goals for these items be included in the specification as soon as possible. On the basis of these goals, the maintainability program plan is developed; this is the master list of tasks and milestones to be monitored during the entire life cycle.

4.2.2 Design Review for Maintainability

Maintainability review starts in preliminary design. Before a design is approved, the maintainability features should be established. Especially to be considered are:

How much downtime is acceptable? (Specify availability.)

At what level will the repair be made? At component level? At circuit board level? At black box level?

Who will do the repair? The contractor? The customer? A technician? An engineer? Or should the failed product be thrown away? (The answers will affect the levels of skill required.)

Does the equipment have to be designed for inspectability?

Should the equipment be designed for ease of accessibility?

Can any built-in test features be added to reduce downtime?

Is modular design helpful? For hardware? For software?

Is fault-diagnostic software helpful?

Is there a need to isolate the faults?

What kinds of skill do the repair technicians require?

What kinds of inspection and test methods are required for crack detection (aircraft, nuclear plants)?

What are the provisions for replacing safety-related components?

What are the allowable crack sizes?

What are the corrosion-prevention measures?

What kinds of maintainability tests are required?

Are components easy to remove and replace?

Answers to such questions as these determine the design concepts for maintainability. All questions are aimed at reducing or eliminating downtime.

It is very important at this stage to study the weaknesses and the strengths of competitors. A dramatic example shows why. A very famous North American maker of business machines lost its entire 90% market share of desktop copiers to Canon, a Japanese company. The Canon copier did not require preventive maintenance, while the North American copier required a $1200 per year maintenance contract. Besides, the Canon copier sold for only $1100 in 1981 while the North American manufacturer charged $3000.

One of the authors bought the Canon copier and used it for more than 20 years without any downtime or preventive maintenance, except for changing the ink cartridge (which took less than a minute). The copier was still in excellent condition at the time of this writing: that's maintainability—a practically maintenance-free design! Granted this is one example; the fact that Canon increased its market share from 9% to 90% indicates that other customers had similar experiences.

If competitors' equipment requires much maintenance and many spares, maintainability by design is a newcomer's opportunity to offer a better value. The manufacturer that thrives on selling spare parts will make profits in the short term but market death is almost certain in the long run, as the fate of the formerly dominant desktop copier manufacturer testifies.

4.2.3 Maintainability Analysis

As design progresses, it becomes important to analyze such requirements as human factors, hierarchy of repair, diagnostics, fault isolation, and testability. For example, consider the level of repair. There are at least seven options:

Repair at component level

Repair at subassembly level (e.g., circuit board level)

Repair at assembly or black box level

Replace at subassembly level and repair later

Throw away the subassembly

Throw away the assembly

Throw away the entire product

Maintainability analysis may include decisions such as preventive maintenance hours required per operating hour and the cost of downtime. There is a right way to approach the design. First, decide how you can do better than the competition and be able to market maintainability as a selling feature. Second, evaluate the life-cycle costs for each option. If the life-cycle costs are too complex to estimate, one can compare the option on some meaningful measure such as cost per circuit mile for a power company or cost per million instructions for a computer network. The final decision should be such that it benefits both the supplier and the customer.

4.2.4 FMECA for Maintainability

One of the best-known tools for maintainability analyses is failure-mode, effects, and criticality analysis. Although FMECA was originally applied to reliability analysis, it can easily be tailored to maintainability by assessing the criticality of a failure in terms of downtime costs. The goal is to minimize the risk of unavailability. The desirable corrective action therefore is to modify the design so it requires the least amount of downtime.

Figure 4.1 shows a portion of a maintainability FMECA for a large electromechanical system. As an example, consider the failure modes for pipe in this figure. Two possible causes of failure are corrosion and cracks in the solder at a joint. The best way to prevent downtime from these causes is to eliminate them. If this is not feasible, one can design for ease of maintenance. Changing the pipe material from copper to stainless steel can eliminate failure from corrosion. Substituting a flexible joint for the stiff one can eliminate failure at the solder joint; a flexible joint can withstand much larger structural loads.

Unfortunately, most design engineers do not perform FMECA for maintainability. Even if they do, their corrective action usually is to perform more preventive maintenance instead of improving the design. This is usually costly to the user. If

Part name/ part number	Potential failure modes	Causes (failure mechanism)	Effects	Expected downtime	Frequency per year	Recommended improvement	Maintenance requirements
Pipe	Leakage in pipe	1. Corrosion	Loss of Freon	8 h	2	Use stainless steel pipe	None
		2. Temperature cycling	Loss of Freon	8 h	5	Monitor temperature with thermocouples	Replace probes every 120 days
	Leakage at the joint	1. Cumulative fatigue	Loss of Freon	14 h	4	Monitor vibration with accelerometers	Calibrate accelerometers semiannually
		2. Poor soldering	Loss of Freon	2 h	2	Provide flexible plaster coupling at the joint	None
Valve	Sticky, intermittent	1. Dirt or foreign objects	Loss of control of temperature	5 h	2	Electronic redundant valve action and fault identification	Clean every 15 days
	Stuck open	1. Component wearout	Loss of control of temperature	4 h	1	Electronic redundant valve action and fault identification	Check every 3 months
	Stuck closed	1. Component failed	Loss of control of temperature	4 h	2	Electronic redundant valve action and fault identification	Check every 3 months
		2. Component expansion and contraction	Loss of control of temperature	5 h	3	Electronic redundant valve action and fault identification	None

Figure 4.1   An example of FMECA for maintainability.

users have to pay for the high cost of downtime, they can take their business to a better supplier.

One reason maintainability has not received proper recognition is that project managers readily see the increased design cost it incurs but fail to see the benefits. Increases in customer satisfaction and in market share are rarely perceived as major components of the life-cycle cost model. There is yet another roadblock. The engineer is heavily criticized for increasing product design costs and rarely credited for the tenfold reduction in support costs that maintainability brings.

Remember that any product that can advertise reduced downtime is a highly desirable product. Most customers are willing to pay extra if they are convinced it means less maintenance; Sony televisions, Honda Accord cars, and IBM 3061 computers attest to this fact.

4.2.5 Maintainability Prediction

Maintainability prediction is done for the indexes chosen in the specification. As an example, consider the prediction of inherent availability,

$$A_i = \frac{\text{MTBF}}{\text{MTBF} + \text{MTTR}}$$

This requires prediction of two parameters: MTBF and MTTR. An equipment MTBF and MTTR prediction will depend on what components are likely to fail. The values will be based primarily on information from the FMECA. The MTBF is computed as the reciprocal of the total failure rate. The MTTR is computed as

Category	Elemental activity	Activity no.
Preparation	System turn-on, warm-up; setting dials and counters as necessary.	1
	Activity 1 plus time awaiting particular component stabilization.	2
	Opening and closing radome.	3
	Gaining access and reinstalling covers (other than radome).	4
	Obtaining test equipment and/or tech orders.	5
	Checking maintenance records.	6
	Procuring components in anticipation of need.	7
	Setting up test equipment.	8
Malfunction verification	Observing indications only.	1
	Using test equipment to verify malfunctions inherently not reproducible on ground.	2
	Performing standard test problems or checks.	3
	Testing for pressure leaks.	4
	Attempting to observe elusive or nonexistent symptom(s).	5
	Using special test equipment designed specifically for this equipment.	6
	Making a visual integrity check.	7
Fault location	Fault self-evident from symptom observation.	1
	Interpreting symptoms by mental analysis only (from knowledge/experience).	2
	Interpreting displays at different settings of controls.	3
	Interpreting meter readings.	4
	Removing unit(s)/subunit(s) and checking in shop.	5
	Switching and/or substituting unit(s)/subunit(s).	6
	Switching and/or substituting part(s).	7
	Removing and checking parts.	8
	Making a visual integrity check.	9
	Checking voltages, continuity, waveforms, and/or signal tracing.	10
	Consulting tech orders.	11
	Conferring with tech reps or other maintenance personnel.	12
	Performing standard test problem(s).	13
	Isolating pressure leak.	14
	Using special test equipment designed specifically for this equipment.	15
Part procure-ment	Obtaining replacement component from aircraft spares or tool box.	1
	Obtaining replacement(s) from bench, shop, or preissue stock.	2
	Obtaining replacement component(s) by cannibalization.	3
	Attempting to obtain replacement component(s). Unavailable.	4
Repair	Replacing unit(s)/subunit(s).	1
	Replacing parts.	2
	Correcting improper installation or defective plug-in connection(s).	3
	Making adjustments in aircraft.	4
	Making adjustments in shop.	5
	Baking magnetron.	6
	Precautionary repair activity (includes so-called fault location, part procurement, and repair times spent when symptom not verified).	7
	Repairing wiring or connections.	8
Final mal-function test	Function checkout following completion of repair.	1

Figure 4.2   FMECA for maintainability.

the average of expected repair times for various breakdowns; an industrial engineer can roughly estimate the hours required for the repair, and Mil-Hdbk-472 gives 26 generic steps required for repairing electronic equipment (Fig. 4.2). For new technology, the FMECA can help by pointing out the additional failure modes.

Predictions of other appropriate indexes suggested in Section 4.2.1 can be made from historical data. The predictions pinpoint the opportunities for improving the design.

4.2.6 Life-Cycle Cost Analysis

During maintainability analysis, life-cycle costs are used frequently to assess the savings in support costs for various options. Maintainability usually has the highest influence on downtime costs and therefore the LCC. In general, life-cycle costs are costs to society. Specifically, they are costs to (1) the supplier,

(2) the user, and (3) innocent bystanders who might get killed or injured or whose property might be damaged from such mishaps as aircraft accidents. As far as maintainability is concerned, bystanders also include operating personnel and maintenance and repair personnel. One can write the LCC model as

$$LCC = \text{nonrecurring costs} + \text{recurring costs}$$

Examples of nonrecurring costs are:

- Basic engineering
- Test and evaluation
- Program management and control
- Manufacturing and quality engineering
- Nonrecurring production costs
- Experimental tooling

Examples of recurring costs are:

- Spares
- Repair labor cost
- Inventory entry and supply management
- Support equipment
- Technical data and documentation
- Training and training equipment
- Logistics management
- Operators
- Utilities

Other versions of the model are

$$LCC = (\text{initial price} + \text{warranty costs to the company})$$
$$+ (\text{repair, maintenance, and operating costs to the customer})$$

and

$$LCC = \text{acquisition cost} + \text{cost of operation, repair, and maintenance}$$
$$+ \text{cost of downtime} + \text{cost of liabilities} + \text{cost of lost sales}$$

The last model is most appropriate, but many companies find it very difficult to quantify the cost of downtime and the cost of lost sales. They use the other two models, which reveal only some of the facts.

The real world has many hidden costs. They are usually unaccounted for but are huge. They include costs of excess spare parts inventory, idle personnel, missed schedules, promises not kept to customers, time spent in meetings and solving problems, engineering changes, loss of goodwill, and penalties.

These costs are not easy to estimate. At the same time, ignoring them will lead to a very unrealistic model. Therefore, a reasonable estimate based on a consensus should be included. Other cost factors, if they are related and significant, should also be included: for example, increases in interest rates, fuel costs, power costs, and labor costs.

One of the best uses of a life-cycle cost model is to evaluate the merits of several options. Assume two options for a desktop copier. Design A is a traditional design, which requires periodic replacement of the toner (black ink powder, cost: $50 per year). A maintenance technician must clean the parts, and limited-life components must be replaced when they fail. A technician is needed to do the repairs. A maintenance contract for the period not under warranty costs $1200 per year. The selling price is $2500. The warranty cost, paid by the manufacturer, is $750.

Design B has a sealed cartridge containing ink and copying components. The cartridge has to be replaced twice a year to ensure a fresh supply of ink; the user can easily replace it in less than a minute. The cartridge itself is reliable and costs $75. Since most of the technology is in the cartridge, there are few other critical components, and they are designed not to fail during the expected 5-year life of the copier. Just in case an external component should break down, the lightweight copier can be hand-carried to the dealer, who can fix it in minutes.

By using a simple LCC model, the two options can be compared. Assume that design B costs $2800, somewhat more than design A because of new technology. Assume also that design A, like design B, has a 5-year life.

Design A:

$$\text{LCC} = (\text{manufacturer's cost})$$
$$+ (\text{maintenance costs and downtime costs to customer})$$
$$= (\text{selling price} + \text{warranty costs during first year})$$
$$+ (\text{cost of maintenance for 4 years}$$
$$+ \text{cost of toner} + \text{cost of downtime})$$
$$= (\$2500 + \$750) + (\$1200 \times 4 \text{ years} + \$200 + \$1100)$$
$$= \$9350$$

Design B:

$$\text{LCC} = (\text{selling price} + \text{warranty costs at 5\% of selling price})$$
$$+ (\text{expected cost to repair for 4 years} + \text{cost of cartridges})$$
$$= (\$2800 + \$140) + (\$600 + \$75 \times 10)$$
$$= \$4290$$

The conclusion is in favor of design B even if its initial cost is higher. The distribution of costs makes design B even more attractive to the customer:

	Supplier Cost ($)	Buyer Cost ($)
Design A	3250	6100
Design B	2940	1350

The supplier will save $310 per unit with design B ($3250 − $2940), but the buyer will benefit far more, saving $4750 ($6100 − $1350). If the gain in market share that is likely for design B is taken into account, the unit cost of design B will be far lower.

4.2.7   Design for Accessibility

Accessibility means having sufficient room around a component to diagnose, troubleshoot, and complete maintenance work in a safe and effective manner. Provision must be made for movement of necessary tools and equipment with consideration for various body positions [3].

Detailed plastic scale models can be used in assessing accessibility. If these are not possible, then the volume available around the equipment for activities must be estimated from the drawings. The former approach is much more effective, since it provides an opportunity to test for accessibility and ease of maintenance.

4.2.8   Design for Ease of Maintenance

Ease of maintenance means making activities at the human–equipment interface easier. It can be assured in many ways.

Minimize Maintenance in the First Place   In automobiles, the fuel injection method of ignition has eliminated the need to check the condition of the distributor.

Allow Repairs with Least Handling   A person changing a component on an automobile should not be required to remove a major assembly.

Design for Reliability of the Interface of Mating Parts   If an integrated circuit (chip) on a circuit board is going to be removed and replaced a few times, it should not be soldered to the circuit board. It should be mounted on a gold socket (aluminum sockets may not provide a good contact after about three replacements).

Design for Off-line Repair   A repair on a very large bearing in an automated process plant was expected to result in 2 days of downtime. However, the material handling equipment was designed so that the bearing could be removed and repaired off-line. A spare bearing was installed to reduce the process downtime to less than 1 hour.

Provide Fault Tolerance   This approach makes use of an alternate means already installed so that the downtime is practically zero. The failed primary unit can be repaired while the secondary line is substituted. Fault-tolerant equipment often permits preventive maintenance while operation continues. The IBM 3081 mainframe computer, for example, performs self-maintenance without the customer being aware of it.

Plan Tools and Equipment in Advance   Pillar [3] suggests useful rules of thumb: "If a service device or a tool is required to construct a piece of equipment, it will also be needed later to repair it .... When considering features for ease of maintenance major emphasis should go to the weight of components that must be moved." This means additional equipment such as a forklift truck or a crane may be required.

Design for Remove-and-Replace Instead of Repair   It may be cheaper to replace the whole circuit board than to have a technician isolate the faulty component.

Use Automated Test Equipment   The automated test equipment (ATE) can be a portable intelligent system which can generate thousands of test patterns to isolate a faulty path.

Design for Testability   This design approach has become very important for reducing the cost of troubleshooting. With microprocessors doing most of the work, the equipment can be designed to self-test and make highly skilled technicians unnecessary for troubleshooting complex electronics. Technicians with less skill can make the repair in minutes instead of days. Microprocessors built into the product can be programmed to test the product and isolate the fault. Some are designed to heal themselves, switching over to an alternate means or a backup device.

Testable designs incorporate two features. First, the microprocessor itself must be capable of being tested to assure its own performance. Second, the product (in which the microprocessor resides) should also lend itself to ease of testing involving many features other than the microprocessors.

Microprocessors with as many as 240,000 gates and millions of combinations of paths require millions of test patterns for testing. With physical limits on the number of input and output pins, it is almost impossible to perform all the tests. Some paths cannot be tested at all unless appropriate test points are monitored. In the circuit in Fig. 4.3 there are two input pins, x_1 and x_2, and one output pin z. Columns x_1 and x_2 in the figure show all the possible inputs. The output for each combination of inputs is shown in column z. Columns A/0, B/0, C/0, D/0, E/0, and F/0 show the output if any of the points A through F is faulty, being stuck at zero. The remaining columns show the output if any of the points is faulty, being stuck at digital one.

It is not easy to isolate the faults. If all the inputs at x_1 and x_2 produce the outputs in column z, one may think the circuit is functioning correctly. This, however, is not necessarily a correct conclusion; fault condition C/1 also gives the same output for the identical inputs. Under these conditions, the test will indicate a good circuit.

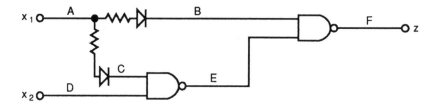

All single stuck faults

x_1	x_2	z	A/0	B/0	C/0	D/0	E/0	F/0	A/1	B/1	C/1	D/1	E/1	F/1
0	0	1	1	1	1	1	1	0	0	0	1	1	1	1
0	1	1	1	1	1	1	1	0	1	0	1	1	1	1
1	0	0	1	1	0	0	1	0	0	0	0	1	0	1
1	1	1	1	1	0	0	1	0	1	1	1	1	0	1

Figure 4.3 A simple circuit with only two gates.

Columns A/0, B/0, E/0, D/1, and F/1 produce identical outputs. This means that these faults cannot be isolated in the circuit test if only x_1, x_2, and z are monitored. From a maintainability viewpoint, however, technicians should be provided with the means to distinguish and isolate important faults. If these paths are known to be safety-critical during the design stage, the design engineer is often able to design for testability. Recently, several books have been published on this subject [4, 5].

Modular Designs This approach requires designs to be divided into physical and functionally distinct units to facilitate removal and replacement. It enables subassemblies and assemblies to be designed as removable and replaceable units for an enhanced design with minimum downtime. These units, called modules, can be a black box, a circuit board, or a major assembly. A few advantages of modularity [6] are the following:

- New designs can be simplified and design time can be shortened by making use of standard, previously developed building blocks.
- Only low-level technician skills are required.
- Training is easier.
- Engineering changes can be done faster, with fewer side effects.

Design for Disposal at Failure This approach is modular also, except that the failed unit is not repaired. This can sometimes be costly when many electronic devices are thrown away because of temporary faults. In one automotive company, as much as 40% of returned circuit boards were found to have no fault. On the other hand, a few advantages of this approach are:

• Savings in repair time, tools, facilities, and personnel
• Improved reliability because of reduced maintenance-induced errors
• Improved interchangeability because of standardization

Additional disadvantages are:

• Increased inventory cost because more replacement parts must be stocked, since repairs cannot be made
• Higher costs of engineering changes because of larger inventory

Design for Working Environment Where possible, the environmental factors affecting the ability of personnel to perform work should be considered. Environments such as temperature, radiation, humidity, chemicals, dust, and inadequate illumination increase the chance of maintenance-induced errors. Figure 4.4, for example, shows the effect of temperature on the average number of mistakes.

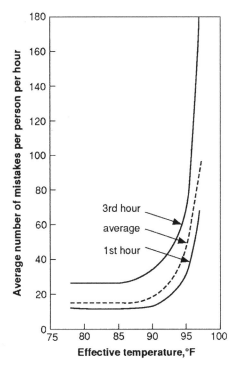

Figure 4.4 Effect of temperature on number of mistakes. (From Ref. 6, p. 70-12.)

Standardization Components such as nuts, bolts, lock washers, and fuses can easily be standardized. This makes the right part much easier to obtain and cheaper. Inventory costs are reduced drastically, and reliability is improved. In one organization, more than 360 part numbers were reduced to only 45 part numbers by standardization.

Design for Interchangeability A product can be designed for functional or physical interchangeability. In the former case, two or more items serve the same function. In the latter case, the same fastening, connecting, and mounting means can be used for more than one item.

Caution: Watch out for safety concerns. Interchangeability can result in accidents. See Chapter 5.

Design for Technician Transparency The U.S. Air Force program for the R&M 2000 process [7] requires a revolutionary approach: standard operating, maintenance, and testing characteristics across similar subsystems. The same technician can then work on all subsystems. This technician "transparency" will reduce personnel requirements, because specialists for certain subsystems are not needed. Any technician can isolate a fault to the lowest field-replaceable module. It also reduces the skill level required by the technicians.

4.2.9 Design for MM of Testing

There are many strategies, that can make testing easier and more efficient, thus enhancing *testability*. Mil-Std-2165 [8] includes a large list of strategies and scoring guidelines to assess design effectiveness for testability. These strategies are in the following checklist.

Mechanical Design

Is a standard grid layout used on boards to facilitate identification of components?

Is enough spacing provided between components to allow for clips and test probes?

Are all components oriented in the same direction (pin 1 always in same position)?

Are standard connector pin positions used for power, ground, clock, test, and other signals?

Are the numbers of input and output (I/O) pins in an edge connector or cable connector compatible with the I/O capabilities of the selected test equipment?

Is defeatable keying used on each board to reduce the number of unique interface adapters required?

When possible, are power and ground included in the I/O connector or test connector?

Have test and repair requirements affected decisions on conformal coating?

Is the design free of special setup requirements (e.g., special cooling), which would slow testing?

Test Access

Are unused connector pins used to provide additional internal node data to the tester?

Are signal lines and test points designed to drive the capacitive loading represented by the test equipment?

Are test points provided that let the tester monitor and synchronize with on-board clock circuits?

Are test access points placed at those nodes that have high fan-out?

Are buffers employed when the test point is a latch and susceptible to reflections?

Are buffers or divider circuits employed to protect those test points that may be damaged by an inadvertent short circuit?

Are active components, such as multiplexers and shift registers, used to make necessary internal node test data available to the tester over available output pins?

Are all high voltages scaled down within the item before the test access point so as to be consistent with tester capabilities?

Is the measurement accuracy of the test equipment adequate for the tolerances established for the item being tested?

Parts Selection

Is the number of different part types the minimum possible?

Have parts been selected which are well-characterized in terms of failure modes?

Are the parts independent of refresh requirements? If not, are dynamic devices supported by sufficient clocking during testing?

Is a single logic family being used? If not, is a common signal level used for interconnections?

Analog Design

Is one test point per discrete active stage brought out to the connector?

Is each test point adequately buffered or isolated from the main signal path?

Are multiple interactive adjustments prohibited for production items?

Are functional circuits (amplifiers, regulators, etc.) of low complexity?

Are circuits functionally complete, without bias networks or loads on another unit?

Is the minimum number of multiple phase-related or timing-related stimuli required?

Is the minimum number of phase or timing measurements required?

Is the minimum number of complex modulation or unique timing patterns required?

Are stimulus frequencies compatible with tester capabilities?

Are stimulus rise time or pulse width requirements compatible with tester capabilities?

Do response measurements involve frequencies compatible with tester capabilities?

Are response rise time or pulse width measurements compatible with tester capabilities?

Are stimulus amplitude requirements within the capability of the test equipment?

Are response amplitude measurements within the capability of the test equipment?

Does the design avoid external feedback loops?

Does the design avoid or compensate for temperature-sensitive components?

Does the design allow testing without heat sinks?

Are standard types of connectors used?

Digital Design

Does the design contain only synchronous logic?

Are all clocks of differing phases and frequencies derived from a single master clock?

Are all memory elements clocked by a derivative of the master clock? (Avoid having elements clocked by data from other elements.)

Does the design avoid resistance–capacitance one-shot circuits and dependence on logic delays to generate timing pulses?

Does the design support testing of "bit slices"?

Does the design include data wraparound circuitry at major interfaces?

Do all buses have a default value when unselected?

For multilayer boards, is each major bus laid out so that current probes or other devices can be used for fault isolation beyond the node?

Is a known output defined for every word in a read-only memory (ROM)? Will the improper selection of an unused address result in a well-defined error state?

Is the number of fan-outs for each internal circuit limited to a predetermined value?

Are latches provided at the inputs to a board in those cases where tester input skew could be a problem?

Is the design free of wired ORs?

Does the design include current limiters to prevent domino-effect failures?

If the design incorporates a structured testability design technique (e.g., scan path, signature analysis), are all the design rules satisfied?

Are sockets provided for microprocessors and other complex components?

Built-in Testing

Can built-in testing (BIT) in each item be exercised under control of the test equipment?

Is the test program set designed to take advantage of BIT capabilities?

Are on-board BIT indicators used for important functions? Are BIT indicators designed so that a BIT failure will give a FAIL indication?

Does the BIT use a building-block approach (e.g., are all inputs to a function verified before that function is tested)?

Does building-block BIT make maximum use of mission circuitry? Is BIT optimally allocated in hardware, software, and firmware? Does on-board ROM contain self-test routines?

Does BIT include a method of saving on-line test data for the analysis of intermittent failures and operational failures that are nonrepeatable in the maintenance environment?

Is the failure rate contribution of the BIT circuitry within stated constraints?

Is the additional weight attributed to BIT within stated constraints?

Is the additional volume attributed to BIT within stated constraints?

Is the additional power consumption attributed to BIT within stated constraints?

Does the allocation of BIT capability to each item reflect the relative failure rate of the items and the criticality of the items' functions?

Are BIT threshold values, which may require changing as a result of operational experience, incorporated in software or easily modified firmware?

Is processing or filtering of BIT sensor data performed to minimize BIT false alarms?

Are the data provided by BIT tailored to the differing needs of the system operator and the system maintainer?

Is sufficient memory allocated for confidence tests and diagnostic software?

Does mission software include sufficient hardware-error-detection capability?

Is the failure latency associated with a particular implementation of BIT consistent with the criticality of the function being monitored?

Are BIT threshold limits for each parameter determined as a result of considering each parameter's distribution statistics, the BIT measurement error, and the optimum fault detection versus false alarm characteristics?

Test Requirements

Has level-of-repair analysis been accomplished? Should the replacement be done at component level or circuit board level, or at assembly level?

For each maintenance level, has a decision been made for each item on how built-in testing, automatic test equipment, and general-purpose electronic test equipment will support fault detection and isolation?

Is the planned degree of test automation consistent with the capabilities of the maintenance technicians?

For each item, does the planned degree of testability design support the decisions on the level of repair, test mix, and degree of automation?

Are the test tolerances established for BIT consistent with those established for higher-level maintenance tests?

Test Data

Do state diagrams for sequential circuits identify invalid sequences and indeterminate outputs?

If a computer-aided design system is used for design, does the CAD database effectively support the test generation process and test evaluation process?

For large-scale integrated circuits used in the design, are data available to accurately model the circuits and generate high-confidence tests for them?

For computer-assisted test generation, is the available software sufficient in terms of program capacity, fault modeling, component libraries, and postprocessing of test response data?

Are testability features included by the system designer documented in terms of purpose and rationale for the benefit of the test designer?

Is a mechanism available to coordinate configuration changes with test personnel in a timely manner?

Are test diagrams included for each major test? Is the diagram limited to a small number of sheets? Are intersheet connections clearly marked?

Is the tolerance band known for each signal on the item?

4.3 MAINTAINABILITY IN THE MANUFACTURING STAGE

This section covers minimizing the downtime on manufacturing equipment. Companies such as M&M Mars, Firestone, General Motors, Intel, Caterpillar, and Eastman Chemicals (Kodak) have applied maintainability techniques to save millions of dollars. One company estimated that a 2% reduction in downtime will result in a savings of $36 million over 5 years on an annual sales volume of $800 million.

4.3.1 Maintainability for Existing Equipment

When the equipment has already been procured, major design changes are usually out of the question. However, many modifications can still be made. Sometimes life-cycle cost analysis will show that it may be cheaper to scrap the equipment and buy new.

The first step is to identify the 20% of problems that are responsible for 80% of the downtime (a famous principle). The next step is to analyze the causes of downtime by a systematic analysis such as fault-tree analysis. The authors use process analysis maps (PAMs), which are simplified fault trees. The details are covered in Chapter 6. Any maintenance technician can be trained to construct PAMs. In fact, a company should train all maintenance technicians to use such tools. Technicians can use analysis tools effectively and come up with many useful ideas—but, of course, management must listen in order to benefit from them.

Example 4.1 A manufacturing process used two rollers, jacketed in 1-in. thick rubber (Fig. 4.5) to squeeze chemicals from thin, wet sheets of steel at high speed. The rubber jacket blew into pieces roughly once a week, like a tire blowing out on a semi truck. It tore the steel sheets apart, resulting not only in very poor quality control but also in a severe safety problem. One worker had his neck lacerated. The total downtime cost to recover from failure was roughly $250,000 per week.

The engineers tried to improve the design by substituting better materials for the rubber jacket. They did not make significant progress. They tried better bonding compounds between the steel and the jacket, again with no progress. They also tried making rough surfaces and grooves on the steel so that the rubber would be held more tightly. Two years went by and $24 million were surrendered to downtime.

Solution A systematic analysis solved the problem in a few hours! The process analysis map in Fig. 4.5 was constructed with the help of the maintenance technicians. Five major causes were identified. The technicians assessed the percent contribution of each cause. The figure shows that the highest contributor to the failure was heat buildup in the steel roller (70% contribution), which was unable to transfer heat except through the rubber jacket. Since the technicians were trained to eliminate the root cause of the downtime, one of them took no time in coming up with the solution. The following questions and answers led to the solution:

Q. What is the most probable cause of downtime?

A. Excess heat buildup in the steel roller.

Q. If the cause of downtime is excess heat, can we eliminate excess heat?

A. Yes. We can circulate refrigerated drinking water inside the steel roller. That should remove a lot of heat. I can bring a pump from the stockroom and circulate water from the drinking water fountain.

Q. Excellent idea. Why haven't you tried?

A. I have been telling management about this for 5 years, but no one cares.

Maintainability is a gold mine—if someone is willing to mine. The technician was allowed to use his idea. It worked. After the change, not a single incidence of downtime in 2 years saved the company $12 million per year!

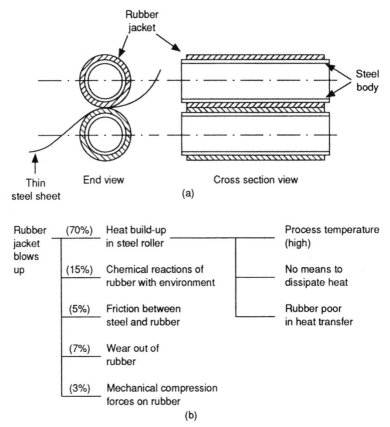

Figure 4.5 The chemical rinse process. (a) End and cross-section views of roll and (b) partial process analysis map.

4.3.2 Maintainability for New Equipment

Maintainability for new manufacturing equipment is the same as maintainability for a new product, except that the supplier may team up with the customer in analyzing maintainability if it is a custom design. For off-the-shelf equipment, maintainability information may not be available to the user. We have worked with several suppliers of large capital equipment items priced in millions of dollars and found that, although the approach is still in its infancy, the two most effective tools are failure-mode, effects, and criticality analysis (FMECA), covered in Chapter 3, and fault-tree analysis (FTA), covered in Chapter 5. Other valuable techniques are quality function deployment, design reviews, and maintenance engineering safety analysis (Chapter 5).

Example 4.2 A pattern-recognition optical system for manufacturing very large scale integrated circuits was to be procured by a large semiconductor manufacturer. A fault tree was constructed by a team of engineers (Fig. 4.6). An independent team

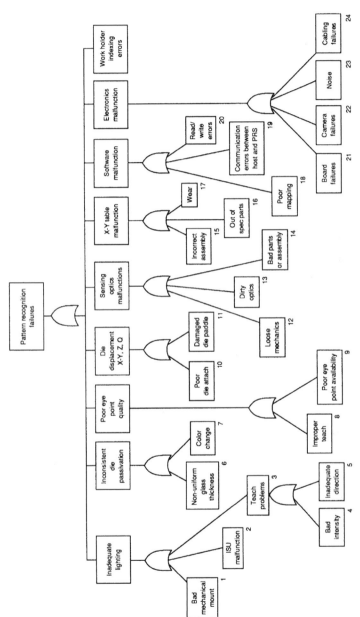

Figure 4.6 Fault-tree analysis for pattern-recognition equipment.

Failure mode	Causes	Effects	Criticality	Design action	Fault verification	RCM action
Optics malfunction	Ambient heat	Permanent deformation	II A	Provide fan	Warn of fan failure	Check fan tolerances every 2 months
	Dirt	Erroneous output	II B	Add filter	Not required	Replace filter monthly
Circuit parameters drift	High leakage current	Parameters out of control	II D	Qualify critical components	Not required	Install software to monitor parameters
	Dirt on circuit	Intermittent performance	II B	Conformal coat	Not required	Not required
	High junction temperature	Degraded performance	II B	Derate parts below 50%	Not required	Use infrared inspection
X–Y table innaccurate	Supplier design	False output	II A	Perform FMEA with supplier	To be determined	To be determined
	Horizontal position drift	False output	II A	Software control	Not required	Check eccentricity during routine maintenance

Figure 4.7   FMECA for pattern-recognition equipment.

performed FMECA. (If the same team performs FTA and FMECA, the results are likely to be the same.) A sample sheet of the FMECA is shown in Fig. 4.7. Both teams came up with different problems, with few exceptions. By the time the analysis was complete, over 100 design concerns were identified. At least 50 design changes were determined to be important. The time of about eight engineers for 3 days was a small investment compared to the value of 60 design improvements. Since the system was still on the drawing board, the cost of engineering changes was insignificant.

4.4   MAINTAINABILITY IN THE TEST STAGE

Maintainability tests are often called *maintainability demonstration tests*. At least four types of tests can be conducted:

1. Test for inherent equipment downtime
2. Test for human variation in performing the maintenance work
3. Maintenance level tests, which are done to verify the ability to repair at organization, intermediate, and depot levels
4. Tests for maintenance during emergencies

Maintainability tests start during product development. The tests can be done with the help of portions of the system, mockups, and models. In some cases, a detailed analysis may be the only option. For example, the analyses may suggest that a forklift truck is required. If the access area is too small, this is the best time to redesign the access area. It is too expensive to make changes later.

4.4.1 Prerequisites for Maintainability Tests

Before expensive tests are done, some strategy is needed. The tests should not be based only on guesswork. No maintainability test for complex equipment should be performed without a FMECA. This analysis reveals many important failure modes—those that are likely to result in downtime. Ignoring this analysis will result in many design changes at a huge cost later.

The other pertinent pieces of information required may be maintainability specifications, where all required parameters are specified (if the specification does not contain maintainability information, then it should be developed), past histories on similar equipment so that such problems are removed from design, and interviews with repair and maintenance crews.

4.4.2 Tests for Inherent Equipment Downtime

Demonstrations of inherent downtime may consist of tests for testability, fault diagnosis, fault isolation, interchangeability, simulation, and any automated repair such as switching over to a backup device. Two types of verification may be done: (1) to verify if the downtime for each failure mode is within specified limits and (2) to assure that the allocation of maintainability goals is adequate.

4.4.3 Tests for Human Variations

Human variation tests are statistical in nature. A probability distribution is constructed to determine the downtime distribution. Tests are conducted for such parameters as mean time to repair (MTTR), mean downtime (MDT), mean worker-hours per operating hour (MWH/OH), and maximum time to repair at a certain percentile.

Several technicians should test for each task to construct the distribution of downtime. The variation in repair times for mechanical systems is very large. Besides, this information will help in workforce planning. If it would be too expensive to test all the tasks, then they may be selected. We follow the strategy of selecting tasks related to safety and those that can result in high downtime. Usually repair times are distributed in a log normal manner. Chapter 2 contains an example of the construction of this distribution.

4.4.4 Maintenance Level Tests

These tests are conducted to see if the maintenance tasks designed to be done at the operational level (organizational level) can indeed be performed by the operations technician. Similar verification can be done for the ability to perform maintenance and repair tasks at the intermediate level (sometimes called *unit level*) and at the depot level (in an automotive environment, depots are equivalent to *remanufacturing centers*).

4.5 MAINTAINABILITY IN THE USE STAGE

Maintainability analysis is a continuous improvement process. Many causes of downtime are not known until the system is in the field. There are at least three major areas where costs can be reduced through design improvements:

1. Prediction and reduction of limited-life items
2. Monitoring and predicting operational availability
3. Minimizing maintenance and support costs through maintainability engineering

4.5.1 Prediction and Reduction of Limited-Life Items

Limited-life items are those that require repair or replacement during the operating life of the system. Several texts contain models based on the assumption that the failure profile of the limited-life item is a normal distribution or an exponential distribution. If these assumptions are not justified, which is usually the case, a better tool is Weibull analysis. This reliability analysis can give insight to the number of spares required by the system. It includes exponential as well as the normal distribution cases. This distribution applies to several families of data. It uses actual data (even partial data will help) in making an assessment of the distribution. Chapters 2 and 3 cover several examples of this analysis. An example is shown here to demonstrate the application of this technique for spares prediction.

Example 4.3   A sample of 25 bearings out of a lot of 400 was tracked for time to failure in a brewery. These bearings are used in similar equipment system-wide. At the time of analysis, only 8 bearings had failed. Their time to failure in equipment hours is converted to weeks below. How many spares are required between weeks 30 and 40?

Bearing Serial Number	Weeks to Failure
1	6.0
2	11.0
3	2.2
4	7.8
5	12.2
6	4.3
7	6.2
8	8.4

Solution The data are ranked below and plotted against median rank (see Chapter 2) as they coordinate:

Weeks to Failure	Median Rank Value
2.2	2.73
4.3	6.62
6.0	10.55
6.2	14.49
7.8	18.44
8.4	22.38
11.0	26.32
12.2	30.27

The growth for the data is shown in Fig. 4.8. The Weibull line is projected from the plot of the first eight points, based on a sample size of 26.

From the graph:

Percent expected to fail in 30 weeks—89.5%

Percent expected to fail in 40 weeks—97.0%

Percent expected to fail during the period 30–40 weeks—97.0% − 89.5% = 7.6%

A conclusion can be made that 7.5% of the bearings from the 400 procured can be expected to fail during the period between 30 and 40 weeks. Therefore, $400 \times 0.075 = 30$ bearings are expected to fail during this period.

4.5.2 Monitoring and Predicting Operational Availability

Availability is the ultimate objective of maintainability. Tracking data are important information that leads to design improvements. One method of obtaining such data is to keep records on downtime.

Then availability is

$$A = \frac{\text{uptime}}{\text{total scheduled time}} = \frac{\text{total scheduled time} - \text{downtime}}{\text{total scheduled time}}$$

If there is significant amount of downtime, the causes may be analyzed. Often, minor design changes will yield as much as a 100-fold reduction in costs.

Another method of assessing availability is to use Monte Carlo simulation. The steps and an example are given below.

1. Establish the model for availability.
2. Construct statistical distributions for each element of the model.
3. Randomly draw a value of each element from its probability distribution.
4. Compute availability from the data obtained in step 3.
5. Construct the distribution of availability.

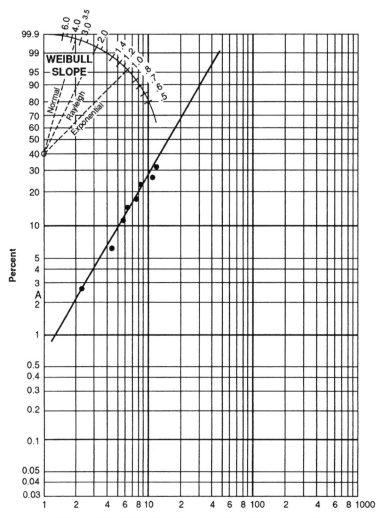

Figure 4.8   Weibull graph for bearing failure, in weeks.

The simulation can be done manually, but a computer is normally used to handle the large amount of data.

Example 4.4

Step 1: Consider this model for availability:

$$A = \frac{\text{MTBF}}{\text{MTBF} + \text{MDT}}$$

Step 2: Construct Weibull plots for MTBF and MDT from actual data.

Step 3: Randomly pick a value of MTBF from the Weibull plot. Also randomly pick a value of MDT from the Weibull plot.

Step 4: Compute availability using the equation in step 1.

Step 5: Repeat the first four steps several hundred times using a computer program. This will result in hundreds of values for the expected availability. Construct the probability distribution of availability (see Chapter 2). The distribution will give minimum and maximum values of availability. It will also give the probability associated with each availability target.

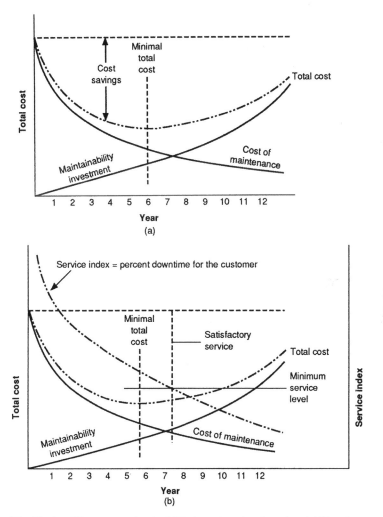

Figure 4.9 Trade-off between maintainability investment and service. (a) Year-by-year cost performance with minimum cost goal achieved in sixth year. (b) Minimum cost goal is achieved but the service goal is not achieved.

4.5.3 Minimizing Support Costs

Minimizing support costs is an important goal, but customer satisfaction is just as important. The two goals go together.

Support costs consist of costs of an ongoing maintainability effort, preventive maintenance and corrective maintenance, and many other logistics costs. An investment in maintainability is required to reduce the cost of preventive and corrective maintenance.

The investment should continue as long as the total of the investment and the support cost is lower. The difference between the initial total cost and the total cost in any year is the direct savings. Figure 4.9a shows year-by-year performance of total cost. It shows that the minimum cost is achieved at the sixth year. Should the investment in maintainability be stopped? The answer is no, because the customer service goal, as shown in Fig. 4.9b, has not been achieved (the downtime goal). Some may argue that extra cost in achieving a customer service goal is not justified; the total cost will no longer be the minimum. Actually, the cost will be lower when the customer is satisfied. A satisfied customer brings more customers and more business volume. This should lower the unit cost of the product and boost profits.

4.6 MAINTAINABILITY AND SYSTEM SAFETY

Although the objective of maintainability is to minimize maintenance, complex systems still will need some maintenance, reconfiguration, remote control, upkeep, revision, field change, or monitoring. Unfortunately, lots of accidents occur during these efforts, and latent hazards may also be introduced into the system. Recently, hybrid vehicles stalled because a wrong version of software was downloaded during maintenance. In a Japanese car, the latest version of software could not be downloaded due to lack of resources in the memory. Inappropriate maintenance actions have inadvertently shut down processes or systems. Errors have been made when operators have assumed that maintenance processes, tasks, or functions must be safe since related regulations, codes, and standards have been met. This may not be a prudent assumption to make. In the consideration of system safety, it is important that safety-related risks have been identified and eliminated, or controlled to an acceptable level. To accomplish this, hazard analysis and risk assessment are needed and activities related to the maintenance must be evaluated.

4.6.1 Remote Maintenance Safety and Security

Automatic remote maintenance systems provide the means of monitoring, calibrating, and adjusting processes or systems. Maintenance is now going wireless. Maintenance engineers and technicians can assess the health of complex systems remotely and make adjustments, conduct system reconfiguration, or command switchover to redundant lags or paths. Malfunctions or errors can occur, which can adversely affect the on-line system and result in inadvertent operation and accidents. The design of remote maintenance systems must be evaluated from a system safety perception. There may be hazards associated with accessing a system remotely. Consideration should be given to the following hazards:

- Inadvertent system shutdown
- Inadvertent automated action
- Inadvertent process function
- Inappropriate reconfiguration of alarm parameters
- Altering of built-in test functionality
- Errors associated with teaching of robotic subsystems
- Maintenance design errors associated with human access
- Exposed energy sources
- Uncontrolled potential energy
- Incomplete or inaccurate maintenance instruction
- Inappropriate system response
- Incomplete or inaccurate cautions and warnings
- Laborious or confusing maintenance tasks
- Hazardous misleading information (HMI) communicated
- Inadvertent release of hazardous gases or materials
- Software error, malfunction, or logic error
- Failure of remote maintenance system when needed
- Hacking, spoofing, or jamming
- Loss of important data/information
- Inadvertent altering of data/information
- Bypassing of interlocks, guards, or protective devices
- Miscommunication associated with maintenance instruction
- Language barriers
- Conflicting literature, manuals, or booklets
- Excessive display of information
- Display masking

Security Threats Information security threats are also a concern where an intruder gains access through a remote access maintenance port and intentionally alters safety-related data or information. Consider a hacker entering a control system via a back door and reprogramming a safety-critical function, or altering parameters within a security system, shutting down monitors or changing alert triggers. Such instances can result in harm. Hackers have frequently stolen credit card numbers and social security numbers.

4.6.2 System Health Monitoring and Maintenance

It is important to monitor the health of complex systems from a system safety view. In some cases, when these systems are automated, built-in testing (BIT) is designed into the system. Initial built-in test (IBIT) may be designed to operate when an automated system is first powered up. Continuous background built-in test (CBBIT) may be on during system operation. When the system is configured during shutdown, an additional BIT may operate to monitor and test the reconfiguration of the system for power shutdown. The automatic health monitoring of a complex system is apparently needed when malfunction, failure, and anomalies can present hazards. BIT capabilities drive alerts, warnings, or cautions, when the system is not functioning according to designed parameters. When a BIT is not adequately designed or should a BIT fail when needed, accidents can happen.

Hardware Health Monitoring For high-reliability physical hardware designs such as building or bridge structures, ships, or aircraft, visual inspection and nondestructive testing may be conducted to monitor and determine the health of such systems. However, there are instances when it is not totally possible to determine wear, degradation, or damage. Depending on the material selection of metals, composites, chemicals, or aggregates, it may not be apparent if there has been excessive wear, degradation, or damage. Consequently, maintenance of hardware structures is a subject of continuous study in the fields of materials science, fracture mechanics, physical science, and chemical, civil, nuclear, aerospace, and mechanical engineering.

4.6.3 Using Models to Develop Maintenance Diagnostics and Monitoring

Models can be stochastic and deterministic. They can be derived from previous statistics such as loss or failure information or prior data, or they can be developed experimentally, such as in burn-in tests. Models can be derived from both prior statistics and estimated future distributions in the application of Bayesian analysis.[1] From a system safety view, an analyst should closely evaluate information and data. The accuracy and the assumptions made concerning the use and applications of models can affect the determination of the diagnostics, monitoring, and maintenance status of a system. Here are examples of some stochastic and deterministic approaches.

4.6.3.1 *Stress–Strength Analysis* To understand concepts of safety factor (SF) and safety margin (SM) the relationship between stress and strength should be discussed. Stresses are the result of abnormal energy effects from electrical,

---

[1]There are many texts that discuss Bayes' theorem. For concise information consult the following: T. B. Sheridan, *Humans and Automation: System Design and Research Issues*, John Wiley & Sons, Hoboken, NJ, 2002, pp. 190–191.

thermal, mechanical, chemical, and radiation interactions. Stresses can be environmental or operating; they include electrical loads, temperature, vibration, and humidity [9, p. 161]. There are two design approaches for controls against stress-induced hazards: select parts and material with sufficient strength to withstand the maximum possible load, or protect the part against excessive stresses.

Two hazard controls that are applied against stress-related hazards consider the application of design safety margins or safety factors. The safety factor (SF) is defined to be the ratio of the capacity of the system to the load placed on the system, or the ratio of minimum strength and maximum stress. The safety margin (SM) is the difference between the system capacity and the load. Failure will occur if the safety factor is less than 1 or the safety margin becomes negative.

4.6.3.2 *Safety Factor and Safety Margin Variability*   Caution must be followed when using SF and SM concepts. Tests are conducted to measure yield strength, tensile strength, and fatigue life, often indicating "average" or "best-fit" curves. Random variables with specific statistical distributions are derived from these tests. Consider all possible variability that can occur—differences in material quality, contamination, manufacturing and process deviations, assumptions in calculations, or errors in testing. As a protection against variability, safety factors and safety margins should be designed into the system. Safety factors can range from 1 to 4. To calculate SF and SM, let X and Y be random variables representing the stress placed on the system and the strength of the system, respectively. Then

$$SF = Y/X \quad \text{and} \quad SM = Y - X$$

4.6.3.3 *Safety Margin as a Hazard Control*   There are many techniques in the application of SM [10, pp. 166–167]. The safety analyst and designers should collaborate on which approach is most appropriate for a particular system.

Overdesign approach was an early application of safety margin. The Brooklyn Bridge is classic example of overdesign. Provide very strong structure that will withstand at least 10 times the expected load. Grumman had built aircraft with this overdesign approach toward constructing airframes. Grumman aircraft were highly survivable against battle damage.

Statistical approach considers the development of random variable distributions, for example, random stress and random strength distributions [9, pp. 130–131], where reliability remains the probability that stress is less than strength (or equivalently, that strength is greater than stress).

The worst-case safety margin for normally distributed strength is defined as [10, p. 167]

$$SM = \frac{\text{maximum stress} - \text{mean strength}}{\text{standard deviation of strength}}$$

Rule-of-thumb approach is used in many construction projects, such as dams, bridges, and highways. Allowances for applied and impact loads provide for an increase in strength. The analyst should evaluate this approach with caution. All assumptions made in the application of rules-of-thumb should be evaluated.

Damage-tolerance approach considers the monitoring of the system to determine degree of degradation, that is, extensiveness of corrosion, erosion, crack growth, and wear. Judgment is applied to estimate system failure. In order to understand degradation, the physics-of-failure models [9, pp. 137–141] are mathematically derived, which are deterministic models based on knowledge of the failure mechanisms. A failure is not viewed as a stochastic event. Time to failure is determined for each failure mode and failure site based on the stresses, material properties, geometry, environmental conditions, and conditions of use.

Developing physical failure models involves the following.

- Identification of failure sites and mechanisms of failure
- Construction of mathematical models
- Estimating reliability for a given operating and environmental profile and for specific component characteristics
- Determining service life
- Redesign of system to increase design life
- Consideration of the total system life cycle
- Consideration of real world applications

4.6.3.4 Integration Considerations Between Safety-Related Models and Hazard Analysis

An important consideration involving the use of safety-related models is to fully understand all the assumptions made in the development of the graphic or mathematical models. The following questions should be considered:

Was the model developed by existing reference or by experimentation?

What were the steps taken?

How was data collected?

Are calculations accurate?

What are the physical models used?

What are the empirical models used?

What statistical approach was used?

Who conducted the test?

Will the test reflect real world application?

Where was the test conducted?

Can the experiment be reproduced?

As discussed, consider all possible variability that can occur—differences in material quality, contamination, manufacturing and process or test deviations, assumptions in calculations, or errors in testing. These situations could introduce latent hazards that could be initiators or contributors. If there are concerns, appropriate hazard scenarios should be developed and controls introduced to eliminate or reduce these risks.

4.6.3.5 Real World Verification

When a hazard control has been validated based on modeling and test assumptions, the verification of the control involves answering the following question: Does the actual field data compare with the estimated test results? It remains very important to verify all the assumptions made with real world data and observation. It is apparent that this effort should be conducted as soon as possible. There may be current risks associated with the maintenance diagnostics and monitoring due to the assumptions made within the design. If the real world data does not adequately compare with assumptions made, the system may have to be recalled or modifications may have to be made in the field. The earlier in the design that hazard controls can be validated and verified, the less will be the safety-related risk and cost.

4.6.4 Hazard Analyses in Support of Maintenance

It can be estimated that there are hundreds of safety-related methods and techniques that can be applied in assurance technologies. Almost any type of system analysis technique can be adapted for application to system safety, if properly implemented by an experienced safety professional. This logic holds for maintainability and safety. A brief discussion of some maintenance applicable methods is addressed below. This listing is not all-inclusive.

Accident Analysis Any accident or incident should be formally investigated to determine the initiators and contributors of the unplanned event. Accidents are a sequence of unplanned events that result in harm. They are the result of unsafe acts or conditions that form adverse event flows. Incidents are also unplanned sequences of events that result in minor or minimal harm. Incidents could have been accidents if harm had occurred. Many methods and techniques can be applied to conduct investigations. Accidents or incidents associated with maintenance activities should be investigated. Maintenance staffs can be briefed on safe operating procedures, safety program requirements, and special mitigations. Examples of past accidents or incidents generally can be topics of toolbox safety talks, safety briefs, or safety meetings. Accident analysis is a reactive activity whereas hazard analysis is a proactive effort. Within scenario-driven hazard analysis, potential accidents are hypothesized. This analysis is considered proactive in nature.

Action Error Analysis The analyst evaluates interactions between humans and machines. It is used to study the corollary of human errors in tasks related to directing automated functions. Any automated interface between a human and automated process can be evaluated, such as maintainer–equipment interactions. Errors can be made anywhere within the life cycle of a maintenance effort. Design errors can introduce latent hazards (e.g., errors related to the design of a guard or inter-lock). Errors can also be made within a particular procedure or task.

Barrier Analysis   This method is employed by identifying uncontrolled energy flow(s) that may be hazardous and then identifying or developing the barriers that must be in place to prevent the unwanted energy flow from damaging equipment and/or causing system damage and/or injury. Any system is comprised of energy; should this energy become uncontrolled, accidents can result. Barrier analysis is an appropriate tool for systems analysis, safety reviews, and accident analysis.

Barrier analysis can also be conducted by evaluating failure or hazard propagation throughout a system and identifying the mitigations or hazard controls that will hinder or stop the adverse event flow. Each mitigation or control can be shown as a barrier within a diagram, flowchart, or sequence flow.

Bent Pin Analysis   This method enables the evaluation of the effects should connectors short as a result of bent pins and inappropriate mating or demating of connectors. Any connector has the potential for bent pins to occur. Connector shorts can cause system malfunctions, anomalous operations, and other risks. Diagrams of connector footprints are used and each pin is labeled or identified as to its input or output electrical function. Pin-to-pin shorts are hypothesized and the sneak circuit assessed. If connector pins are not properly assigned, single point failures may occur should a pin inadvertently bend. Such situations can have catastrophic outcomes.

Cable Failure Matrix Analysis   Within this technique, the risks associated with any failure condition related to cable design, routing, protection, and securing are evaluated. Should chafing, creeping, corrosion, or vermin damage cables, system malfunctions can happen. Inadequate design of cables can result in faults, failures, and anomalies, which can be hazards. The analyst assesses the layout of cables within particular cable runs, cable trays, and cable ducts, bends, cut-through, channel-ways, fire divisions, and hazardous locations. It may be important to isolate particular safety-related signals, power runs, and high-voltage sources. Cables should be protected from physical damage. Consideration concerning system redundancy requirements may also be needed. An evaluation of the overall placement of cable runs is needed. Further consideration should also be given to multiple effects. Fire, explosion, battle, flood, earthquake, and other physical damage effects should be included within this analysis.

Change Analysis   The analyst examines the effects of modifications from a starting point or baseline. Any change to a system—such as new equipment procedure, software upgrade, minor adjustment, new operation, new task, new instruction, and manual revision—should be evaluated from a system safety view. Changes can inadvertently introduce latent hazards. Almost any change within a system should be evaluated. A mix of other techniques can be used to conduct change analysis. Requirements cross-checking, visual inspection, modeling, markup, prototyping, and flowcharting are examples of other related techniques that can be used within change analysis.

Checklist Analysis This method can be used to make comparison to criteria, or a checklist can be used as a memory jogger. The analyst uses a list to identify items such as requirements, hazards, or operational deficiencies. Checklist analysis can be used in any type of safety analysis, safety review, inspection, survey, or observation. Checklists can enable a systematic, step-by-step process. They can provide formal documentation, instruction, and guidance.

The purpose of comparison-to-criteria is to provide a formal and structured format that identifies safety requirements. Comparison-to-criteria is a listing of safety criteria that could be pertinent to any system. This technique can be considered in a requirements cross-check analysis. Applicable safety-related requirements are reviewed against an existing system.

Confined Space Safety Evaluation The purpose of this analysis technique is to provide a systematic examination of confined space risks. Any confined areas where there may be a hazardous atmosphere, toxic fumes, gas, or lack of oxygen could present risks. Confined space safety should be considered at tank farms, fuel storage areas, manholes, transformer vaults, confined electrical spaces, and raceways—anywhere a human can gain access to maintain, inspect, construct, or diagnosis a problem. Specific techniques are used to test and evaluate a confined environment for toxic or otherwise hazardous atmosphere. Mitigations for monitors, alarms, two-person buddy system, repelling equipment, and ventilation/breathing apparatus may be needed.

Contingency Analysis This is a method of minimizing risk in the event of an emergency. Potential accidents are identified and the adequacies of emergency procedures and controls are evaluated. Contingency analysis should be conducted for any system, procedure, task, or operation where there is the potential for harm. Contingency analysis lists the potential accident scenario and the steps taken to minimize the situation. It is an excellent formal training and reference tool. Other supportive techniques can be applied for contingency analysis, such as makeup, simulation, drills, alerts, readiness reviews, role-playing, and dry runs.

Critical Incident Technique This technique [11] is based on collecting information on hazards, near misses, and unsafe conditions and practices from experienced personnel. It can be used to investigate human–machine interfaces in future designs or in existing systems, or for the modification and improvement of systems. This technique is a method of identifying unsafe acts or unsafe conditions that contribute to potential or actual accidents within a given population by means of a stratified random sample of participant-observers who are selected from within the population. Observers are selected from specific departments or operational areas so that a representative sample of the operational system existing within different contributory categories can be acquired.

Damage Modes and Effects Analysis   Similar to failure-modes and effects analysis, this method enables the evaluation of the damage potential (rather than failure potential) as a result of an accident. Risks can be eliminated or controlled by evaluating damage progression and severity. Designs can be implemented to decrease the damage in the event of harm, thereby reducing severity of risk. Engineering controls that can decrease damage involve the control of abnormal energy releases. Examples of such designs include pressure or explosion venting, shunting power to ground, and decreasing the energy requirements within a particular design.

Energy Analysis   There are many methods of evaluating energy from a system safety view. These techniques can be applied to all systems that contain, make use of, or store energy in any form (e.g., potential or kinetic mechanical energy, electrical energy, ionizing or nonionizing radiation, chemical energy, and thermal energy). These techniques can be conducted in conjunction with barrier analysis. Energy trace and barrier analysis combines energy analysis and barrier analysis. The analysis can produce a consistent, detailed understanding of the sources and nature of energy flows that can or did produce accidental harm, (also see Chapter 3, page 97).

Event and Casual Factor Charting   This method utilizes block diagrams, charts, or models to depict cause-and-effect propagation. The technique is effective for solving complex situations because it provides a means to organize the information, produces a summary of what is known and unknown about the event(s), and results in a detailed sequence of functions, facts, and activities. Many different types of charts, models, and diagrams can be used in combination, such as flowcharts, event trees, fishbone diagrams, or network diagrams.

Event Trees   An event tree models the sequence of events that results from a single initiating event. The tool can be used to organize, characterize, and quantify potential accident sequences. The analysis is accomplished by selecting initiating events, both desired and undesired, and developing their consequences through consideration of system/component failure-and-success alternatives. Nodes within event trees can be represented by probabilities, which have been developed by fault-tree inputs.

Flow Analysis   This analysis enables the evaluation of confined or unconfined flow of fluids or energy, intentional or unintentional, from one component/element/subsystem/system to another. The technique is applicable to all systems that transport or control the flow of fluids or energy.

Health Hazard Assessment   This method is used to identify health risks associated within any system, subsystem, operation, task, or procedure. The method considers the routine, planned, or unplanned use and releases of hazardous materials or physical agents. The technique is applicable to all systems that transport, handle, transfer, use, or dispose of hazardous materials or physical agents.

High Elevation/Fall Hazard Analysis This method is used to evaluate risks associated with conducting operations at elevations, where the risk of falling personnel or objects can exist. The physical controls related to fall protection are evaluated along with all operations and tasks where personnel may be exposed to falling from heights.

Human Error Analysis This method can be used to evaluate the human interface, and error potential within the human–system interaction and to determine human-error-related hazards. Many techniques can be applied in this human factors evaluation. Initiating and contributory hazards are the result of unsafe acts such as errors in design, procedures, and tasks. Human error analysis is appropriate to evaluate any human–machine interface related to maintenance activity.

Human Factors Analysis There are many techniques that represent an entire discipline that considers the human engineering aspects of design. There are many methods and techniques to formally and informally consider the human–engineering interface of the system. Human factors analysis is appropriate for all situations were the human interfaces with the system and human-related hazards and risks are present. The human is considered a main subsystem. There are specialty considerations such as ergonomics, biomachines, and anthropometrics.

Human Reliability Analysis The purpose of the human reliability analysis is to assess aspects that may impact human reliability in the operation of the system. The analysis is appropriate were reliable human performance is necessary for the success of the human–machine systems. Human reliability is important in high-risk operations such as weapon system manufacturing and decommissioning, hazardous materials handling and disposal, weapon delivery, nuclear power, and transportation. The two-person concept is a administrative control in high-risk operations to arm and launch weapons, combat operations, medical procedures, and dealing with high-energy equipment.

Job Safety Analysis This method is one of the first formal safety analysis techniques, which was conducted before World War II. This technique is used to assess the various ways a task may be performed so that the most efficient and appropriate way to do a task is selected. Each job is broken down into tasks, or steps, and hazards associated with each task or step are identified. Controls are then defined to decrease the risk associated with the particular hazards. Job safety analysis can be applied to evaluate any job, task, human function, or operation. The results of job safety analysis can be posted near a particular workstation and can be used as a reference. With appropriate coordination and training, the personnel conducting the particular maintenance task can conduct job safety analysis. However, this analysis can be somewhat complex due to the diversity of the particular maintenance tasks. It is important that the personnel conducting such complex task-related activities be involved in the analysis.

Management Oversight and Risk Tree (MORT)   The MORT technique is used to systematically analyze an accident in order to examine and determine detailed information about the process and accident contributors. This is an accident investigation technique that can be applied to analyze any accident. In some cases, a MORT-like process can be developed to be used as a checklist or guide for conducting hazard analysis. The MORT concept considers three major aspects: the safety program elements, risk understanding, and the control of energy sources associated with the system under evaluation. A fault-tree structure is used with defined criteria to evaluate.

Operating and Support Hazard Analysis   This analysis is performed to identify and evaluate risks associated with the environment, personnel, procedures, and equipment involved throughout the operation of a system. The analysis is appropriate for all operational and support efforts. At the hands of an experienced analyst, this analysis can be fairly extensive, addressing the human element within any system. Risks that address human factors, ergonomics, toxicology, environmental health, and physical stressors can be evaluated from a system view, throughout the life cycle.

Point of Operation Analysis   This analysis evaluates mechanical risks at the point of operation. The analysis assesses the adequacy of machine guarding, interlocking, cutoffs, safety devices, two-hand controls, and automated protective equipment. During maintenance, such safety devices may be bypassed. In specific instances, these risks must be analyzed and additional engineering controls are needed to provide appropriate access for maintenance, teaching of robots, programming, adjustment, or recalibration. When machines are accessed for maintenance, engineered safety precautions must be designed in. For example, stored power should be shunted to ground, physical potential energy sources must be at zero energy state, and personnel must be properly isolated to protect against uncontrolled energy.

Procedure Analysis   This is a step-by-step analysis of specific procedures to identify risks associated with procedures. The technique is universally appropriate. Any procedure must be evaluated from a system safety view. Many related supportive techniques can be applied to conduct procedure analysis, such as task analysis, link analysis, what-if analysis, scenario-driven hazard analysis, flowcharting, prototyping, simulation, and modeling.

Prototype Development   These methods provide a means of modeling/simulation analyses that are constructed in early preproduction of a system, project, or product so that the developer may inspect and test an early version. These techniques are appropriate during the early phases of preproduction and test. Prototyping a system can save costs and provide a capability to evaluate potential subsystem designs, interactions, and interfaces. Consequently, subsystem, interactive, and interface risks can be identified early within a particular complex system.

Root Cause Analysis This set of methods identifies causal factors for an accident or incident. This technique goes beyond the direct causes to identify fundamental reasons for the adverse sequence. Any accident or incident should be formally investigated to determine the initiators and contributors of the unplanned adverse event(s). The root cause is the underlying initial hazard that started the adverse event flow. Combinations of methods have been applied to conduct root cause analysis, such as event trees, fault trees, checklists, and energy and flow analysis. Fault-tree analysis is considered the best tool for identifying root causes. The Cadillac repair and diagnostics procedures make use of this tool, covered in Chapter 5.

Formal and Informal Safety Review Depending on system complexity, system risk, and the extensiveness of the assessment, reviews can range from an informal review by an experienced analyst to a formal board review with expert judgment.[2] The types of reviews conducted will depend on the extensiveness of the system, available resources, regulatory requirements, and current practices [12]. The results of the reviews are to be recorded and documented. Periodic inspections of a system, operation, procedure, or process are a valuable way to determine their safety integrity. A safety review might be conducted after a significant or catastrophic event has occurred or prior to a particular milestone.

Sequentially Timed Events Plot Investigation System (STEP) This method is used to define systems and system flows; to analyze system operations; to discover, assess, and find hazards; to find and assess options to eliminate or control risks; to monitor future performance; and to systematically investigate accidents. In accident investigation, a sequential time of events may give insight into documenting and determining sequential causes of an accident. The technique is universally appropriate.

Task Analysis This technique can be used to evaluate a task performed by personnel from a system safety standpoint in order to identify hazards, to develop notes, cautions, or warnings for integration into procedures, and to receive feedback from operating personnel. Task analysis can be used to evaluate any process or system that has a logical start or stop point or intermediate segments. This methodology is universally appropriate to any operation when specific tasks are conducted. Task analysis can also be used as an input to develop learning objectives for maintenance and safety-related training.

Test Safety Analysis Many methods can be applied to conduct test safety analysis. When there is risk of injury or damage associated with a system, testing safety analysis and risk assessment are required. The risks associated with the initial system (hazardous) exposure are to be identified, eliminated, or controlled to an

---

[2]Expert judgment can be applied to enhance the qualitative assessment. Concurrence is reached on the scenario logic, risk parameters, models, and matrixes. A true consensus review is appropriate and there are formal ways of conducting consensus by the application of *organizational theory*.

acceptable level. Consider initial hazardous exposures involving the first power-up, flight, engine ignition, weapon test, launch, shakedown, or deep dive. During testing, additional system hazards may be discovered, which have to be eliminated or controlled, and safety lessons to be incorporated into the design. Lessons learned from testing can be applied to enhance the design. This approach is applicable to the development of new systems, particularly in the engineering/development phase. It is also important to conduct contingency analysis and planning in the event that the test introduces real-time hazards inadvertently.

Walk-Through Analysis   There are many techniques to conduct a safety walk-through. Generally, a physical inspection or survey is conducted and unsafe acts or unsafe conditions are noted and corrected. Any unusual occurrences, noise, vibration, leakage, elevated temperature, or physical damage are to be noted and investigated. Checklists with performance criteria may also be used. To a safety professional, this could be referred to as real-time hazard analysis. Note that the personnel conducting the walk-through may be exposed to particular risks and they should be properly trained and protected. This technique is a systematic analysis that should be used to determine and correct root causes (or initiating hazards) of unplanned occurrences related to maintenance.

What-If Analysis   This methodology can be equated to scenario-driven hazard analysis, where potential accidents are hypothesized. Initial, contributory, and primary hazards are identified, which are considered unsafe acts or conditions. The analyst hypothesizes what-if scenarios that can occur in the event of any deviation, failure, malfunction, error, fault, or anomaly associated with the system under study. Generally, hazards related to the human, hardware, software, and environment are identified. Consideration is also given to abnormal energy events, interactions, flows, or interfaces.

4.7   TOPICS FOR STUDENT PROJECTS AND THESES

1. Suggest the best definition of maintainability as far as you are concerned. Suggest measures or models.
2. Write a maintainability specification for a product.
3. Write a maintainability demonstration plan for a product.
4. Describe your approaches to designing maintenance-free products.
5. Conduct a study to assess the impact of maintainability on life-cycle costs for a product.
6. Describe how fault-tree analysis can be used for maintainability analysis.
7. Write a generic procedure for a design review for maintainability.
8. Write your approach to testing for maintainability.
9. Discuss the pros and cons of remote maintenance of a complex system.

10. Identify a specific system equipped with remote maintenance capability and select three methods to conduct hazard analysis and compare your results.

11. Choose a complex automated assembly line and identify and evaluate the maintenance-related risks.

12. Consider that you are to evaluate the safety of an existing maintenance facility. The facility meets all applicable codes and regulations. State your case as to why additional hazard analysis may be needed.

13. Select a complex system with human, hardware, software, and environmental elements. Conduct an appropriate hazard analysis and risk assessment of the maintenance operations.

REFERENCES

1. *Maintainability Prediction*, Mil-Hdbk-472, Naval Publications and Forms Center, Philadelphia, 1984.

2. *Electronic Reliability Design Handbook*, Mil-Hdbk-338, Naval Publications and Forms Center, Philadelphia, 1988.

3. C. S. Pillar, Maintainability in Power Plant Design. In: *Proceedings of the Sixth Reliability Engineering Conference for Electric Power Industry*, American Society for Quality Control, Milwaukee, 1979.

4. H. Fujiwara, *Logic Testing and Design for Testability*, MIT Press, Cambridge, MA, 1985.

5. P. K. Lala, *Fault Tolerance and Fault Testable Designs*, Prentice-Hall, Englewood Cliffs, NJ, 1985.

6. *Engineering Design Handbook*, Pamphlet AMCP 706-134, Headquarters, U.S. Army Materiel Command, 1970.

7. *USAF R&M 2000 Process Handbook SAF/AQ*, U.S. Air Force, Washington, DC, 1987.

8. Mil-Std-2165, Testability Program for Electronic Systems and Equipments, 1985.

9. O. E. Ebeling, *Reliability and Maintainability Engineering*, McGraw-Hill, New York, 1997.

10. D. G. Raheja, *Assurance Technologies—Principles and Practices*, McGraw-Hill, New York, 1991.

11. W. E. Tarrents, *The Measurement of Safety Performance*, Garland STPM Press, 1980.

12. M. Allocco, Consideration of the Psychology of a System Accident and the Use of Fuzzy Logic in the Determination of System Risk Ranking. In: *Proceedings of the 19th International System Safety Conference*, System Safety Society, September 2001.

FURTHER READING

Arsenault, J. E., and J. A. Roberts, *Reliability and Maintainability of Electronic Systems*, Computer Science Press, Potomac, MD, 1980.

Blanchard, B. S., and E. E. Lowery, *Maintainability Engineering, Principles and Practices*, McGraw-Hill, New York, 1969.

Patton, J. D., *Maintainability and Maintenance Management*, 4th edn., ISA: Instrumentation Systems and Automation Society, Research Triangle Park, North Carolina.

Goldman, A. S., and T. B. Slattery, *Maintainability, A Major Element of System Effectiveness*, John Wiley & Sons, Inc., Hopoken, NJ, 1967.

Mil-Std-471A, *Maintainability Verification/Demonstration/Evaluation*, 27 March 1973, Superseding Mil-Std-471.

Mil-Std-470B, *Maintainability Program for Systems and Equipment* (current version).

CHAPTER 5

SYSTEM SAFETY ENGINEERING

5.1 SYSTEM SAFETY PRINCIPLES

System safety is defined as the application of engineering and management principles, criteria, and techniques to optimize safety within the constraints of operational effectiveness, time, and cost throughout all phases of the system life cycle [1]. This definition suggests that merely a good design is not the goal; rather, the goal is a successful design, one that anticipates unknown problems in advance, is implemented on time, and minimizes life-cycle costs.

Often, tests are not done under conditions that duplicate field environments, and many human errors, such as miswiring in an aircraft, are easy to make. The safety profession, which deals with preventing loss of human lives, cannot wait for reports on accidents. By that time too many may have died! A safety professional may accept the probability of loss of human life to be less than one per million per year. Since no scientist or engineer can predict such a low probability with any data, attempts are made to eliminate the product hazards altogether. This concern for preventing accidents without the help of much data is responsible for expanding the concept of product safety to system safety. To understand system safety, one must understand the definition of a system as "a composite, at any level of complexity, of personnel, procedures, materials, tools, equipment, facilities, environment and software. The elements of this composite entity are used together in the operational or support environment to perform a given task or achieve a specific mission" [1].

Assurance Technologies Principles and Practices: A Product, Process, and System Safety Perspective, Second Edition, by Dev G. Raheja and Michael Allocco
Copyright © 2006 John Wiley & Sons, Inc.

This definition has a special meaning in system safety. It implies that it is the mission safety that is important. It also implies that all the hardware in the system, including the interfacing hardware, since it can cause an accident, must be compatible. Similarly, all the internal and external software that affects mission safety must be compatible. One must look beyond one's own product to assure safety of the mission. This concept applies even to simple products such as toys. If a certain toy has small parts that can come loose, then the toy design is unsafe; the manufacturer cannot rely on parents to prevent a child from swallowing the parts. Since it is not easy to predict the behavior of parents, the goal should be to design a toy whose parts cannot be swallowed, even if the parents are totally careless. System safety uses such an approach.

To assure mission objectives, the system boundaries must be defined. The boundaries for system safety are subjective. They include interfaces among hardware, software, environments, and humans. The extent to which these factors influence mission safety will determine the boundaries on systems. For example, an electronic system in an automobile would include a mechanical system in its boundary if the mechanical parts degrade the performance of the electronics. Consider the early warning system on an aircraft called engine condition monitoring (ECM). The mission of this equipment is to report on the condition of the jet engines before they malfunction. System safety deals with predicting as many lifetime hazards as possible and controlling them by sound design. The following may be considered as boundaries for ECM.

Hardware Associated with the Mission

- Sensors that may measure incorrect exhaust temperature, compressor pressure, engine speed, fuel consumption, and quietness of operation
- A computer that may malfunction during data analysis
- A printer that prints out wrong reports for maintenance technicians
- Display equipment in the cockpit that may display misleading information

Software Associated with the Mission

- Analytical software in the ECM computer that may have logic faults
- Software for cockpit displays that may have coding errors
- Fault isolation software that may give wrong indication

Humans in the System

- A pilot who may misunderstand warnings
- Technicians who may not follow procedures
- Software engineers who may introduce errors while modifying the software
- Maintenance personnel whose actions may give wrong warnings
- Manufacturing personnel who may not install the sensors right, may miswire, or may mount components backward

Procedures in the System

- Preventive maintenance procedures that may be ambiguous
- Configuration control procedures that may not get implemented
- Software change procedures that may be lax

Materials in the System

- Fuel for engines
- Lubricants
- Deicing agents
- Chemicals used during cleaning, maintenance, and repair
- Deodorants used in passenger cabins

Environments of the System

- High altitude, which may cause erroneous readings in some measurements
- Hot-and-cold cycles that may degrade the accuracy and precision of instruments
- A large number of landings, which may cause cracks
- Vibration, which can cause loose connections

Looking at the partial list above, it is easy to see that there are many road-blocks to system performance and hence safety. Not to be forgotten is peripheral equipment, which may introduce additional hazards. Unless hazards in all the areas above are controlled by design, it is impossible to feel confident about safety.

Example 5.1 Consider a baby who has a heart defect. The parents want to monitor the breathing of the baby so they can immediately take corrective action when needed. The objective here is to monitor breathing performance 24 hours a day. Ordinarily, the parents would consider as the boundary an item of medical equipment that can sound an alarm. If this equipment is truly the boundary, then it should be enough to accomplish the mission safely, and the safety analysis should be limited to it. But the fact is that even if the equipment works reliably, the mission can be unsafe. The system boundary contains some additional factors: the parents may not be able to hear the alarm, or the alarm may not be sounded because of electromagnetic interference from a videocassette recorder or because the remote controller of a TV may interfere with the execution of software in the equipment. The boundary is now extended to the analysis of the parental patterns, the electronic equipment used in the house, and the surroundings, which may accidentally cut off the equipment. Even the baby may accidentally disconnect the equipment from the power source. If the parents are going to be outside the house, a remote alarm system should also be included in the system boundary.

5.1.1 System Safety Process

The system safety process is a risk management process developed to the highest degree. The steps in the process are:

1. Identify the risks using hazard analysis techniques as early as possible in the system life cycle.
2. Develop options to eliminate, control, or avoid the hazards.
3. Provide for timely resolution of hazards.
4. Implement the best strategy.
5. Control the hazards through a closed-loop system.

System safety is not only a function of engineering but is an integral part of top management activities. Participation of management can assure the timely identification and resolution of hazards. Therefore, a major requirement of system safely is that it must be institutionalized. The management tasks as well as the engineering tasks are well identified in Mil-Std-882 series B through E. They are listed in Table 5.1.

5.1.2 Risk Assessment

Risk is a function of the severity of an accident and its probability of occurrence. Typically, various types of analyses will produce hundreds of hazards for complex equipment. No organization can afford to react to all the hazards. The hazards are assigned priorities so that at least the catastrophic and critical accidents are prevented. The example severity categories in Mil-Std-882 are Catastrophic (I)—death or catastrophic system loss; Critical (II)—severe injury, severe occupational illness, or major system damage; Marginal (III)—minor injury, minor occupational illness, or minor system damage; and Negligible (IV)—less than minor injury, occupational illness, or system damage. The example probability (or likelihood) classifications are listed in Table 5.2.

To determine which risks should receive priority, a system of making decisions should be established. Suggested guidelines from Mil-Std-882, based on combinations of severity and frequency, or likelihood, are given in Table 5.3. This approach offers a good structure for risk control. In practice, however, it is difficult to implement. Management participation in resolution of unacceptable risks has been very meager. See Chapter 12 for additional detailed discussions on system risks.

5.1.3 Technical Risk Analysis

Since severity categories I and II deal with catastrophic and major risks, they usually comprise over 80% of the risk. This does not, however, mean category III and IV risks

TABLE 5.1 Management and Engineering Tasks According to Mil-Std-882

Task	Title	Type[b]	Conceptual	Validation	Full-Scale Engineering Development	Production
					Applicability[a]	
100	System safety program	Mgt	G	G	G	G
101	System safety program plan	Mgt	G	G	G	G
102	Integration/management of associate contractors, subcontractors, and AE firms	Mgt	S	S	S	S
103	System safety program reviews	Mgt	S	S	S	S
104	System safety group/system safety working group support	Mgt	G	G	G	G
105	Hazard tracking and risk resolution	Mgt	S	G	G	G
106	Test and evaluation safety	Mgt	G	G	G	G
107	System safety progress summary	Mgt	G	G	G	G
108	Qualification of key system safety personnel	Mgt	S	S	S	S
201	Preliminary hazard list	Eng	G	S	S	N/A
202	Preliminary hazard analysis	Eng	G	G	G	GC
203	Subsystem hazard analysis	Eng	N/A	G	G	GC
204	System hazard analysis	Eng	N/A	G	G	GC
205	Operating and support hazard analysis	Eng	S	G	G	GC
206	Occupational health hazard assessment	Eng	G	G	G	GC
207	Safety verification	Eng	S	G	G	S
208	Training	Mgt	N/A	S	S	S
209	Safety assessment	Mgt	S	S	S	S
210	Safety compliance assessment	Mgt	S	S	S	S
211	Safety review of ECPs and waivers	Mgt	N/A	G	G	G
212	Software hazard analysis	Eng	S	G	G	GC
213	Government furnished equipment/government furnished property system safety analysis	Eng	S	G	G	G

[a]Applicability codes: S—selectively applicable; G—generally applicable; GC—generally applicable to design changes only; N/A—not applicable.
[b]Task type: Eng—system safety engineering; Mgt—management.

TABLE 5.2 Probability Classifications from Mil-Std-882

		Frequency of Occurrence	
Description	Level	Individual Items	Fleet or Inventory
Frequent	A	Likely to occur frequently	Continuously experienced
Probable	B	Will occur several times in life of an item	Will occur frequently
Occasional	C	Likely to occur sometime in life of an item	Will occur several times
Remote	D	Unlikely but possible to occur in life of an item	Unlikely but can reasonably be expected to occur
Improbable	E	So unlikely, it can be assumed occurrence may not be experienced	Unlikely to occur, but possible

should be neglected, but much less attention is required. In deciding whether to accept the risk from unresolved hazards, an organization should develop a model for assessing their impact. The most practical model, which can be translated into dollars, is

$$\text{Risk} = F \times S \quad \text{summed for all category I and II risks}$$

where F is the frequency of the risk and S is the severity of the hazard in dollars. The severity of a hazard in dollars is computed from the value of the personnel liability and property damage, as estimated for that risk. A common mistake is that engineers do not define whether the frequency is per year or for the entire life of the system.

If one wishes to assess total technological risk, the risk model must contain not only the hazards but also the failures not related to safety. This can be written as

$$\text{Technological risk} = F \times S$$

summed for all critical and major failures or hazards whether or not the failures or hazards result in an accident

Typical publications quantify risk as a product of severity and probability or frequency, but they fail to state a period for the probability or frequency. Is it 1 year,

TABLE 5.3 Hazard Priority Guidelines from Mil-Std-882

Hazard Risk Index	Suggested Criteria
IA, IB, IC, IIA, IIB, IIIA	Hazard unacceptable
ID, IIC, IID, IIIB, IIIC	Hazard undesirable (higher management decision is required)
IE, IIE, IIID, IIIE, IVA, IVB	Acceptable with review by management
IVC, IVD, IVE	Acceptable without review

2 years, or the lifetime? They also fail to clarify when to use probability and when to use frequency. The word probability refers to one accident, but in the lifetime of a system there can be multiple accidents. Hence, frequency should be used if the risk is considered over the entire life cycle. The relation between frequency and probability is

$$F = PNn$$

where P = probability of an accident during a period, N = number of equipment items, and n = number of periods over the intended life.

5.1.4 Residual Risk

If some risks are uncontrolled and management cannot or does not want to control them, then someone must assume the risk. At least the following options are available:

1. Avoid risk by not marketing the product.
2. Transfer risk to a contractor or an insurance company and pay for it in advance.
3. Accept the risk and keep doing what comes naturally. This is the most undesirable approach but also the most common.
4. Prevent risk by switching to another technology or different procedure.
5. Control risk by eliminating the hazards of high priority.

For additional discussions on risk management see Chapters 1 and 12.

5.1.5 Emergency Preparedness

Management must ensure that safety is inherent and there are no operational risks. This can be verified by conducting emergency drills, covered in Chapter 8, and by institutionalizing a closed-loop reporting and correction system for all risks. A major task in emergency preparedness is the amelioration plan: viscinity of hospitals, spontaneous response by doctors, and quick fire rescue squad. Operating personnel must report all "incidents" that have not but could result in accidents. This is referred to as incident analysis; see Chapter 4 for additional detail.

5.2 SYSTEM SAFETY IN DESIGN

Since system safety engineering is the process of applying scientific and engineering principles, criteria, and techniques, its goal is to develop products that are immune to component failures and human errors. To develop robust designs, it is important to understand potential causes (initiators and contributors) of accidents.

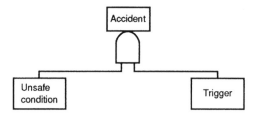

Figure 5.1 Generic causes of an accident.

For hazards to result in mishap, something else also must happen. This is called a trigger event or an initiator. If either the hazard or the trigger event is prevented, the mishap can be prevented, as Fig. 5.1 shows. Examples of hazards and triggers are the following:

Hazard	Trigger Event
High temperature	Temperature is above flash point
Nuclear gas leakage	Operator does not notice
Train driver is under the influence of drugs	Train on wrong track close to an oncoming train
Cracks in aircraft fuselage body	Too many landings and takeoffs

5.2.1 Criteria for a Safe Design

A safe design is one that accomplishes at least the following four goals:

1. It minimizes the gravity of danger in normal use: there are some gray areas. For example, if an automobile bursts into flames when accidentally hit by a driver who could not stop at the traffic light, the automobile may be considered inherently dangerous. One can argue that the accident was caused by human error and the product did not contribute to the accident. A jury may not agree; it may decide that such accidents are normal events and the automobile should not go into flames in normal use.

2. It minimizes the likelihood of danger: in other words, it minimizes the frequency of accidents. This can be achieved by designing products with higher reliability.

3. It uses state-of-the-art technology for minimizing risks as much as possible. In the example in goal 1, a steel plate between the fuel tank and the driver should prevent an explosion.

4. It minimizes life-cycle costs. Consider the cost of a steel plate at $11 per automobile. If millions of cars are produced, the total cost will run into many millions of dollars. If the cost is prohibitive, one should evaluate alternatives

such as adding a plate made of a tough plastic like Lexan or lining the gas tank with plastic. Usually a life-cycle cost analysis (Chapter 4) will reveal that a costly design change will produce lower life-cycle costs for at least one of the alternatives. Cost/benefit analysis is required for a product that is needed by the society such as X-ray or MRI machines, which save millions of lives. A particular utility product such as an automobile should be safe regardless of the cost. Society can do well without unsafe cars.

Caution: Safety decisions are not based on minimizing life-cycle costs alone. They should be based on all four goals above.

Example 5.2   What are the benefit-to-cost ratios for each design option below. The data apply to prevention of the exploding car accident described above.

Option 1: Install a steel plate: cost $14. This will prevent all explosions.

Option 2: Install a Lexan plastic plate: cost $4. This will prevent 95% of explosions.

Option 3: Install a plastic lining inside the fuel tank: cost $2. This will prevent 85% of explosions.

The following data apply to the current design:

Possible fatalities from vehicles already shipped: 180
Expected cost per fatality: $500,000
Number of injuries expected (no fatality): 200
Cost per injury: $70,000
Expected number of vehicles damaged (no injury): 3000
Cost to repair the vehicle: $1200
Number of vehicles shipped: 6,000,000

Solution   The cost for each option is the cost of implementing the change. The benefits are in terms of lives saved and avoidance of injury and damage.

Option 1:

$$\text{Cost} = \$14 \times 6{,}000{,}000 \text{ vehicles} = \$84{,}000{,}000$$

$$\text{Benefits} = (180 \text{ lives saved})(\$500{,}000) + (200 \text{ injuries prevented})$$

$$\times (\$70{,}000) + (3000 \text{ damaged vehicles prevented})(\$1200)$$

$$= \$107{,}600{,}000$$

$$\text{Benefit/cost ratio} = \frac{\$107{,}600{,}000}{\$84{,}000{,}000} = 1.2809$$

Option 2:

$$\text{Cost} = \$4 \times 6{,}000{,}000 = \$24{,}000{,}000$$

$$\text{Benefits} = (95\% \text{ accidents prevented})$$
$$\times [(180 \text{ fatalities})(\$500{,}000) + (200 \text{ injuries})$$
$$\times (\$70{,}000) + (3000 \text{ vehicles})(\$1200)]$$

$$\text{Benefits} = 0.95 \times \$107{,}600{,}000 = \$102{,}220{,}000$$

$$\text{Benefit/cost ratio} = \frac{\$102{,}220{,}000}{\$24{,}000{,}000} = 4.2592$$

Option 3:

$$\text{Cost} = \$2 \times 6{,}000{,}000 - \$12{,}000{,}000$$

$$\text{Benefits} = (85\% \text{ accidents prevented})[(180 \text{ fatalities})(\$500{,}000)$$
$$+ (200 \text{ injuries})(\$70{,}000) + (3000 \text{ vehicles damaged})(\$1200)]$$
$$= 0.85 \times \$107{,}600{,}000 - \$91{,}460{,}000$$

$$\text{Benefit/cost ratio} = \frac{\$91{,}460{,}000}{\$12{,}000{,}000} = 7.6217$$

Option 3 has the highest benefit/cost ratio. As noted earlier, however, the decision cannot be based on this figure of merit alone. From a moral standpoint, option 1 is the best. Actually, option 3 has some hidden costs, which were not considered here, such as loss of future customers. But this option is better than not taking any action at all.

5.2.2 Safety Engineering Tasks

Several analyses can be applied as the design progresses. A typical chart of tasks is shown in Fig. 5.2. It is important to recognize that as much as possible the tasks should be performed according to principles of concurrent engineering. This assures that system safety is consistent with reducing costs and enhancing schedules. Safety must always make business sense. Therefore, hazard analysis techniques form the core of system safety methodology. They allow designers to make engineering changes very early when they are least costly to implement.

5.2.3 Preliminary Hazard Analysis

Preliminary hazard analysis (PHA) is done at the concept stage so that safety considerations are included in trade-off studies during early design. The purpose is to obtain initial risk assessment of a concept and to make sure the conceptual framework is sound. Time and money are wasted if the concept has to be

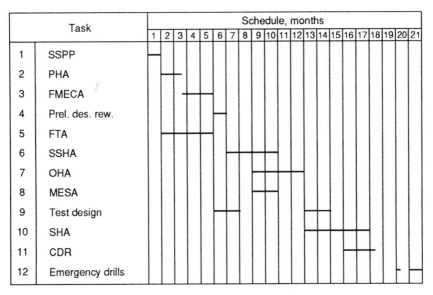

	Task	Schedule, months
		1 2 3 4 5 6 7 8 9 10 11 12 13 14 15 16 17 18 19 20 21
1	SSPP	
2	PHA	
3	FMECA	
4	Prel. des. rew.	
5	FTA	
6	SSHA	
7	OHA	
8	MESA	
9	Test design	
10	SHA	
11	CDR	
12	Emergency drills	

Figure 5.2   A chart for safety task.

changed later. Besides, it may be too late to change the concept. Many analysts find it difficult to conceptualize the hazards, but a systematic process should make it easier. The depth of analysis always depends on the objectives and on the severity of the accidents. The objective in PHA is to identify the hazards and initial risks. As a minimum, the hazards that will impede the mission objectives must be identified. Such hazards are not limited to hardware and at system level include:

Hardware hazards
Software hazards
Procedural hazards
Human factors
Environmental hazards (including health hazards)
Interface hazards

The PHA process can be demonstrated with a chemical plant example (Fig. 5.3). The plant in this example stores a chemical in an underground storage tank from which the daily requirements are delivered to a unit tank. The chemical, if exposed to humans, is deadly. In addition, it can react within itself and cause an explosion; it also reacts violently if it comes in contact with moisture or water. The plant contains some safety features. Temperature and inventory sensors monitor the heat and the quantity of the chemical, respectively. A pressure monitor is included as a precaution. If the pressure is too high, a pressure-relief device vents small amounts of gas, which are piped through a scrubber and neutralized. In addition, a flare unit can burn off most of the gas from a major leak.

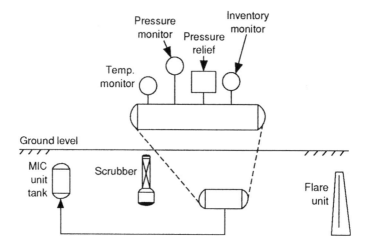

Underground storage, or "MIC"

Figure 5.3 A chemical plant.

PHA is basically a brainstorming technique, but some kind of organized approach helps in starting the process. Some system-level questions can be:

Approach A: System Engineering Approach

Are there any hardware hazards?

Are there any software hazards?

Are there any human-induced hazards?

Are there any procedure-related hazards?

Are there any hazards to the public environment such as noise, radiation, or contamination?

Are there hazards from the existing environment?

Are there any obvious interface hazards from the existing software, hardware, and humans?

Are there hazards due to materials such as lasers, propellants, and fuels?

Approach B: Controlling the Energy

According to approach B, if certain energies in the system are controlled, there will be no accident. The following energies can be present in the chemical plant in Fig. 5.3:

Chemical Energy: too much reaction, chemical reacting with the tank material, moisture in the tank, pipe corrosion, tank corrosion, gas leaks

Thermal Energy: high temperature in the tank, high temperature outside the tank
Pressure Energy: high pressure in the tank, high pressure in the pipes
Mechanical Forces: vibration in the pipes, vibration in the tank
Electric Energy: high voltage in the cables on control panels

Once these energies are identified, the following checklist can be applied. The checklist is a 10-point precedence test proposed by Dr. William Haddon [2]:

1. Prevent the marshaling of the energy in the first place.
2. Reduce the amount of energy marshaled.
3. Prevent release of energy.
4. Modify the rate or spatial distribution of energy.
5. Separate the energy release in space or time.
6. Do not separate the energy release in space or time, but interpose a material barrier.
7. Modify the contact surface or basic structure.
8. Strengthen the structure.
9. Move rapidly in detecting energy release and counter its extension.
10. If energy is beyond control, then stabilize the altered condition.

Approach C: Exploring the Life-Cycle Phases

This method uses systematic prediction of possible hazards during the entire life cycle of a product. For each life-cycle phase, each hazard is assessed and controlled, if possible.

Design: Are there hazards in the present design? Could there be any design mistake to consider?

Manufacturing: Are their any hazards if the product is made as intended? Could any manufacturing mistakes be made that will result in a hazard?

Testing: Are there any hazards during testing the product? Could any hazard be created during testing?

Storage: Are there any hazards in the way we plan to store the product? Could anyone make mistakes in storage leading to unsafe conditions?

Handling and Transportation: Are there any hazards if the product is handled and transported as intended? Could someone introduce hazards using unintended procedures?

Operations and Use: Are there any hazards during intended use? Could someone introduce hazards by misuse or use in a different manner?

Repair and Maintenance: Are there any hazards present during normal repair and maintenance? Could someone introduce hazards by mistake?

Preliminary hazard analysis					
Hazard	Cause	Effect	Criticality	Recommendations	Final action
1) Temperature too high	Gauge malfunction	Gas leak	I	Provide cooling jacket on tank or redundant gauge	Provide cooling jacket
2) Too much inventory	Sensor malfunction	Explosion	I	Install redundant sensor	Install redundant sensor with automated control
3) Leaks through pipe joints	Vibrations, corrosion	Gas leak	I	Install stainless steel pipes	Install stainless steel pipes with flexible joints

Figure 5.4   PHA matrix for the chemical plant of Fig. 5.3.

Emergency Situations: Are there hazards in a situation such as emergency shut-down of an operating system? Can someone make mistakes because of unusual work pressure?

Disposal: Are there hazards in disposing of the product as intended? Could someone use different disposal procedures and introduce hazards?

An example of PHA format is provided in Fig. 5.4. It should be noted that at PHA level major design improvements can and should be made. These improvements are shown in the last column of all hazard analyses. In the PHA in Fig. 5.4, the design changes are such that the whole design philosophy has to be changed. The existing design is a series structure (see Chapter 3). The improved design is a fault-tolerant structure involving redundancies.

A fault-tree analysis (explained later in this chapter) for the system level failures can also help in identifying potential hazards. Other techniques are cause–effect diagrams (Chapter 6), design reviews, and simulation.

Hazard Control Criteria   Figure 5.5 shows the design criteria for controlling potential hazards. They apply to all hazard analyses. They are presented in order of precedence and can be summarized as follows:

1. Design the hazard (or risk) out of the product. If the hazard cannot be eliminated, minimize the residual risk.
2. Design for a fail-safe default mode by incorporating safety devices or fault-tolerant features.
3. Provide early warning through measuring devices, software, or other means. The warning should be clear and should attract the attention of the responsible operator.
4. Implement special procedures and training when the above means are unable to eliminate the hazard.

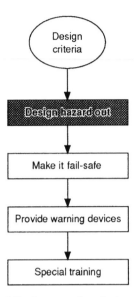

Figure 5.5 System safety design criteria.

5.2.4 Subsystem Hazard Analysis

The literature is very scarce in the important area of subsystem hazard analysis (SSHA). It emphasizes the importance of SSHA, but rarely the "how" of it. SSHA is a detailed extension of PHA. For less complex equipment, where it is not necessary to break down the system into subsystems, it may simply be called hazard analysis. SSHA is based on the premise that an accident is a result of many events (hazards and trigger events) and can be prevented as long as the events are prevented. If the events can be prevented, then it is prudent to perform a fault-tree analysis to reduce the number of initiators and contributors forming the accident.

To perform subsystem hazard analysis, the entire system is divided into subsystems. The subsystems themselves usually are complex enough to be considered systems. As an example, consider a mass transit subway system. The system mission is to transport the public safely underground. The subsystems may be defined as follows:

Trains

Station structures, including escalators and elevators

Tunnels and underground system

Central control room

Track and controls

Provisions for the elderly and handicapped

Communications

Fare collection

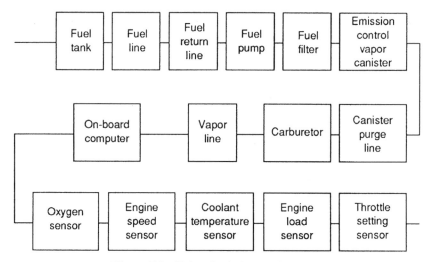

Figure 5.6 Units of a fuel control system.

For each of these subsystems, a separate analysis should be performed. Each component or unit of the subsystem should be analyzed. In Fig. 5.6, the units of a fuel control subsystem of an automobile are shown. The subsystem hazard analysis for this subsystem is given in Fig. 5.7. One should consult the bill of material to determine what to analyze. It is easy to miss important components such as adhesives, lubricants, and cables; high-voltage cables have been known to cause

Subsystem Fuel control System 'x' vehicle	Subsystem Hazard Analysis			Date _____ Page ___ of ___ Analyst _____	
Part Name/ part number	Hazard description	Accident (trigger event)	Criticality	Recommended control	Revised criticality
Fuel tank	Tank rupture	Differential in pressure between the inside and outside of the tank	I B	Qualify tank for 20-year ability to withstand rupture forces. Use ductile material.	I
	Failure of tank seal	Any kind of flame or spark	I B	Provide double seal.	I C
Fuel pump	Fire	Spark in the area	II A	Mount fuel pump inside the tank.	II E
Fuel line	Rupture	Corrosion sufficient to penetrate through the wall thickness	II C	Use stainless steel lines or corrosion-resistant material.	II E
Filter	High buildup of pressure	Dirty or clogged filter	I C	Design for pressure relief. In addition, design filter for twice the current life.	I C
Carburetor	Fuel leakage	Vehicle rollover	II D	Use electronic fuel injection.	III D

Figure 5.7 Subsystem hazard analysis.

deformities in animals and farm products, while adhesives can be toxic. The analysis steps for each unit are:

1. Identify hazards and risks.
2. Identify initiators and contributors.
3. Classify the severity.
4. Identify potential accident sequences.

The corrective action is determined by the severity of risk. For category I and II risks, the mitigation is accomplished by improving the design. Inspection, testing, warnings, and procedural actions are not acceptable unless one is willing to accept the risk. If a design improvement is not obvious, it is necessary to perform fault-tree analysis (Section 5.2.5) or the equivalent to justify why one cannot change the design. Many times the FTA will lead to a reduction in frequency or may provide a complete solution.

5.2.5 Fault-Tree Analysis

Fault-tree analysis (FTA) is a deductive process especially useful for analyzing category I and II risks, when the risks have not been resolved. If the risks can be resolved by design, the fault-tree analysis may not be necessary. The purpose is to identify the initiators and contributors so that an effort can be made to eliminate as many causes as possible.

FTA can also be used for analyzing any problem: technical, management, or administrative. It can be used for manufacturing process control, human factors, and software mishaps. Some perform FTA at system level to identify many risks.

FTA is a cause-and-effect diagram, which uses standard symbols developed during the Minuteman missile program. It is also called root cause analysis. A sample fault-tree analysis is shown in Fig. 5.8. The definitions of symbols commonly used are shown in Fig. 5.9.

Construction of the tree in Fig. 5.8 started with a logical question: How can electromagnetic interference (EMI) cause an accident? Remembering that an accident is always a result of many initiators and contributors, an engineer constructed the second level of FTA. Here the hazard is "excessive EMI" and the trigger event is "EMI able to interfere with navigation." The AND gate symbolizes that both of these events must occur for an accident to occur.

The branch "excessive EMI" shows four elements under an OR gate. Any of the four elements (electric motor operations, equipment transients, power generation noise, other causes (secondary faults)), or any combination of them, can produce excessive EMI. The process of analyzing causes for each event is carried out until further analysis is not meaningful. In a hardware tree, the analysis may stop when failure (or hazard) causes are traced to component faults, which are represented by circles. Similarly, circles may be used for the human or software details when the root causes have been postulated.

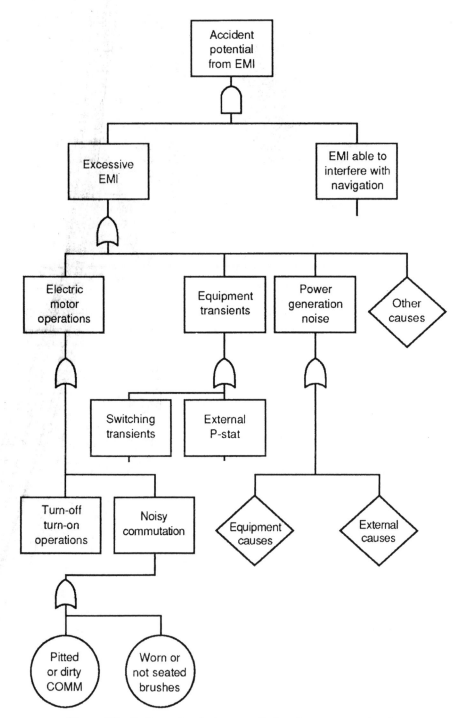

Figure 5.8   Fault tree for electromagnetic interference on an aircraft.

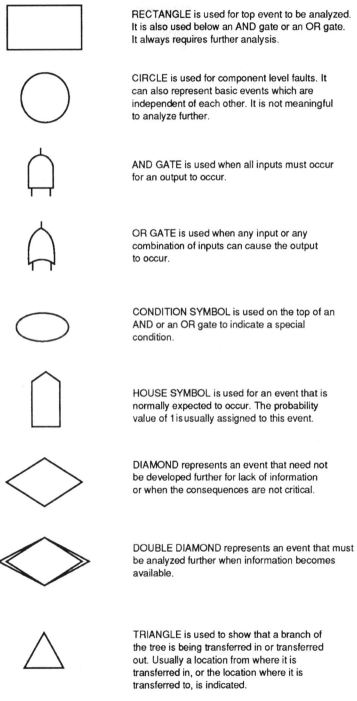

RECTANGLE is used for top event to be analyzed. It is also used below an AND gate or an OR gate. It always requires further analysis.

CIRCLE is used for component level faults. It can also represent basic events which are independent of each other. It is not meaningful to analyze further.

AND GATE is used when all inputs must occur for an output to occur.

OR GATE is used when any input or any combination of inputs can cause the output to occur.

CONDITION SYMBOL is used on the top of an AND or an OR gate to indicate a special condition.

HOUSE SYMBOL is used for an event that is normally expected to occur. The probability value of 1 is usually assigned to this event.

DIAMOND represents an event that need not be developed further for lack of information or when the consequences are not critical.

DOUBLE DIAMOND represents an event that must be analyzed further when information becomes available.

TRIANGLE is used to show that a branch of the tree is being transferred in or transferred out. Usually a location from where it is transferred in, or the location where it is transferred to, is indicated.

Figure 5.9 Common fault-tree symbols.

Usually a tree cannot be fitted on one page. A triangle is then used to denote that a portion of the tree is somewhere else. The location of this portion is indicated in the triangle. A triangle is also used for a branch, which is repeated in several places.

Identifying Causes (Adverse Sequences) Within a Fault Tree   Conceptually, one needs to have a systematic method of developing the tree. At least eight factors, shown below, can provide clues about the sources:

1. Can any equipment failures contribute to this effect?
2. Can any material faults contribute?
3. Can human errors contribute?
4. Could methods and procedures contribute?
5. Could software performance be a problem?
6. Could maintenance errors or the absence of maintenance contribute?
7. Could inaccuracies or malfunction of the measuring devices contribute?
8. Could environments such as chemicals, dust, vibration, shock, and temperature contribute?

Computing Probability or Likelihood of an Accident   If the fault tree contains the estimates of component or event probabilities, then the probability of accident can be computed by applying two fundamental rules to AND and OR gates starting at the bottom of the tree:

1. *Multiplication Rule*: When an output results because of two or more input events occurring at the same time, as in an AND gate (Fig. 5.9), the output probability is computed by multiplying the input probabilities:

$$P(\text{Output}) = P(1) \times P(2) \times P(3) \times \cdots \times P(N)$$

This rule assumes the input probabilities are independent of each other. For example, carburetor and radiator failures in an automobile are independent of each other because if one fails, the other is not affected. If a component fails as a result of the failure of another component, its probability of failure is not independent. Then a new equation has to be written on the basis of the relationships. Reference 3 covers this situation. However, most fault-tree components are statistically independent of each other. This computation will rarely be required.

2. *Addition Rule*: For an OR gate, where the occurrence of any input causes the output to occur, the probability calculation is based on whether the events are mutually exclusive or not. If two events are mutually exclusive, such as high pressure and low pressure (if one occurs, the other cannot occur at the same time), the probabilities are simply added up:

$$P(\text{Output}) = P(1) + P(2) + P(3) + \cdots + P(N)$$

If the events can occur at the same time (if they are not mutually exclusive, such as high temperature and high pressure), then

$$P(\text{Output}) = 1 - \{[1 - P(1)][1 - P(2)][1 - P(3)] \cdots [1 - P(N)]\}$$

If the probabilities are small (less than 0.001), they can be added together like probabilities of mutually exclusive events. The answer will be very close. Since probabilities in the fault trees are very low, most practitioners disregard the distinction between mutually exclusive and not mutually exclusive.

FTA is an excellent tool for approving design for safety. If an accident scenario has an OR gate at the top, reject the design. Either the analysis is flawed or the design is dependent on a single-point flaw.

5.2.6 Cut Set Analysis

Fault trees can be put to many uses. One such use is identifying design weaknesses and strengths. A single event or component that can cause the system to fail is a potential weakness in the system. A tool called cut set analysis allows identification of all single-point failures and other paths that lead to a failure.

A cut set is a set of basic events whose occurrence causes the system to fail. For a hardware system, it is a set of component failures leading to a system failure. A cut set is called minimum cut set if it cannot be reduced and can still cause the system to fail. Barlow and Proschan [4] developed an algorithm, which is available in several commercial computer programs, for reducing a cut set to the minimum.

This algorithm is illustrated in Fig. 5.10. Gates are labeled G0, G1, G2, ..., GN. Numbers 1, 2, 3, ..., N represent the other basic events. The algorithm begins with the top event, G0. The elements (G1, G2) of this gate are connected by an OR gate and are entered in separate rows in the matrix in the figure. (If the elements were connected by an AND gate, they would be entered in the same row.) All the gates are entered in the same manner until no more are left. This completes the analysis. The final analysis shows all the possible cut sets. A step-by-step walk-through follows.

The matrix in Fig. 5.10 shows eight steps. The first step shows gate G0 replaced by G1 and G2 in separate rows because they are connected by an OR gate, as noted earlier. In step 2, gate G1 is replaced by its elements G3, G4, G5, and 4. (Gate G2 is not replaced yet.) Step 3 shows G3 replaced by its elements G6 and G7. (Other elements are untouched.) In step 4, gate G6 is replaced by its elements 11 and 12 in the same row because they are connected by an AND gate. (The remaining elements are untouched.) The analysis continues in this way until step 8, when no gates are left. Each row is then a cut set. In this case, since the basic events are not repeated in the other sets, each row is a *minimum* cut set. Events (or components) 3, 7, 8, 9, 10, and 4 are single-point failures, which can cause the system to fail. It is here a designer should consider redundancy or fail-safe features.

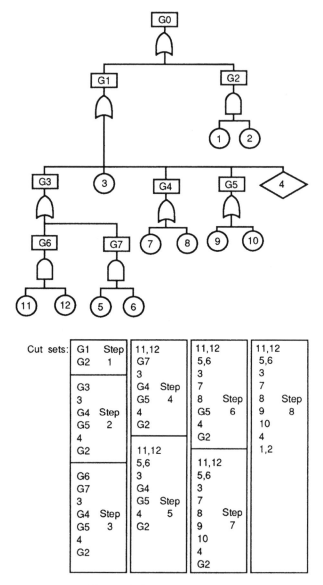

Figure 5.10   Cut set analysis.

A similar tool called minimum path set analysis is an extension of this analysis. A path set is a set of components (or events), which, if it works, assures the system will work. A path set is a minimum set if it cannot be reduced further. The procedure is to change the OR gates to AND gates, and vice versa, in a fault tree and use the cut set algorithm. Each row in the final step is a path set. If any of these paths work, the system will work. An example of this analysis is given in Reference 4.

5.2.7 Failure-Mode, Effects, and Criticality Analysis

Failure-mode, effects, and criticality analysis (FMECA) evaluates the effect of component failures and is usually done by reliability engineers. (For details, see Chapter 3.) This is insufficient. For system safety, a FMECA should be done first for the failures of the system functions. That will lead to many changes in the system specification. As a result, many component designs will change. For safety-critical components, the safety margin should be at least 3.[1] Only after implementing these changes does it make sense to do FMECA at the component level. Safety engineers, reliability engineers, and design engineers working together would be more effective, however. The reliability engineers may not pay detailed attention to safety-critical failure modes, and they often will not consider a failure critical when redundancies in design are present. A safety engineer may still consider such a failure a dangerous situation. For example, if an aircraft device has a redundant circuit board, a failure may not be critical if only one circuit board fails. But if the failed board is not detected and replaced immediately, another failure could be catastrophic. The safety engineer in this case has to assure that the repair action will be performed immediately. Assessing the consequences of failure requires special training. Many reliability engineers do not go to proper depths. For example, if a motor for an escalator on a train station fails, a reliability engineer may not relate the stopping of the escalator to a risk, but a safety engineer knows that this can contribute to accidents. If an escalator comes to an abrupt stop, people fall on each other. A good design would require the escalator to come to a gliding stop. The least a safely engineer should do is let reliability engineers complete the FMECA and then review the effects of failures independently.

5.2.8 Maintenance Engineering Safety Analysis

Maintenance engineering safety analysis (MESA) should first be done in the design phase, even though some may claim that maintenance procedures are not known until the system is in place. If one waits for the system to be installed, it is too late. If maintenance hazards are to be prevented, the system must be designed accordingly.

The difficulty of not having procedures is overcome in the following manner. A rough draft of the potential procedures is constructed with the help of experienced maintenance technicians who have worked on similar systems. Usually, at least 80% of the tasks are similar. This knowledge is supplemented by a method called technique of human error rate prediction (THERP) [5]. This consists of analyzing maintenance safety on the basis of interviews with operating and maintenance personnel on similar systems. The analysis is updated as the design details become available. The best time to start this analysis is when preliminary design is approved. It must be

---

[1]See the following for further discussion: D. Raheja, There Is a Lot More to Reliability than Reliability, SAEWorld Congress paper, 2005.

complete before the critical design review. This is the last opportunity to make design changes.

MESA consists of writing the procedure in logical tasks and then treating each task as a component for hazard analysis. As an illustration, take three tasks in sequence for a proposed preventive maintenance procedure for a high-voltage transformer.

Task 1: Shut off the power.

Task 2: Check the oil level with a dipstick (a metal rod).

Task 3: Remove an oil sample for chemical analysis.

In task 1, the following hazards are possible:

- The technician may think the power is off, but it is really on because of feed-back loops.
- The power may be off, but there is a residual high voltage.
- Someone may turn on power at the source.

The task 2 hazards are

- The technician can be electrocuted through the metal dipstick if the power is accidentally on.
- Contact with transformer oil can cause toxic effects.

The analysis is shown in Fig. 5.11. It is interesting to note the design improvement for task 2. To prevent exposure to high voltage, use of a sight gauge is suggested (instead of a dipstick) so that oil levels can be observed remotely. This action also prevents an accident from task 1, since it is no longer necessary to shut down the power. The oil sample can be removed through an insulated pipe without shutting down the power. Figure 5.12 shows a format suggested by the U.S. Navy [3].

5.2.9 Event Trees

Event trees are complements to fault trees. They trace the effect of a mishap and lead to all possible consequences. This information is required to develop training for emergency drills. In large systems, it is imperative that a mock accident be created periodically to see if people know what to do during emergencies and disasters. This is an excellent opportunity to check whether the evacuation plans are realistic. Such drills reveal many weaknesses in the system, which can result in multiple fatalities.

The procedure for constructing event trees is to visualize two sides, positive and negative, for each event. For example, the tree in Fig. 5.13 shows that an alarm may

Procedure title	Transformer preventive maintenance	Maintenance Engineering Safety Analysis			Date 3/16/89 Page 1 of 25	
System	Power distribution				Analyst John Doe	
Task number	Task description	Hazard	Cause of hazard	Criticality	Recommended control	Revised criticality
001	Shut off power.	Power may not be cut-off.	Someone may accidentally turn on the switch, or the circuit may malfunction.	I C	Provide postive cut-off switch locally within the equipment.	Hazard no longer exists.
002	Check the oil level with a dipstick.	High voltage shock through metal dipsticks.	Same as in task 001.	I C	Provide a sight gauge so the need for this task is eliminated (Note: task 001 is no longer required.)	Hazard no longer exists.
003	Remove an oil sample for chemical analysis.	Electrocution if power is on accidentally.	Oil conducts electricity.	I C	Provide a sight gauge which may indicate contamination. Alternatively, provide an insulated pipe through which oil can be removed without shutting off the power.	Hazard no longer exists.

Figure 5.11 Simplified maintenance engineering safety analysis.

work or it may not. If it works, there is no damage done. If it does not, then there is a chance of manual detection; a person may detect the leak or may not. The analysis is carried on until all the consequences are considered. Event trees can also be used for estimating the probability of each consequence, since the sequence of events is known.

System Surface Missile			Maintenance Hazard Analysis (LIHA)		Analyst _____ Sheet ____ of ___	
Subsystem _____					Maintenance level ____ Date ____	
Equipment	Maintenance type	Maintenance function	Hazard	Hazard class	Safety feature or recommendation	Remarks/ recommendation
1	2	3	4	5	6	7
CW transmitter	Corrective maintenance	Remove and replace high voltage power supply	Access to HVPS will activate safety interlocks causing HV to be discharged. Activation of battle short during this maintenance action bypasses the interlocks, creating a serious personnel hazard.	IV	Cabinet has an inadequate safety interlock design capable of being bypassed during battle short mode operation. Should have method positively assuring no voltage when desired.	Recommended power interrupt switch located on the cabinet to inhibit all power including battle short, during maintenance. A holding relay circuit could also be considered for this purpose.
Launcher	Preventive and corrective maintenance	All maintenance or work performed on or near the launcher.	Personnel working in or near the launcher are susceptible to injury from the launcher rails during remotely controlled train and elevation.	III IV	Locking pins are adequate when not in an alert status. A warning alarm is installed at the launcher to warn personnel of movement. Sound power telephone outlet for safety watch.	A launcher train circle should be provided to indicate danger zones. Ref. ship installation specification.

Figure 5.12 Maintenance hazard analysis from a U.S. Navy example.

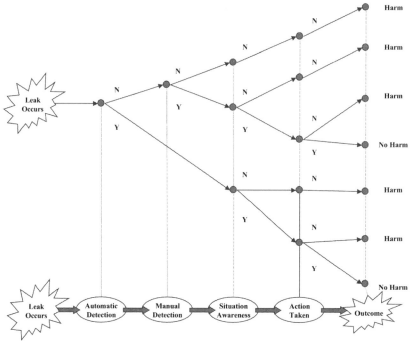

Figure 5.13 Event tree for a gas leak.

5.2.10 Operating and Support Hazard Analysis

Operating and support hazard analysis (O&SHA) is done for operating procedures and support functions (see Chapter 4 for further information on operating and support hazard analysis) such as:

- Production
- Testing
- Deployment
- Handling
- Servicing
- Storage
- Modification
- Demilitarization
- Disposal
- Emergency action

The technique requires identification of activities by a procedure similar to MESA. Maintenance activities can also be covered in this analysis. Since maintenance activities usually are too numerous, MESA is done separately.

The inputs for this analysis come from previously mentioned analyses in which many operating hazards are identified. Other inputs that can help are prototype tests, mock installations, emergency procedures, and interviews with operating and maintenance personnel.

The analysis should include [1] the following:

- Activities that occur under hazardous conditions, their time periods, and the actions required to minimize their risk
- Changes needed in functional or design requirements for system hardware/ software, facilities, tooling, or support/test equipment to eliminate hazards or reduce associated risks
- Requirements for safety devices and equipment, including personnel safety and life-support equipment
- Warnings, cautions, and special emergency procedures (e.g., egress, rescue, backout)
- Requirements for handling, storage, transportation, maintenance, and disposal of hazardous materials
- Requirements for safety training and personnel certification

Analysis shows the weaknesses and critical tasks where a hazard is possible. Such tasks can be highlighted by darker print, and flowcharts can be constructed with darker blocks depicting safety concerns and dotted lines for nonsafety activities. Similarly, dark arrows can be used to show that certain sequences must be adhered to. A format for reporting operating hazards is shown in Fig. 5.14.

Subsystem (1) _____

Operation Safety Matrix   Activity (2) _____

System safety engineering analysis

OSM Item No. (3)	Task Description (4)	Hazardous element (5)	Potential accident (6)	Effect (7)	Hazard class. (8)	Safety requirements (9)	Procedure reference (10)

Prepared by _____

Date _____ Issue ___ Rev. ___

Figure 5.14 Operating and support hazard analysis.

5.2.11 Occupational Health Hazard Assessment

Occupational health hazard assessment (OHHA) identifies health hazards so that engineering controls can be implemented, rather than making short-term fixes or depending entirely on persons to protect themselves. Items to be considered are the following:

- Toxic materials such as poisons, carcinogens, and respiratory irritants
- Physical environments such as noise, heat, and radiation
- Explosion hazards such as concentrations of fine metal particles, gaseous mixtures, and combustibles
- Adequacy of protective gear such as goggles and protective clothing
- Facility environment such as ventilation and combustible materials
- Machine guards and protective devices

5.2.12 Sneak Circuit Analysis

Often a complex product is designed in portions by several design engineers. It is easy to overlook certain paths and conditions, especially with electronics and software. The sneak circuit analysis (SCA) technique was standardized by Boeing in 1967 as a result of its effort on the Minuteman missile [4]. It requires a lot of effort and should be applied selectively.

A sneak circuit is a latent path or condition in a system which inhibits a desired condition or initiates an unintended or unwanted action. This can happen in even a mechanical system. It is not a result of a component failure, nor is it a problem created by electromagnetic interference. It is mainly a result of lack of visibility of all the details. The methodology [6] consists of identifying and analyzing topological patterns from detailed manufacturing drawings, installation schematics, and engineering schematics. A network of trees is constructed with the help of a computer program. Five basic patterns, shown in Fig. 5.15, are vulnerable. Of these, the H pattern is most vulnerable; about 50% of sneak conditions are found in it. Design engineers should avoid the H pattern. The analytical portion consists of asking questions related to each pattern. For example:

Can a switch be open when a load is required?

Can a switch be closed when a load is not required?

Can the clocks be unsynchronized when they are required to be synchronized?

Does the label indicate the true condition? Is anything else expected to happen which is not indicated by the label?

Can a current flow in the wrong direction?

Can a relay open prematurely under certain conditions?

Can power be cut off at a wrong time because of an energized condition in the interfacing circuit?

Can a circuit lose its ground because of interfacing circuits?

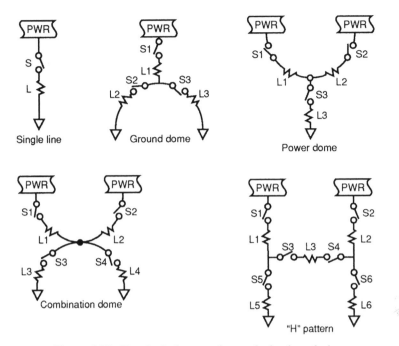

Figure 5.15 Topological pattern for sneak circuit analysis.

Clardy [6] points out that there are four broad categories of sneak conditions.

1. *Sneak Path*: Allows current or energy to flow along a suspected path or in an unintended direction.
2. *Sneak Timing*: Causes functions to be inhibited or to occur at an unexpected or undesired time.
3. *Sneak Label*: On a switch or control device, causes incorrect actions to be taken by operators.
4. *Sneak Indication*: Causes ambiguous or incorrect displays.

Since Boeing software is proprietary, some companies have contracted work through Boeing. Others have developed their own software. Some perform analysis informally. One company put the design engineers of various circuits together in a room and encouraged them to trace the effects of each circuit on interfacing circuits. Another very large company traced all the effects of failures by FMECA in a space program and determined whether a circuit was fault-tolerant. It discovered several sneak conditions.

Caution: A question such as this may come up. What will happen to the interfacing circuits if a relay is closed suddenly because of electrical overload? If the effect is not easy to trace, the best approach is to try it out during a scheduled shutdown or on a simulator. Do not ignore the question because no one has the answer.

5.2.13   System Hazard Analysis

System hazard analysis (SHA) is the analysis of interface effects and interface integration. It should never be done without doing hazard analysis of internal and external interface functions. Results of other subsystem hazard analyses are evaluated to assess the impact on other subsystems and on the total system. Interfaces are of several kinds: hardware to hardware, hardware to software, and software to software, and all the interfaces in a system of systems. Human interfaces also belong in this analysis. There is no standard method, and the format for reporting varies according to the needs of the system.

In the beginning of a project, PHA serves as a rough SHA. PHA is then replaced by complete and detailed analysis. Techniques similar to SSHA can be used for SHA. Inputs from FMECA and SCA are especially valuable because they affect the entire system. Fault-tree analysis can also be a very efficient tool for SHA.

5.3   SYSTEM SAFETY IN MANUFACTURING

Manufacturing processes are vulnerable. They can introduce hazards into products and can be a hazard to working personnel. Analysis for occupational hazards created by processes is covered in Section 5.3.2. Section 5.3.1 covers hazards introduced into a product by manufacturing.

5.3.1   Determining Safety-Critical Items

To assess the hazards and risks being introduced during manufacturing, the method of producing the product should be analyzed as in maintenance engineering safety analysis. The steps of the manufacturing process are reviewed to determine where safety-critical error can occur. Each risk is classified by severity rating. All category I and II risks are called safety-critical items. If a product is shipped with these items uncontrolled, costly recalls can be expected. In one case, a component used in a smoke detector required chemical cleaning during manufacturing, and the cleaning solution had to be changed every 8 hours. A production operator erroneously thought that the solution was clean enough and did not change it. As a result, components rusted in the field. The rust caused some alarms not to function. Under pressure from the Consumer Product Safety Commission, the company recalled all its smoke detectors. The company is no longer in business.

Safely-critical items are not limited to manufacturing processes. They can be a result of handling, packaging, and shipping. Vibration and shock during transportation can cause a product to malfunction. These processes should also be reviewed. Some organizations conduct process FMECA. This is one of the best sources of hazard and risk identification.

5.3.2   Manufacturing Controls for Safety

Control of safety-critical items is generally the responsibility of the process engineer or the production foreperson. But system safety engineers must make them aware of their responsibility. Some organizations label such items critical or major on drawings. Others highlight them in the process sheets.

The following general procedure is a guide for controlling hazards. It is illustrated in Fig. 5.16 and the steps below for a welded joint.

Step 1: Identify the safety-critical item (strength of welded joint).

Step 2: Identify the inputs to the process, which, if not right, will result in a weak joint. (Time allocated for welding; chemistry of weld; material, especially the carbon content, which influences temper qualities; equipment qualified for environment; quality of welding rods; welder certification; surface cleanliness; and storage environment of electrodes. *Note*: If electrodes are not dry, they will absorb oxygen, resulting in hydrogen bursts.)

Step 3: Identify characteristics to control in the process itself (see CAUTION in the figure); other important characteristics such as cooling rate can also be identified.

Step 4: Verify the output characteristics. If the output is not as expected, then either the inputs or the process need reevaluation. (The outputs for a welded joint are good weld penetration, continuous weld, strong weld, freedom from residual stresses, and ductility measured as Brinell or Rockwell hardness number.)

An example of what can go wrong in welding is the new aircraft hangar at a Chicago airport that collapsed when a commercial airliner flew over it. The failure analysis revealed that the welding was done on a very cold day. The faster cooling rate resulted in very brittle joints.

Generally, once quality assurance engineers are aware of the safety-critical items, they can help manufacturing in controlling them. See Chapters 4 and 6 for more techniques.

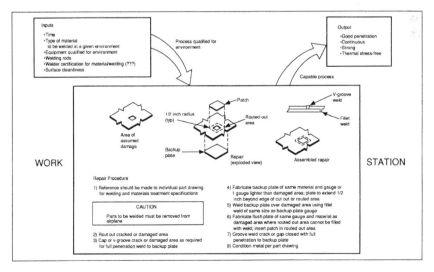

Figure 5.16 Manufacturing control of weld strength.

5.4 SYSTEM SAFETY IN THE TEST STAGE

Most of this section is based on Reference 7. The important principles and applications are covered here.

5.4.1 Testing Principles

Accidents may result from unsafe conditions and unsafe events. In order to design effective safety tests, the analyst must know the potential unsafe conditions and events associated with foreseeable product uses. The purpose of testing is to assure that the product is capable of performing under these foreseeable situations without causing property damage or personal injury. The primary analytical tools used to assess potential accidents are the preliminary hazard analysis, failure-mode, effects, and criticality analysis, and fault-tree analysis. In addition to the tools listed above, products must be analyzed for their ability to withstand the environments in which they are going to be used so that no unforeseen product or health hazard is created. To evaluate foreseeable environments of use, consideration must be given to conditions such as humidity, temperature, chemical vapors, dust, radiation, vibration, and shock. Safety tests are used to simulate the effects of long-term exposure to such environments. For example, products such as automotive seat belts degrade after such exposure and lose their ability to restrain passengers safely.

Some environments can create long-term health hazards. Other environments, such as a lack of oxygen in canned foods, are known to cause health hazards such as botulism. A recent controversy has arisen over blindness due to infection caused by lack of oxygen between the cornea and extended-wear contact lenses. Where issues of toxicity to humans exist, tests on laboratory animals are used. Drug companies routinely conduct laboratory experiments on test animals to determine the safety and efficacy of drugs before use by humans.

In addition to the analytical tools of system safety, special techniques have been developed by human factors specialists to evaluate the likelihood of human behavior and the effect of human error on safety. One technique is THERP (Section 5.2.8). FMECA can also be applied to each use and abuse situation to determine the impact of product component malfunctioning on product safety. Human factors analysis requires analysis of foreseeable misapplication and misuse. Tests can be developed to determine what human behavior will be under these situations and whether safety hazards will result because of human actions. Warning label verification testing, such as the program conducted by the National Electrical Manufacturers Association to develop the Mr. Ouch label to warn children of the hazard of electrical shock from electrical equipment, is an example of this type of testing. For some products, additional analytical tools may be developed or special tests may be required.

Assuming that hazards are not preventable by design, the criterion for effective safety tests is that the tests should demonstrate that, if the product can fail, it will fail in a safe manner and will not expose users or bystanders to dangers during its lifetime. Life tests can be developed to demonstrate safety over the life of the product. Make sure you conduct fault injection tests, not just simulation of fault injection. Simulation tests may make no provision for loose connections, intermittent faults because of vibrations, or high temperature.

5.4.2 Prerequisites for Developing Appropriate Tests

Tests may be required to evaluate the safety impact of human actions, use, misuse, abuse, and repair and maintenance activities. The selection of appropriate tests depends on the potential hazards created during these actions. Therefore, thorough safety analyses are a prerequisite to effective safety testing. The safety analyses highlight areas in the product and human–product interactions that require testing. Depending on the nature of the product, preliminary hazard analysis, failure-mode, effects, and criticality analysis, and fault-tree analysis may be performed.

Many times, PHA reveals the relationship between the design and the manufacturing process. A manufacturer of consumer products discovered that certain safety tests could be avoided if the tooling in manufacturing was changed. The improvements in tooling prevented production of out-of-tolerance parts that could have led to hazards. Such design and process improvements minimize the necessity for future safety testing. Because it may not be possible to eliminate some generic hazards by design changes, safety tests must be developed to evaluate the residual hazardous characteristics of the product. These tests will determine the likelihood of product failure under foreseeable use conditions, and the possibility of failure resulting in a hazardous condition. The tests also measure the severity of the hazard. Safety tests are also used to determine the effectiveness of safety devices, warnings, and interlocks in modifying user behavior and minimizing the likelihood of an accident resulting in injury or property damage. A study of water sprinklers showed that 3% to 5% of sprinklers, depending on the design, are not effective in actual fires [8].

Before safety tests are developed, the causes of hazards should be minimized. Fault-tree analysis is one system safety technique that is used to systematically study the possible causes of failure. Each cause of failure is analyzed to determine if it is possible to eliminate the cause or change the product's design to decrease the likelihood of a failure. Tests to determine the response of the product to the various causes of failure are developed. Improving the design reduces the number of tests required to verify the safety of the product. The FMECA identifies many failures that are considered hazards not identified in the preliminary hazard analysis or the fault-tree analysis. Effective design of product safety tests depends on knowledge of those components whose failures are most likely to result in hazards.

A procedure similar to preliminary hazard analysis can be used to evaluate the safety of maintenance and repair operations. At a minimum, human–equipment interface hazards must be analyzed. The working procedure rather than the product is tested in this instance. The designated procedures should create no hazards during product use or foreseeable abuse and misuse.

Design review is another tool used to evaluate the safety of a product. During the design review, the results of the safety analysis are discussed. Often, additional hazards are found to exist during these design review discussions. In order to practice reasonable care in design, the designer must try to eliminate these hazards. Durability tests should be designed for unpreventable hazards. A technique, which can be used to evaluate the effects of faults, is the event tree.

Cost is an important consideration in determining the type and number of tests to be performed. A complex product may require many tests. The decision is based on the severity of the danger, its frequency, the cost of the improvement, and the availability of economically and technologically feasible alternative designs.

5.4.3 Product Qualification Tests

After the design review team accepts the final design of a product, products are usually tested to determine the type and frequency of failure over time, before production and sale to the consumer. These tests are known as qualification tests; they are designed to qualify the product for production. If the product exhibits unacceptably high failure rates, the design or production process is evaluated to determine what changes are necessary. Qualification testing is also performed when major changes are made to existing products, to determine how the changes affect the safe useful life of the product. A change meant to correct one product deficiency may result in other safety-critical failures.

Flaws appear in three stages of any product life cycle. Together they are known as the bathtub failure distribution. Every engineer should be familiar with the details of the distribution (see Chapter 3).

To simulate service life in the laboratory under controlled conditions, accelerated tests are required. Without higher stresses, the time required to test a new product might be many years. With accelerated testing, years of service life can be simulated economically in a much shorter time. Accelerated testing must be designed and performed with care. The engineering and physical properties of the test specimens must not change except at the point of failure. Stresses may range from 150% to 400% over those anticipated in actual service of the product. There are some precautions to be followed when performing accelerated testing. The failure modes at normal stress levels and those at accelerated stress levels should be the same. The approximate relationship between test time and field life of the product can be predicted analytically by the design engineer. Chapter 3 contains more details. These tests must be performed periodically to assure that the changes in manufacturing are not detrimental.

5.4.4 Production Tests

Production tests are designed to detect variations in the manufacturing process. These variations include those caused by human error. Deviations in the manufacturing processes can introduce safety-critical flaws in the product. The product may be improperly heat-treated. A brittle material is more likely to break in service, possibly injuring the user. A worker may forget to put a lockwasher or gasket in place. If these components are critical to the safe use of the product, production tests should be designed to detect their absence. Statistical design of experiments is useful in resolving many common problems. Special tests may be required for unique situations.

Production tests should, at a minimum, assure that the product conforms to engineering specifications, regulations, and standards. Absence of these tests may constitute negligence. The proper design of these tests is important to assure that they are able to find defects. Other production tests consist of testing for proper handling, packaging, and shipping. Tests should be conducted to evaluate the product's capability to withstand shock, temperature, and vibration during these activities. As with other tests, a thorough understanding of the potential hazards that can result from improper handling, packaging, and shipping is required to develop appropriate tests.

5.4.5 Tests for Human-Related Errors

Human actions can introduce safety-related risks. These can arise during intended use, abuse, or misuse. Tests should be designed to determine how the product would react under these foreseeable situations. It is often difficult to predict how the product will behave without sufficient testing under these conditions. One example involves the location of the horn in automobiles. Most automobiles used to have the honking mechanism located in the center of the steering wheel. Drivers become accustomed to pressing the center of the steering wheel in an emergency. However, some late-model automobiles locate the mechanism on the periphery of the steering wheel or on a separate lever behind the steering wheel. A person renting different automobiles in different cities, and thus unfamiliar with the location of the horn mechanism, may be unable to react appropriately in an emergency. Tests are performed to determine how suitable the nonstandard feature is under foreseeable conditions of use. Human factors specialists through the observation of test drivers can evaluate mockups of automobile interiors.

Tests are often required to determine foreseeable misuse and abuse. One toy manufacturer invites children to a company facility to use its products and observes how the children misuse or abuse the products, through one-way mirrors. A controlled test may not be adequate. Therefore, the company also provides toys for children to use at home. Parents are asked to record how the child uses the toy and to note possible safety problems. On the basis of these tests, the manufacturer can decide to alter the design of the product or to recommend parental supervision during product use. The manufacturer may also determine that only children above a certain age should be allowed to use the product. Without adequate testing, a jury may find that injuries caused by foreseeable use of the unmodified product are attributable to product defects.

For large, complex systems such as aircraft or chemical plants, companies use simulators to evaluate the effectiveness of the arrangement of controls and instrumentation and to train new operators. One method of testing deliberately introduces defects into a product to determine how the user reacts. Where testing is too dangerous for actual users, carefully designed dummies that simulate human responses can be used. Automotive crash testing of new vehicles makes extensive use of this method. A number of computer models have been created to simulate the motion of occupants involved in crashes to determine the crashworthiness of automotive designs.

5.4.6 Testing the Safety of Design Modifications

Often the solution of one safety problem creates another. Design modification made to reduce costs without testing for the effect of the modification can affect the safety of the product. Usually it is not necessary to test the complete product. Safety analysis is necessary to determine the extent of the required testing. The analysis can be in the form of a mini-design review to critique the effects of the change on durability, assembly clearances, chemical properties, physical properties, and failures during shelf life. The following example is not unusual. A roof disintegrated into an almost muddy material because a supplier had increased the sulfur content of the roofing material. The sulfur reacted with the moisture in the air to cause an undesirable change in the finished product, the roof. Changes in the production process should also be evaluated. The production process may affect the safe performance of the product.

5.4.7 Testing Procedures

A product may not necessarily be hard goods. A repair manual containing procedures is also a product. These procedures must be tested for safety the same way that products are tested. As a product is made up of components, a procedure is made up of steps. Some steps can result in accidents; a wrench falling from a scaffold or ladder can injure a person below. Methods for testing procedures include job sampling, emergency drills, change analysis, and THERP.

Job sampling is generally applied to inspection processes. It can also be applied to the evaluation of procedures. Several users are observed to measure adherence to correct safety procedures. Observations may reveal unsuspected hazards in the procedures and may result in improvements to the procedures.

Emergency drills are used to evaluate complex procedures. For example, to determine whether passengers can evacuate an aircraft in 90 seconds, an emergency is simulated and the ability of the passengers to react properly is evaluated. This is a powerful tool because it evaluates whether people can respond to unforeseen hazards safely.

5.4.8 Analyzing Test Data—The Right Way

Engineers who do not have formal training in statistics are likely to make mistakes when evaluating test data. For example, the following test data were collected relating to the strength of a material (in pounds): 110, 120, 150, 150, 150, 160, 170, 170, 180, 190, 200, 400. The minimum strength required for this product is 100 pounds.

Should the analyst accept this product on the basis of the test data? Each of the 12 test samples had strength higher than the specification. If the data are plotted according to the Weibull distribution, it is discovered that there is 50% chance that approximately 4% of the units can fail below 100 pounds. Knowing these facts may alter the decision to accept the material. Because a sample size of 12 units was not large enough, the defective units did not show up in the test. When a larger sample of this product was tested, some defective units did show up. The lesson is that a manufacturer should not use data casually when dealing with safety. The probability distribution must be adequately analyzed before a decision is made. Chapter 2 contains details of this methodology.

5.4.9 How Much Testing Is Enough?

A manufacturer should test 100% of its products for safety if possible. Some companies inspect a product twice during the production process using two different inspectors. If the test is destructive, 100% testing cannot be performed, of course. In such cases, a statistician can determine the necessary sample size.

5.5 SYSTEM SAFETY IN THE USE STAGE

Once a system is installed, more hazards are discovered. They will require engineering changes. These changes must be reviewed for safety. Several tasks outlined below must be performed on an ongoing basis.

5.5.1 Closed-Loop Hazard Management (Hazard Tracking and Risk Resolution)

The effort to resolve hazards permanently must go on. Hazards should be reported even if they did not result in an accident. For example, a nurse who trips on a wire in a hospital may not get injured. But later someone else may trip on the same wire and sustain an injury. Such occurrences are usually reported on an incident report. A closed-loop system is one that ensures that the wire will never again be in someone's path anywhere in the hospital. Someone must verify this in future audits.

The safety audit is a powerful tool to assure that hazards are removed. If possible, an independent person from outside the audited area should conduct the audit. The auditor should also check for hazards reported in previous audits.

5.5.2 Integrity of the Procedures

In no case should anyone be allowed to violate safety procedures. A separate audit should be conducted to check their validity and those that were adhered to. Job sampling (Chapter 8) is a good technique for procedure audits.

5.5.3 Control of Changes

Make sure you have a system of traceability. When we change either hardware or software, we must know what other subsystems are affected. We must also know what versions of software are affected. There are two basic types of changes: (1) changes in the product design and (2) changes in the manufacturing process. These two changes cause many other changes in test procedures, inspection procedures, training material, and disposal procedures, for example.

5.5.3.1 *Changes in Product Design*   To control changes in product design, configuration control procedures must be followed. Configuration control means preserving the functional integrity of the product; it does not mean merely controlling the documents. Every change must be reviewed to see that the following product configurations are not affected:

- Functional configuration, which requires that the functional baseline will be maintained after a change.
- Allocated configuration, by which certain safety and reliability requirements are allocated to certain portions of the system to be compatible with another system. Integrity from system to system must be assured after a change.
- Product configuration, by which a product or a subassembly is required to have certain features to assure its form or its fit in the overall system.

Three major steps are carried out in a configuration control system. First, decide which of the three configuration controls above is appropriate (configuration identification). Second, assure that the changes are evaluated and implemented in all the affected documents (configuration control). Third, record and report the status of all the changes (configuration accounting).

5.5.3.2 *Changes in Manufacturing Processes*   Changes in methods of production should be carefully evaluated. One efficient procedure [9] is change analysis. In this method, the new procedure is written and compared side by side with the old procedure. The differences are identified. They are evaluated for their effects. Other techniques that can be used are fault-tree analysis and event trees.

5.5.4 Accident/Incident Investigation

If a mishap has already occurred, failure analysis laboratories are utilized for establishing the failure mode. The tools described in Section 5.5.3 can also be used for accident analysis, especially change analysis procedure. For example, a mushroom packaging factory went from a manual assembly line to an automated line. Surprisingly, automatically packaged mushrooms were found to have botulism after two consumers died. Change analysis revealed that the only difference between the old and new processes was that the mushrooms were cut into smaller, flatter pieces in the new process. Investigation showed that when the flat surfaces of two mushroom pieces were joined, oxygen was driven out between them, and this caused botulism.

5.6 ANALYZING SYSTEM HAZARDS AND RISKS

Products, processes, and systems are becoming more complex and extensive. An apparent minor failure or malfunction can propagate throughout the system. Accidents associated with extensive products, processes, and systems can be comprised of many errors, failures, or malfunctions. These situations present system risks. The scenario-driven hazard analysis[2](SDHA) was developed to enable an overall process to systematically analyze system and synergistic risks. The technique relies on the understanding the dynamics of an accident. Accidents are unplanned sequences of events that usually result in harm. Accidents are never the result of a single cause or hazard. There has to be at least a hazard present and a trigger event as shown earlier in the theory of accidents. Accidents are the result of many contributors. In reconstructing an accident or hypothesizing a potential accident, the analyst thinks of a scenario. The SDHA process involves constructing scenarios by identifying initiators, subsequent contributors, and defining the harm.

---

[2]A number of papers have been presented concerning the scenario-driven hazard analysis process. Within the last 15 years many analysts have been applying a scenario-based approach toward hazard analysis. The following papers are provided for additional reference:

Allocco, M., Computer and Software Safety Considerations in Support of System Hazard Analysis. In: *Proceedings of the 21st International System Safety Conference*, System Safety Society, August 2003.

Allocco, M., and R. P. Thornburgh, A Systemized Approach Toward System Safety Training with Recommended Learning Objectives. In: *Proceedings of the 20th International System Safety Conference*, System Safety Society, August 2002.

Allocco, M., W. E. Rice, and R. P. Thornburgh, System Hazard Analysis Utilizing a Scenario-Driven Technique. In: *Proceedings of the 20th International System Safety Conference*, System Safety Society, August 2002.

Allocco, M., Consideration of the Psychology of a System Accident and the Use of Fuzzy Logic in the Determination of System Risk Ranking. In: *Proceedings of the 19th International System Safety Conference*, System Safety Society, September 2001.

Allocco, M., W. E. Rice, S. D. Smith, and G. McIntyre, Application of System Safety Tools, Processes, and Methodologies, *TR NEWS*, Transportation Research Board, Number 203, July–August 1999.

Allocco, M., Appropriate Applications Within System Reliability Which Are in Concert with System Safety; The Consideration Complex Reliability and Safety-Related Risks Within Risk Assessment. In: *Proceedings of the 17th International System Safety Conference*, System Safety Society, August 1999.

Allocco, M., Automation, System Risks and System Accidents. In: *Proceedings of the 17th International System Safety Conference*, System Safety Society, August 1999.

Allocco, M., G. McIntyre, and S. Smith, The Application of System Safety Tools, Processes, and Methodologies Within the FAA to Meet Future Aviation Challenges. In: *Proceedings of the 17th International System Safety Conference*, System Safety Society, August 1999.

Allocco, M., Development and Applications of the Comprehensive Safety Analysis Technique. In: *Proceedings of the 15th International System Safety Conference*, System Safety Society, August 1997.

5.6.1 SDHA Process Development

The SDHA process has been in development and refinement since the 1980s. Many papers have been presented, and many discussions and graduate classes have been conducted concerning this concept. Students successfully applied the technique within term projects, and SDHA is being applied in the government and industries in the United States and abroad.

Systems Analyzed with SDHA SDHA has been applied to analyze many different types of systems during term projects and in actual applications. For illustrative purposes, here are some examples of systems analyzed.

Aircraft
Aircraft subsystems
Aircraft ground systems
Airport operations
Automobiles
Automated roadways
Automobile subsystems
Communication systems
Computer networks
Compressed gas processing facilities
Explosives manufacturing
Fire protection systems
Food processing systems
Hazardous waste
Laboratories
Medical equipment
Medical approaches
Metal working
Nuclear reactors
Nuclear waste
Petroleum distribution and processing systems
Power distribution systems
Public facilities
Spacecraft, missiles
Ships
Trains
Radar systems
Railroad systems
Robots
Weapon systems

5.6.2 Designing Accidents

Determining potential event propagation through a complex system can involve extensive analysis. Specific system safety methods such as software hazard analysis, human interface analysis, scenario analysis, and modeling techniques can be applied to determine system and synergistic risks, which are the inappropriate interaction of software, human, machine, and environment. All of these factors should be addressed when conducting hazard analysis and accident investigation.

Consider that hazard analysis is the inverse of accident investigation. An analyst should be able to design prospective accidents, which are potential system accidents—system risks. In order to design a robust system, all the potential and past accidents associated with the system must be determined. To apply this method, the analyst has to be able to identify all potential safety-related risks. The overall objective once these risks are identified is to eliminate or control these risks to an acceptable level. Thinking is not confined to logic of a single hazard and linear cause and effect, but multicausal progression. The analyst should reflect in terms of being able to design potential accidents.

5.6.2.1 *SDHA-Related Concept* The scenario concept first came to mind after study of Willie Hammer's books and material on system safety and in latter discussions with Hammer. Hammer [10] initially discussed concepts of initiators, contributors, and primary hazards in the context of hazard analysis. Hammer noted that determining exactly which hazard is or has been directly responsible for an accident is not as simple as it seems. Consequently, Hammer discussed a set of hazards that form sequences within the potential or actual accident. The sequences are comprised of initiating, contributory, and primary hazards. Initiating hazards define the start of the adverse sequence. They are latent design defects, errors, or oversights that under certain conditions manifest or trigger the adverse flow. Contributory hazards are unsafe acts and/or conditions that contribute within the flow. Primary hazards are the potential for harm. Hammer specifically described an accident sequence involving a series of events that result in the rupture of a high-pressure air tank. The injury and/or damage resulting from the rupture of the tank were considered the primary hazards. The moisture that caused corrosion of the tank is considered the initiating hazard; the corrosion, loss of strength, and the pressure are contributory hazards.

5.6.2.2 *Adverse Event Model* To further reinforce the concepts relating to system risks, system accidents, or scenarios, the *adverse event model* was developed. This model has been designed to show the complexity of hazard control (Fig. 5.17). The model has been used to conduct hazard control analysis. This is when controls are to be validated and verified.

System accidents can be very complex or simple. Initiators could be the result of latent hazards, like software design errors, or specification errors, or oversights involving inappropriate assumptions. If hazard controls are not adequate, they become initiators or contributory hazards. If the control is not verified, it may not function when required. Validation considers the adequacy of the control. It is the

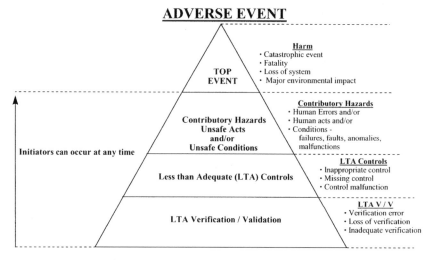

Figure 5.17 Adverse event model.

determination of sufficiency of the control and if it has been appropriately designed or applied. Initiating and contributory hazards are unsafe acts or conditions, which under specific conditions will result in accidents.

5.6.2.3 Life Cycle of a System Accident

The model in Fig. 5.18 addresses and illustrates the concept that a system accident has a life cycle associated with it. Accidents are initiated, they progress, and harm can result. In conducting scenario-driven system hazard analysis, the analyst should consider this concept. A system is in dynamic equilibrium when it is appropriately designed. It is operating within specification and within design parameters. The system is operating within its envelope. However, when something goes wrong and an initiator occurs, the system is no longer in balance. The adverse sequence progresses and the imbalance worsens until a point-of-no-return and harm results. In conducting analysis, consider the accident life cycle and how these adverse sequences progress. By the application of hazard control, the adverse flow can be stopped. It is important that any imbalance is detected and the system is stabilized and brought back to a stable state. Furthermore, should the adverse sequence progress past the point-of-no-return, the resultant harm may be minimized or decreased by the application of hazard control. Should harm result, the system should be brought back to a normal stable state. Consider that additional harm can occur during casualty or contingency. During hazard control application not only are all the initiators, contributors, and primary hazards controlled or eliminated, but also contingency, recovery, damage, and loss control are applied toward the system life cycle.

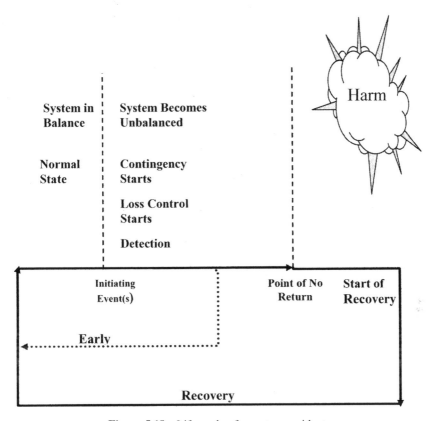

Figure 5.18 Life cycle of a system accident.

5.6.2.4 *Potential Pitfalls in Logic Development* Thinking in terms of a single hazard (or failure) and cause (or failure mode) with linear logic will not be totally appropriate in all situations. The system safety analyst should think in terms of the dynamics that can occur associated with the potential accident. If failures occur, these failures may or may not be hazards; trying to think in terms of reliability and system safety simultaneously is not a good idea. In the United States, these two disciplines have evolved with different objectives. Failures are hazards when they are unsafe conditions within the potential accident. Confining the thought process is also not good when scenarios are being developed.

A classic consideration is to think in terms of system functions and that a functional analysis is required prior to any initial hazard analysis. An assumption is made that in order to do hazard analysis functions have to fail to identify hazards. Analysts should not totally confine their thinking to this mindset. Accidents could still occur during so-called normal functions. Functions can be inappropriately defined or incompletely defined. The logic that drives the function should be confirmed. Consider how the function is to be implemented—an automated process, manual application, or a combination of automation and human involvement. The human will develop software and may input logic manually to a compiler. Once software is compiled, firmware interfaces with hardware within a computer, digital signals may be converted into analog signals, and the human may have to take action based on displayed information. All of the sequential logic should be evaluated.

To avoid isolated approaches, develop potential system accidents from different points of view. For example, consider abnormal energy interaction, oversight, omission, error, anomaly, malfunction, misuse, and any deviation associated with the system under study. Develop the adverse logic throughout the potential accident life cycle.

5.6.2.5 Determining Hazards (Unsafe Acts and Unsafe Conditions)

Most accidents are triggered by human error and it could be argued that many unsafe conditions associated with a design are the result of human error, oversight, omission, poor assumption, or poor decisions. Keeping this human interface in mind when determining hazards, separating the unsafe physical condition from an apparent human error is not simple. Consider that an unsafe act can occur at any time during the life cycle of the system. Decision errors can be made during initial design that may introduce latent hazards. An operator can make a real-time error that affects the system, introducing real-time hazards, which are considered unsafe conditions.

Within the hazard analysis process, initiating and contributory hazards are identified as unsafe acts or unsafe conditions. The determination of unsafe acts is the result of human factors analysis, and the determination of unsafe conditions is the result of abnormal energy analysis, for example. Because of the designation between unsafe acts and unsafe conditions, multilinear analysis can be applied to assess and address the possibility of combinations of contributory events, both unsafe acts and/or unsafe conditions, that are part of the accident scenario.

5.6.2.6 Tabular Worksheet

A typical tabular format heading is displayed in Fig. 5.19. The scenario number in indicated in the first column. The second column contains the scenario description or scenario theme. The initiators are indicated in the third column. All other contributors are listed in the fourth column. The mission phase or operational phase in indicated in column 5. The possible effect or primary hazard is presented in column 6. The recommendations, precautions, and controls are provided in column 7. The next row contains a description of the system state and exposure.

5.6.2.7 Deductive and Inductive Approaches Toward Scenario Development

Depending on training, background, and style of thinking, an analyst may approach scenario development in three ways—deductively,

S #	Scenario Description	Initial Contributors	Subsequent Contributors	Phase	Possible Effect	Recommendations, Precautions, and Controls
System State and Exposure						

Figure 5.19 Typical format for a scenario-driven hazard analysis.

inductively, or from the middle of the model or sequence. Typically, a fault-tree analysis is considered a deductive top–down approach, where a scenario theme is considered the top event. ("What can cause a specific event to occur?") The analyst starts the process with the theme and then conducts an investigation to determine what the contributors and initiators are that will support the scenario logic. Some system safety engineers may approach scenario development inductively from the bottom–up. ("What happens if a specific failure occurs?") This is a more conventional approach. Conducting a fault hazard analysis or failure modes and effects analysis can provide failure logic that is comprised of initiators and contributors—the unsafe conditions. After thinking in terms of accident scenarios, it is quite possible to construct logic starting from midlevel contributors and develop sequential logic in both directions toward the top event, scenario theme, or toward the initiators.

5.7 HAZARD IDENTIFICATION

In order to identify hazards, there are formal and informal methods of analysis—for example, failure mode and effects analysis, fault hazard analysis, and energy flow analysis. Formal methods are applied to identify failures, human error, and malfunctions, which could be initiators or contributory hazards or primary hazards. A highly experienced analyst could directly identify initiators and contributors through more informal methods like brainstorming. During brainstorming sessions, participants develop preliminary hazard lists based on the design. Key words in the form of hazards are listed along with scenario themes.

5.7.1 Scenario Themes

Scenario themes are short concise statements, which describe the potential accident in terms of primary hazards and main contributory hazards. Themes are negative action statements; they are the title of the potential accident. In past vernacular, they are hazard descriptions. Scenario themes can be very specific or general, depending on the type of analysis conducted. Here are some examples of accident scenario themes.

- Loss of engine power as a result of fuel starvation results in aircraft loss of control.
- Failure of valve seal results in inadvertent toxic gas release and consequent fatal injury.
- Software malfunction within engine speed control subsystem results in excessive speed, derailment, and fatal injuries.
- Mechanical failure of the airbag actuator results in loss of airbag function during an emergency and consequently resulting in fatal injury.
- Loss of situational awareness results in human error in conducting the safety-critical process; situation results in overpressure condition in tank and consequent explosion and fatality.

- Human error during weapon disassembly operation results in inadvertent exposure of toxic material and consequent fatal injury.
- Latent software error results in inadvertent shutdown of safety system during an emergency; condition results in release of radioactive gas and fatal exposure to personnel.
- Material degradation results in crack formation and propagation; condition results in leakage of toxic material and exposure to personnel, causing fatal injury.
- Error in software code results in data corruption and hazardous misleading information to be displayed to operator. As a result, operator takes inappropriate action, which causes process failure, fire, and explosion.
- Error in process instruction results in inappropriate human action and consequent injury.
- Less than adequate design of tire results in excessive tire wear and consequent tire failure during high-speed; situation results in loss of control of vehicle and fatal crash.
- Less than adequate design of tire causes inadvertent tire disassembly during high-speed; situation results in loss of control of vehicle and fatal crash.
- Inappropriate material selection results in incompatibility of materials at the interface contributing to excessive corrosion, which causes interface failure and fatal injuries.
- Electromagnetic environmental effects cause single bit upsets, which result in inadvertent system shutdown during a safety-critical time.

Note that in the scenarios above the initiators, contributors, and primary hazards are identifiable in the themes. Scenario statements provide details as to how or why potential accidents can occur. An analysis will consist of the entire scenario set, which is comprised of the individual scenario themes, all of which have initiators, contributors, primary hazards, and the phase in which the accident can occur, along with the associated risk.

5.7.2 Primary Hazards

Other names—catastrophic event, top event, end event, critical event—often are used to describe a primary hazard. The primary hazard is the one that can directly and immediately be the basis of the final harm or end effect, any accidental damage or loss. Examples of primary hazards are listed below (these are descriptions of harm, not the primary hazard such as oil on the floor or the failure of a ground fault interruptor):

- Injury or death
- Loss of system
- Damage
- Property loss
- Environmental damage

Consider any accidental harm and the prior event as a primary hazard:

- Aircraft collision
- Facility fire
- Tank explosion
- Theft of valuables
- Vandalism damage
- Spoofing damage
- Automobile crash
- Loss of valuable data
- Business interruption
- Lightning strike damage
- Earthquake damage
- Flood damage

5.7.3 Initiators

Since humans design systems, it is logical to consider that all accidents associated with the system (excluding acts of nature) are the result of human error, decision error, poor judgment, and omission. Initiators are traced back to system inception—an error, mistake, poor assumption; error in a calculation or theory; misjudgment; less than adequate design; misperception about risk; blunder; poor management decision; inadequate resources; poor planning; or poor engineering judgment. These initiating situations result in latent or real-time built-in hazards when they physically manifest. Consider abnormal energy interaction associated with the system. Other examples of initiators are:

- Inadequate identification and control of safety-related risks>
- Overall less than adequate (LTA) designs
- Inadequate specifications
- Inadequate instructions
- Inadequate life-cycle considerations
- Software errors
- Calculation errors
- Inadequate documentation
- Logic errors
- Algorithm error
- Timing error
- Scheduling error
- Poor approaches
- Process error
- Inadequate procedure

- Manufacturing errors
- LTA quality oversight
- Inadvertent damage
- Inadequate material selection
- Inadequate analysis
- Poor or inappropriate communication
- Oversight
- Omission
- Misjudgment

5.7.4 Contributors

The sequences of other hazards that can or have occurred between the initiator hazards or initiators are contributory hazards or contributors. A contributor is always the direct result of an initiator and eventually results in harm (primary hazard). Generally, these hazards are the result of abnormal energy exchange or unsafe human action or reaction. Examples of contributors are:

- Failures of subsystems
- Failures of components
- Sequential human errors
- Inappropriate system responses
- Inadvertent action due to software error
- Material failure due to error in design
- Confusion
- Fixation
- Display clutter
- Information overload
- Inadvertent release of toxic material
- Inadvertent disassembly
- Loss of situational awareness
- Interface failure
- Blockage of flow
- Adverse environmental effects
- Blowout of seals
- Material deformation
- Inappropriate chemical reaction
- Overpressure
- Loss of control
- Inadequate backup

- Fracture
- Implosion
- Material degradation
- Failure of restraining device
- Excessive decay
- Illness
- Stress affecting humans
- Endothermic reactions
- Excessive heat loss
- Decreased viscosity
- Excessive heating
- Excessive friction
- Incomplete combustion
- Short circuit
- Sneak path
- Fault
- Spontaneous ignition
- Reverse flow
- Adverse physical reactions

5.7.5 Overlapping Hazards

Accident dynamics are not perfectly linear: there may be many initiators, contributors, and primary hazards associated with an accident sequence. In developing scenarios, consider that hazards can overlap. It is not all that important to totally segregate all hazards in the flow. It is, however, important to provide appropriate logic showing the potential accident from inception through the harm, and further to system recovery. Caution should be exercised to assure that safety requirements are independent and not redundant.

5.8 TOPICS FOR STUDENT PROJECTS AND THESES

1. There are several hazard analysis techniques (FMECA, FTA, O&SHA, etc.). They seem to be done independently, and some are not done at all. Users do not seem to integrate them. Develop a better approach so that the hazard analysis is complete, useful, and done in the shortest possible time.
2. Hazard analysis is generally done during the design phase; very rarely do operating personnel make use of it. Discuss how hazard analysis should be used for improving operational safety.

3. Literature is almost completely void as to how one can apply system safety principles in manufacturing. Conduct a literature search and develop recommendations.

4. Perform a literature search to summarize techniques for assessing safety margins. Briefly present your own recommendations.

5. Conduct a literature search to define durability. What definition would you recommend? Justify your views in detail.

6. Propose your own methodology for performing software hazard analysis.

7. Conduct a literature search to define software system safety. Propose and justify your own recommendation.

8. Develop a safety test plan for any product. Justify each test.

9. Perform an O&SHA for a product. Recommend design changes for reducing the risks.

10. Develop a methodology that you feel is appropriate for risk analysis.

11. Discuss why safety in design is a good business practice. Attempt to justify your statements with quantitative examples.

12. Propose a method for considering human factors during product design.

13. Select a complex process and conduct a system hazard analysis; apply the scenario-driven hazard analysis method.

14. Discuss why linear hazard analysis may not be appropriate.

15. Define the differences between reliability and system safety analysis.

16. Indicate how the human, software, firmware, hardware, and environment can be integrated into a system hazard analysis by applying scenario-driven hazard analysis.

17. Select a complex existing system and conduct loss analysis and accident reconstruction and discuss the apparent complexities of system and synergistic accidents.

18. Discuss the differences between systematic hazard analysis and accident investigation.

19. Select a complex process and develop an approach with appropriate worksheet to conduct a system hazard analysis applying scenario-driven hazard analysis.

20. Select a commercial system and develop criteria for risk assessment; devise severity, likelihood, and exposure criteria and conduct hazard analysis and risk assessment.

REFERENCES

1. System Safety Program Requirement, Mil-Std-882, 1984.

2. D. Raheja, A Different Approach to Design Review. In: *Reliability Review*, American Society for Quality Assurance, June 1982.

3. *System Safety Engineering Guidelines*, NAVORD-OD 44942, Naval Ordnance Systems Command, 1992.

4. R. E. Barlow, and F. Proechan, *Statistical Theory of Reliability and Lifting* (Gordon Pledger, 1142 Hornell Dr., Silver Spring, MD 20904), 1981.

5. W. Hammer, *Product Safety Management and Engineering*, Prentice-Hall, Englewood Cliffs, NJ, 1980.

6. R. C. Clardy, Sneak Analysis: An Integrated Approach. In: *International System Safety Conference Proceedings*, 1977.

7. D. Raheja, Testing for Safety. In: *Products Liability: Design and Manufacturing Defeat*, Lewis Bass (Ed.), McGraw-Hill, New York, 1986.

8. Automatic Sprinkler Performance Tables, *Fire Journal*, National Fire Protection Association, Boston, July 1970.

9. W. Johnson, *MORT Safety Assurance Systems*, Marcel Dekker, New York, 1980.

10. W. Hammer, *Handbook of System and Product Safety*, Prentice-Hall, Englewood Cliffs, NJ, 1972, pp. 63–64.

FURTHER READING

American Society for Testing and Materials (ASTM), 1916 Race Street, Philadelphia, PA 19103.

Department of the Air Force, Software Technology Support Center, *Guidelines for Successful Acquisition and Management of Software-Intensive Systems: Weapon Systems, Command and Control Systems, Management Information Systems, Version* 2, June 1996, Volumes 1 and 2 AFISC SSH 1-1, *Software System Safety Handbook*, September 5, 1985.

Department of Defense, AF Inspections and Safety Center (now the AF Safety Agency), *Software System Safety*, AFIC SSH 1-1, September 1985.

Department of Defense, *Standard Practice for System Safety*, Mil-Std-882E, Draft, March 2005.

Department of Labor, *Compliance Guidelines and Enforcement Procedures*, OSHA Instructions CPL 2-2.45A, September 1992.

Department of Labor, *OSHA Regulations for Construction Industry*, 29 CFR 1926, July 1992.

Department of Labor, *OSHA Regulations for General Industry*, 29 CFR 1910, July 1992.

Department of Labor, *Process Safety Management Guidelines for Compliance*, OSHA 3133, 1992.

Department of Labor, *Process Safety Management of Highly Hazardous Chemicals*, 29 CFR 1910.119, *Federal Register*, 24 February 1992.

Department of Transportation, *Emergency Response Guidebook*, DOT P 5800.5, 1990.

Electronic Industries Association, *System Safety Engineering in Software*, EIA-6B, G-48.

Environmental Protection Agency, *Exposure Factors Handbook*, EPA/600/8-89/043, Office of Health and Environmental Assessment, Washington DC, 1989.

Environmental Protection Agency, *Guidance for Data Usability in Risk Assessment*, EPA/540/G-90/008, Office of Emergency and Remedial Response, Washington DC, 1990.

Ericson, C. A. II, *Hazard Analysis Techniques for System Safety*, Wiley-Interscience, Hoboken, NJ, 2005.

FAA Safety Risk Management, FAA Order 8040.4.

FAA System Safety Handbook: Guidelines and Practices in System Safety Engineering and Management, 2001.

Fire Risk Assessment, ASTM STP762, American Society for Testing Materials, Philadelphia, 1980.

Human Engineering Design Criteria for Military Systems, Equipment and Facilities, Mil-Std-1472D, 14 March 1989.

Institute of Electrical and Electronics Engineers, *Standard for Software Safety Plans*, IEEE STD 1228, 1994.

Interim Software Safety Standard, NSS 1740.13, June 1994.

International Electrotechnical Commission, *Functional Safety of Electrical/Electronic/Programmable Electronic Safety-Related Systems*, IEC 61508, December 1997.

Malasky, S. W., *System Safety: Technology and Applications*, Garland STPM Press, New York, 1982.

NASA, *Methodology for Conduct of NSTS Hazard Analyses*, NSTS 22254, May 1987.

National Fire Protection Association, *Properties of Flammable Liquids, Gases and Solids*.

National Fire Protection Association, *Flammable and Combustible Liquids Code*.

National Fire Protection Association, *Hazardous Chemical Handbook*.

Peters, G. A., and B. J. Peters, *Automotive Engineering and Litigation*, Garland Press, New York, 1984.

Procedures for Performing a Failure Mode, Effects and Criticality Analysis, Mil-Std-1629A, November 1980.

Process Safety Management, 29 CFR 1910.119, U.S. Government Printing Office, July 1992.

Reliability Prediction of Electronic Equipment, Mil-Hdbk-217A, 1982.

Roland, H. E., and B. M. Moriarty, *System Safety Engineering and Management*, John Wiley & Sons, Hoboken, NJ, 1986.

Software Development and Documentation, Mil-Std 498, December 5, 1994.

System Safety Program Requirements, Mil-Std 882D, February 10, 2000.

Weinstein, A. S., et al., *Product Liability and the Reasonably Safe Product*, Wiley-Interscience, Hoboken, NJ, 1978.

Yellman, T. W., Event-Sequence Analysis. In: *Proceedings of the Annual Reliability and Maintainability Symposium*, 1975.

CHAPTER 6

QUALITY ASSURANCE ENGINEERING AND PREVENTING LATENT SAFETY DEFECTS

6.1 QUALITY ASSURANCE PRINCIPLES

Most life-cycle costs are locked in by the design philosophy, and quality is no exception. Like reliability, it is built into a product. On the other hand, this does not mean that process control in manufacturing is unimportant. Process control is of the utmost importance to prevent premature failures, which can create catastrophic hazards such as loose fasteners and improper assembly in many automotive recalls.

Quality is conformance to a set of requirements that, if met, results in a product that is fit for its intended use [1]. It is a broad term applied to customer satisfaction. It includes quality of design, quality of processes, quality of services such as on-time product delivery, and the value perceived by the customer. (The "customer" can also be the production operators and managers in the downstream departments.) Since the goal is customer satisfaction (internal as well as external customers), the overall focus should be on customer needs.

Traditionally, companies have aimed at reducing scrap and rework costs rather than satisfying customer needs. Apropos of this philosophy, Dr. W. Edward Deming once said: "Working hard will not help you if you don't do the right things." One of the wrong things done by quality professionals is not conducting any hazard analysis on the manufacturing process. We are referring to the errors in the process that can lead to accidents in the field. Typical hazards are the wrong torque on fasteners, assembling the wrong components, using defective components, and forgetting to assemble a component such as a lockwasher. Of course,

Assurance Technologies Principles and Practices: A Product, Process, and System Safety Perspective, Second Edition, by Dev G. Raheja and Michael Allocco
Copyright © 2006 John Wiley & Sons, Inc.

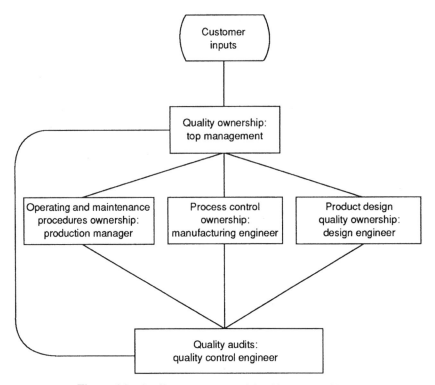

Figure 6.1 Quality assurance model, with suggested help.

we must conduct another analysis on the accidents encountered by the humans in the manufacturing process also: an analysis similar to operations and support hazard analysis covered in Chapter 5. This principle emphasizes that the requirements will result in fitness for intended use only when engineering specifications truly reflect the expectations of the customer. If not, blind adherence to specification will be counterproductive. If customers are offered high-quality products at reasonable prices (the "right things"), then the reduction of scrap and rework costs will automatically be considered.

A good quality assurance system can be represented by the model in Fig. 6.1. The model applies to new products as well as existing ones; in both the cases, knowing the customer's needs is the key to success. For new products, the inputs can come from potential customers and from the design reviews. When the product already exists, the inputs come from field complaints and customer contacts. The model emphasizes the following steps:

Step 1: Use various sources to get customer inputs, including the safety incidence reports.

Step 2: Establish "ownership" (total accountability) for all customer concerns. Specific people must be made clearly responsible and accountable for them.

There should be no "rock throwing," such as manufacturing blaming design and vice versa. Top management must be the leader, not the follower. For efficient quality assurance, ownerships must be coordinated so that product and process design and the maintenance development proceed simultaneously and coherently. This definitely requires not only commitment but also involvement by top management.

Step 3: Make an effort to change the product design so that it is immune to variability and uncontrolled conditions in manufacturing. Alternatively, make an effort to change the process to eliminate these problems permanently. For safety-related errors, do not depend on inspection. Try to design such that inspection is not required. For example, if two parts are welded, we require welding inspection. If they are designed as a single piece, then welding inspection is not required.

Step 4: If a quality problem cannot be removed by step 3, implement process control measures. Similar measures should be worked out with the suppliers. The process controls should be such that there is always an early warning of poor quality.

Step 5: The process should be inherently reliable. Therefore, assess operating procedures and maintenance procedures for their ability to keep the process in a controlled state all the time.

Step 6: Make quality audits to assure that the product meets its ppm (nonconforming parts per million) goal and that the customer is proud of it. The pride shown by the customers for your products is the best measure of quality!

Step 7: Apply these efforts to product design, process design, and operating procedures concurrently and cohesively.

Step 8: All safety-related features such as certain dimensions and material hardness must have six sigma control in the inherent process variability. If such is not the case, either change the component design or go to a new process. If you look at all the life-cycle costs, you can have a very high return on investment.

6.2 QUALITY ASSURANCE IN THE DESIGN PHASE

Quality starts during specification writing with an informal review. Detailed analysis begins as soon as the first draft of the specification is issued. Chapter 10 describes the tasks and important milestones for quality assurance engineering. These are an integral part of the overall system assurance plan.

Two analyses are required at this stage. The first is often referred to as the *product design review*. Quality function deployment, benchmarking, applying the quality loss function, and use of checklists are some tools for the product design review. The second analysis is called the *process design review*, to assure that the anticipated process concept is sound. These reviews are part of the system assurance design reviews in which reliability, maintainability, safety, and logistics support are also determined.

The process must deliver a high yield, which is often defined in terms of parts per million levels. A process producing nonconforming parts at a level less than 100 ppm is looked on as a standard of excellence. Those who achieve this level of performance are those who control quality characteristics closest to the target value.

In some industries, of course, the state of the art is not fully developed and quality is difficult to predict. Nevertheless, during the design phase, even though one may not be able to accurately predict quality levels, one should be able to judge that some processes are more reliable than others. For example, an automated manufacturing process might produce more consistent dimensions than those produced manually by different machine operators.

It is never too early to think about process reliability. Unfortunately, it is this area which industry has failed to recognize as *the* major competitive tool. A process design is done once but the labor and material in manufacturing are wasted hundreds of times by a poor process design.

6.2.1 Product Design Review for Quality

There are several design reviews during product development. They are updated in each stage as the details become available. Their results are used in development of operating procedures, repair and maintenance procedures, and training. As a minimum, the reviews are done at least at the following milestones:

- Preliminary design review
- Detail design reviews (several)
- Critical design review—to approve the final design
- Preproduction design review—after the prototype tests
- Postproduction design review
- Operations and support design review

The involvement of quality assurance starts at the preliminary design review and becomes much larger as design becomes firm. The largest effort should go into the preliminary and the detail design reviews because about 95% of life-cycle costs are established at this time. Lack of a strong effort will increase these costs dramatically. The money saved by preventing losses is at least 10 times the money invested in these design reviews.

The function of quality assurance at the preliminary design review is to make sure the new designs do not have the quality problems of similar designs in the market and that no quality problem will result in accidents in the field. Quality assurance engineers therefore must know the strengths and weaknesses of the competitive products. The review should examine the product's ability to perform its function in all the foreseeable situations. Any problems or concerns that show up might require a change in the concept of the design. For example, we participated in design of a robot that was supposed to pick up and cut thin metal lead frames for a very large scale integrated (VLSI) circuit manufacturing process. There was a good possibility the robot would not be able to perform its function if its timing was not synchronized with another robot, which was to pick up paper packing between two lead frames. The design review team therefore decided to operate both robots from the same clock sequentially. This action changed the design concept.

6.2.1.1 Quality Function Deployment A value analysis tool applied to product and process development, quality function deployment (QFD), can be used for developing test strategies and for converting requirements to specifications. Keep in mind that those who perform the QFD may not be trained to think of accident scenarios and the hazards. The concept originated in Japan. For new product development, it is a matrix of customer needs versus design requirements. The inputs can come from many sources including interviews, market analysis, market surveys, and brainstorming. Vannoy and Davis [2] suggest using fault trees to identify products with negative characteristics and convert them to success trees to identify customer needs (see Chapter 5 for fault-tree methodology). For example, they identify the following needs of the customers for an automobile:

- Price in relation to expectations
- Expectations at delivery
    Performance
    Features
    Attractability
    Workmanship
    Customer service
    Safety
    Serviceability
    Conformance to regulations
    Perceived quality
- Expectations over time
    Reliability
    Effective preventive maintenance
    Less preventive maintenance
    Number of failures below expectation
    Availability of repair parts
    Lower repair expenses
    Expenses covered by warranty
    Reasonable repair time
    Better customer service
    Time to failure within expectation
    Type of failure within expectation
    Durability
    Performance
    Safety
    Customer support

Quality function deployment means turning these needs into design engineering requirements. Similarly, these needs can be translated to manufacturing controls, test programs, and customer service requirements. Reference 3 contains more information.

6.2.1.2 *Benchmarking*

Benchmarking is a process of measuring one's own products and processes against those of leaders in the field and setting goals to gain a competitive advantage.

The procedure [1] starts with identifying items to benchmark during product planning and identifying key engineering features. It continues with these steps:

- Determine what product features to benchmark against. The targets may be companies, industries, or technologies.
- Determine existing strengths by collecting and analyzing data from all major sources, including direct contacts.
- From each benchmark item identified, determine the best-in-class target.
- Evaluate your processes and technologies in terms of benchmarks and set improvement goals.

Actually, the best-in-class target is always a *moving target*. Benchmarking should take this into account and not just be satisfied with the current best-in-class. A leader is the one that arrives first in the market with new targets and establishes its product as the world standard. Examples are IBM mainframe computers, Sony pocket radios, and Intel integrated circuits.

Caution: The benchmark for safety is always zero fatalities.

6.2.1.3 *Quality Loss Function*

The concept of the quality loss function is based on the engineering principle that, if all components are produced very close to their target value, product performance can be expected to be the best, and the costs to society can be expected to be lower. Unfortunately, over the years manufacturing managers have dismissed the principle as too costly to implement, until Dr. Genichi Taguchi modeled it quantitatively. He convinced managers that the cost of quality goes up, not only when the product is outside specifications, but also when the product deviates from the target value within the specifications. The more the deviation, the higher is the expected life-cycle cost. Naturally, the performance over the long term will not be identical for parts produced at the lower and upper ends of the tolerance band. Their performance is not likely to be as good as that of products produced at the target value. Inadequate performance will result in higher warranty costs and higher in-house inspection costs. It will ultimately turn away customers. To understand the quality loss function, it is necessary to understand Taguchi's philosophy, which Kackar [4] has summarized as follows:

1. An important dimension of the quality of a manufactured product is the total loss generated by that product to society.

2. In a competitive economy, continuous quality improvement and cost reduction are necessary for staying in business.

3. A continuous quality improvement program includes incessant reduction in the variation of product performance characteristics about their target values.

4. The customer's loss due to a product's performance variation is often approximately proportional to the square of the deviation of the performance characteristic from its target value.

5. The final quality and cost of a manufactured product are determined to a large extent by the engineering designs of the product and its manufacturing process.

6. A product's (or process's) performance variation can be reduced by exploiting the nonlinear effects of the product (or process) parameters on the performance characteristics. Details of this principle are contained in Section 6.2.3.

7. Statistically planned experiments can be used to identify the settings of product (and process) parameters that reduce performance variation.

The loss function concept is the application of a life-cycle cost model to quality engineering. Taguchi used a simple model of the loss imparted to the supplier, the customer, and society in general. It is oversimplified but serves its purpose in highlighting the fact that a product is cheaper and better if it is consistently produced close to its target value.

Figure 6.2 shows the model. T is the target value of a variable at which the product is expected to perform best. Values of the variable—which may be a dimension, for

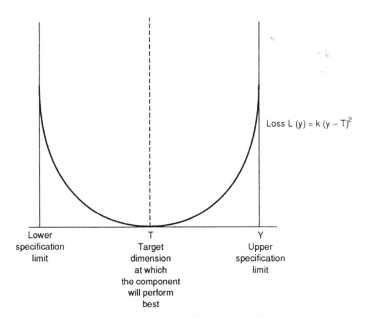

Figure 6.2 Loss function model.

example—are plotted on the horizontal axis and the loss associated with each value of the variable is plotted on the vertical axis. The loss function is

$$L(y) = k(y - T)^2$$

where k is constant and $(y - T)$ is the deviation from the target value. This model assumes the loss at the target value is zero; that is, the customer is satisfied if a component is produced at the target value. It further assumes that customer dissatisfaction is proportional only to the deviation from the target. The value of k may be determined by substituting an estimated loss value for an unacceptable deviation such as the tolerance limit. For example, if a Rockwell hardness number of 56 is the target value and if a customer's estimated loss for Rockwell hardness beyond 59 (the specification limit) is $200, k may be calculated as

$$k = \frac{\text{loss}}{(59 - 56)^2}$$
$$= \frac{200}{(3)^2}$$
$$= 22.22$$

The model may be written as

$$L(y) = 22.22(y - T)^2$$

6.2.2 Process Design Review for Quality and Yield

During the preliminary design review, *all* the customer needs should be considered. As soon as the preliminary design is approved, the proposed flowchart for the process should be prepared. For some processes the analysis can be begun during the concept phase. It is advisable that the quality engineer work with the process engineer and the reliability engineer in a team. The purpose is to assure the process will work properly at the desired level of yield.

In designing for reliability, the team should quantify three factors and include them in the process specification. First, the process equipment must be capable of producing the right output (narrow distribution at the target value) consistently. Second, the process should be in operational condition as much as possible. Third, the process should be utilized to its full capability. These factors in the specifications can be called, respectively, the process yield E_p, process availability A_p, and process utilization U_p. Thus, process effectiveness can be defined as the probability that a process will work during the scheduled hours, that it will deliver the desired yield, and that it will work at an intended speed, or

$$E_p = Y_p \times A_p \times U_p$$

Note: A standard definition of process reliability does not exist.

Process yield performance may be defined as the probability that a process will produce the part or assembly within the specified range of target values during the time until a scheduled maintenance function is performed. It is the ratio of good output to total output. If the specification calls for process yield performance equal to 99.99999%, it means the process deviation outside the target range will be less than 100 ppm. The equipment should demonstrate this capability in a statistical analysis of data over a reasonable length of time. *Process availability* can be defined as the ratio of total uptime to total scheduled time. If a process is used for two shifts a day for 5 days a week, then the total scheduled hours are 8 hours × 2 shifts × 5 days = 80 hours per week. If the equipment is in running condition for 64 hours, then the process availability for the week is $A_p = 64/80 = 0.80$. Process utilization is the ratio of actual production quantity during a given period to its designed capacity. It is sometimes called the *speed ratio*. If the industrial engineering standards require 80 assemblies to be produced in 8 hours, and if only 70 are actually produced, the process utilization is $70/80 = 0.875$ or 87.5%. The following examples demonstrate that quality effort must be concurrent with the design work.

Example 6.1 An integrated circuit manufacturer bought a wire bonding machine. When the parts from a typical production run were analyzed, it was found that about 0.5% (5000 ppm) of the bonds had low strength. A Weibull plot (Fig. 6.3) for 2370 bonds shows not only that 0.6% of the bonds were of poor quality but also that the machine produced two levels of bond quality, as indicated by the two slopes on the plot. (See Chapter 2 for Weibull distribution concepts.) Had the buyer conducted some experiments at the supplier's facility, the mistake of buying such equipment would not have been made. One may say that this kind of study might not be possible during the design phase because the proposed equipment is special. In such a case, demonstrations on similar equipment should be required at a potential supplier's facility.

Example 6.2 A vacuum circuit breaker concept was approved but the design engineers demanded tight tolerances for the parts. One of the authors, having had long experience with the suppliers, knew that the yield of such parts would be less than 60%. In other words, the suppliers would eventually increase the price at least 40% to make up for the scrap and rework. A decision was made to test some suppliers. It was shocking to learn that not only did suppliers raise the prices tremendously but also very few sent bids. The result was that the product idea was dropped before it became a reality. It is better to drop a product at this stage than to waste millions of dollars more and drop it later.

6.2.2.1 *Capital Equipment Analysis*

In capital equipment analysis, an effort is made to minimize downtime by a proper critique of purchased equipment. One of the best-known techniques for this is the equipment failure-mode, effects, and criticality analysis. This technique is essentially the same as design FMECA, but

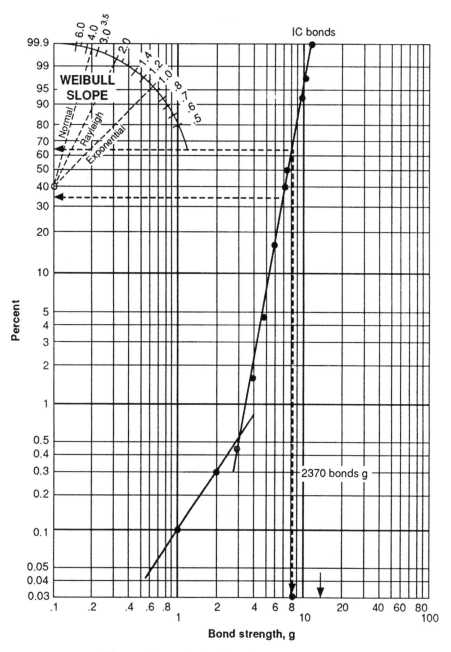

Figure 6.3   Analysis of wire bonding equipment.

emphasis is on minimizing process downtime; the technique is not to be confused with process FMECA, which will be discussed later in this chapter. Make sure the FMECA or hazard analysis considers latent hazards that can be introduced during processing.

Equipment FMECA uses standard design FMECA, covered in Chapter 4, and evaluates the effects on downtime. The main objective is to introduce design changes that will improve process availability. If the design fix is not effective, then reliability-centered maintenance (see Chapter 4) is introduced. As an illustration, consider the equipment FMECA format in Fig. 6.4. It considers the effect of downtime in terms of how long it will take to diagnose the fault and repair the device. It can be noted that the *Action plan* column contains design or preventive maintenance action instead of inspection. Even preventive maintenance should be avoided if a design change is feasible. Inspection should be used only as a last resort.

Many will agree that performing equipment FMECA is a responsibility of the reliability engineer. The authors, however, in over 25 years' experience, have rarely seen reliability engineers get involved in process equipment reliability. Since the process engineer is responsible for the process, the ownership should belong there. If the ownership is assigned to the reliability engineer, the process and the quality engineers should still be involved.

6.2.3 Design Optimization for Robustness

As design work progresses, quality assurance activities should take a foothold in the formal design reviews. Technical analysis should be done to assure that the product design is optimized and the process reliability, yield, and availability are maximized. Design engineers, manufacturing engineers, reliability engineers, and quality assurance engineers should jointly perform experiments, either on hardware or in software simulation, to assure interchangeability, insensitivity to noise, and optimum tolerance for each component. These parameters and tolerance ranges should be

Failure mode	Failure cause	Effects				Criticality			Diagnosis time (h)	Repair time (h)	Expected lost time per month	Action plan
		Local	Sub-assm.	Machine	Other	Downtime	Freq.	S & F				

Figure 6.4 An equipment FMECA format.

scientifically established to optimize the performance so that it is robust under any operating and environmental conditions. Methods proposed by Shainin [5], Taguchi [6], and Hicks [7] should be used. Shainin's and Taguchi's methods are probably the most popular approaches to robust design.

6.2.3.1 Shainin Approach

Shainin uses a variable search pattern technique to isolate the important variables and interactions that govern product performance. The search pattern uses multi-vari charts and/or full factorial (all possible combinations) experiments to zero in on the causes.

Multi-vari charts (Fig. 6.5) are used in stratified experiments to establish whether the variation is within a unit (position variation), from unit to unit (cycle variation),

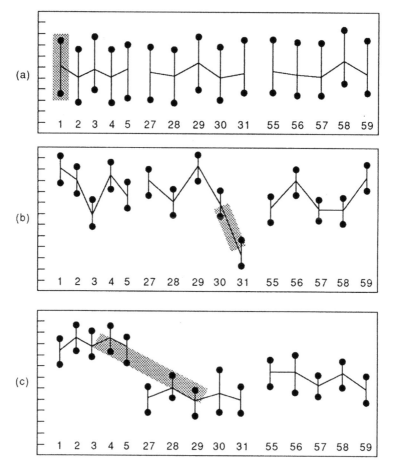

Figure 6.5 Multi-vari charts: (a) position variation (within unit), (b) cycle variation (unit to unit), and (c) temporal variation (time to time). (From Reference 8.)

or from time period to time period (temporal) [8]. A case history of full factorial experimental design is given in Section 6.3.2.

One of the major strengths of the Shainin method is that it gives equal importance to the effects of interactions. Shainin reminds us of many engineering laws governed by interactions [5] such as $PV \gg MRT$, which is a three-factor interaction affecting the pressure and volume. The authors have used the interaction of two variables (gas pressure and voltage) in solving the problems of seemingly mysterious corona failures in high-voltage cables. The study of interactions is the main reason for using full factorial experiments. These may incur more cost, but the analysis is thorough. Also, it is easier for a practitioner to understand just one simple way of experimenting.

6.2.3.2 *Taguchi Approach*[1]

The renewal in the use of design of experiments (DOE) methods in industry since 1982 is primarily a result of the work of Dr. Genichi Taguchi. Classical DOE focuses on identifying factors that affect the *level* (or *magnitude*) of a product/process response, examining the response surface, and forming the mathematical prediction model. Taguchi, however, recognizes that DOE should be applied in both product and process development to identify the factors that affect the *variability* of the response. The combination of different factor levels, identified from the DOE with significant effects on response variability and checked by a confirmation run later, is likely to result in a product/process response insensitive to the change of operating settings, environmental conditions, and material degradation. This pioneering work not only improves the product/process quality, but also enhances the reliability, since the product/process response is consistent over time and different operating conditions.

Taguchi's robust design methodology [8] applies a fractional factorial design technique, which means more factors are studied with less experiments at a loss of some information. It makes use of standard matrices for experimental design setups, called *orthogonal arrays*. Depending on the number of variables and levels of experiments, an appropriate array is chosen. The methodology studies more than the traditional number of effects, takes advantage of nonlinear relations between response and control factors, employs the operating window concept (explained later in this section), and selects appropriate quality characteristics to achieve three objectives: *efficiency* in generating information to make decisions, *economy* in developing a product/process, and *consistency* in product/process performance over time and operating conditions [9].

Role of Orthogonal Arrays and Linear Graphs

Orthogonal arrays are statistically designed experimental layouts. (*Note*: To develop robust designs, one can use any method of design of experiments. Using orthogonal arrays is one approach proposed by Taguchi.) In robust design, an orthogonal array is used to achieve the

---

[1]The authors thank C. Julius Wang of Chrysler Corp. for his generous contribution of material on the Taguchi approach.

addition of the main factor effects. In other words, interactions are practically not welcome. This feature makes the analysis subjective to some extent. Reproducibility from laboratory test results to downstream production and field use is a big concern [9]. By applying orthogonal array experiments and performing confirmation runs to verify the additive prediction, experimenters can get a highly consistent product performance in a timely fashion at a lower cost. One example of an orthogonal array is called the $L_8(2^7)$ array, which means it can be assigned up to seven 2-level factors and that only eight experiments are required (full factorial design would require 128 experiments to study seven variables).

Linear graphs (Fig. 6.6) developed by Taguchi can be used to facilitate assignment of factors to each column. This figure also shows the $L_8(2^7)$ orthogonal array in which the factors A, B, C, D, E, and F are assigned to six columns. Each column shows the levels of each variable in each of the eight experiments. Each row represents a specific experimental run. For example, row 2 indicates that the second experiment will have factors A, B, and C at low level and the factors D, E, and F at high level. Column 7 in this study is left open, since no factor is assigned to it.

The linear graph in Fig. 6.6 shows that each dot and line segment can be assigned a factor and the line segment also is the interaction of the two connecting dots. Therefore, column 3 will be a compound effect of interaction between the factors represented by columns 1 and 2—that is, factors A and B, factor C (the main effect), and some other interactions. This is known as *confounding*. In Taguchi's robust design, seeking main effects is more important than identifying interactions, and this feature makes the Taguchi approach somewhat subjective. The confounding

$$L_8(2^7)$$

		A	B	C	D	E	F	Left open
	Col	1	2	3	4	5	6	7
Row								
1		1	1	1	1	1	1	1
2		1	1	1	2	2	2	2
3		1	2	2	1	1	2	2
4		1	2	2	2	2	1	1
5		2	1	2	1	2	1	2
6		2	1	2	2	1	2	1
7		2	2	1	1	2	2	1
8		2	2	1	2	1	1	2

Figure 6.6 Example of orthogonal array and linear graph.

effect, however, makes it possible to explore the significant main effects and to verify them in a confirmation run [10].

Nonlinearity Application The advantage of using the nonlinear relation between response and control is best illustrated in Fig. 6.7. Suppose the product response (e.g., output voltage) target is $Y(f)$ and the corresponding factor (e.g., resistance) setting under the influence of all other factors should be $X(t)$. The resistance value has a distribution around $X(t)$ because of material inconsistency and manufacturing process variation. The resulting response voltage will have a large variation around the target. However, if the resistance value is changed from $X(t)$ to $X(s)$, the resulting response voltage $Y(s)$ has a more centered distribution, even though the resistance shows more variation than before. Since the response is now off target, adjustment factors identified from DOE—factors having strong impact on the level of response but limited impact on the variability—could be used to adjust the response back to the intended target $Y(t)$. This is sometimes called *two-step optimization* [11–13].

Outer Array Application and Parameter Design Some variables cannot be controlled or are impractical to control without a major investment: for example, eccentricity on an old machine or room temperature and humidity. But their effect on quality may be significant. Such factors are treated as noise but their effect is evaluated by means of an additional experimental setup called an *outer array*. The main orthogonal array is the *inner array*. Figure 6.8 illustrates the experimental layout for handling the outer array. This method is used in parameter design, which is the most critical step in

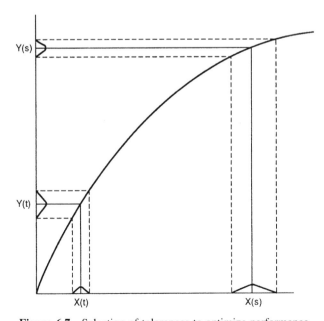

Figure 6.7 Selection of tolerances to optimize performance.

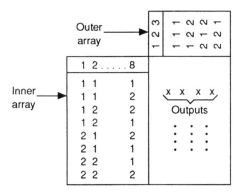

Figure 6.8   A conceptual layout of inner and outer arrays.

Taguchi's three-step development strategy of *system design* (specifying the system requirements), *parameter design* (optimizing factor levels to achieve insensitivity with the least expensive components), and *tolerance design* (upgrading component tolerances when necessary to reach the desired target with the least extra cost).

Operating Window Concept   Taguchi proposes an operating window concept in which the larger the window, the more immune the process is to the production variation. An example [14] from a study of copying machine paper feeders illustrates the concept. Two failure modes, misfeed (no paper) and multifeed (too many papers) were of concern. The paper is fed by using the friction between pressing rollers and the paper, which is pressed by a spring. The spring force is critical, since at a threshold value F_1 the misfeed failure mode disappears and a single sheet feeds. Increasing the pressing force to a second threshold value F_2 introduces the multifeed failure mode. The objective then is to maximize the operating window formed by two threshold values (Fig. 6.9a). Another example is a steel alloy that achieves desirable ductility in a 25 °F temperature range during a hot-rolling process. This temperature range is difficult to maintain. By adding silicon in the steel composition, the range was widened to 75 °F (Fig. 6.9b).

6.2.4   Process FMECA

Design reviews are the best opportunity a quality assurance engineer has for voicing opinions and getting the appropriate action implemented. At this time, the hardware has not been made nor has money been invested in tooling. Therefore, talk is still cheap and has its greatest impact on reducing life-cycle costs. Engineering changes are practically free.

The process failure-mode, effects, and criticality analysis should consider the performance of the entire manufacturing process, which is much more than just the equipment. It includes support services and operating procedures. The purpose is to use brainstorming to develop the best possible process. The product design

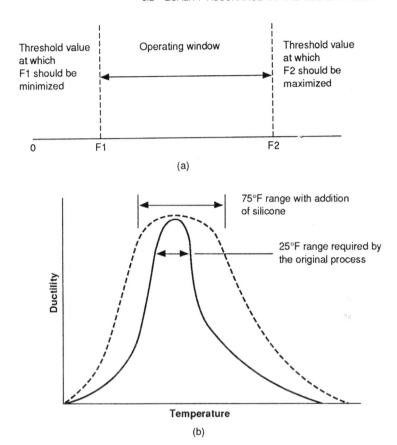

Figure 6.9 Operating window examples: (a) operating window for paper feeding and (b) operating window for steel ductility.

engineer is very much a team member in the review on an as-needed basis. The process or the manufacturing engineer may or may not chair the review team but should assume technical leadership. For a process to work right, it must incorporate the expertise in variation control that comes from quality assurance engineers and the expertise in fabrication methods that comes from manufacturing engineers. The reliability and design engineers should be invited to join in a review at important milestones in the project.

To have the best product, a company must have the best team. It is equivalent to saying that if a company has the best design, best manufacturing, and best quality assurance, then its product must be the best. If any of these elements is not the best, the product will not be a market leader.

One of the major reasons for the failure of the team effort between manufacturing and quality assurance is a lack of integration. Process FMECA allows this integration to take place early. It requires quality assurance and manufacturing engineers to work together and integrate their skills. It is a design review tool for analyzing a new process. It includes analysis of receiving, handling, and storing materials and tools. Whenever possible, suppliers should also participate.

The procedure is as follows:

Step 1: Construct a process flowchart. Include in it all inputs to the process such as the materials; their storage, handling, and transportation; and the tooling. Standard operating and maintenance procedures should also be identified.

Step 2: Using a process FMECA format similar to Fig. 6.10, perform the analysis below.

Step 3: Write down each major process step in column 1.

Step 4: Write down what can go wrong (failure mode) for each process step. For example, if a process step for an assembly is "Insert a lockwasher," some things that could go wrong are: "wrong washer," "cracked washer," and "operator forgets to insert washer."

Step 5: Write the possible causes (failure mechanism) for the discrepancy. For the lockwasher example, the cause may be "human error."

Step 6: Identify the effect of discrepancy on the end product: for example, "Missing lockwasher may prevent product from performing its critical mission."

Step 7: Assign criticality. This can be assigned as a multiplication of severity, frequency, and undetectability ratings, each rated on a scale of 1 to 10. Each organization has to define these three terms and the associated ranking system. The following definitions should apply to most organizations.

Severity may be defined as the impact on safety and functional performance. Multiple-fatality accidents or a failure resulting in enormous financial loss may be rated as 10, while negligible damages will be given the numerical value 1. Sometimes the effects on the process are not severe, but a defect may cause unsafe conditions for the user. In such cases the effects on the user should also be considered.

Frequency can be judged on the probability of occurrence. If the event is certain to occur, it will be assigned a rating of 10. The impossible event will assume the lowest rating. Some companies substitute *probability of failure* for frequency.

Detection can be defined as how late in the process the defect can be caught. If it can never be caught before shipping, then the detection rating is the highest.

Figure 6.11 shows the relationship of these criticality measurements to other FMECA items. It may be noted that severity is a function of the ultimate effect on the customer.

Step 8: If the equipment is not yet fabricated, perhaps the design can be changed to avoid the problems exposed in step 7. For already-built equipment, preventive action may be possible. In most cases, life-cycle costs are lower if the corrective action is taken in design.

As an example of the efficacy of correction in design, consider the failure mode in which an operator forgets to add a lockwasher before putting on a bolt. Inspecting the assembled equipment for the presence of the lockwasher would be monotonous, and inspectors might often think there is a lockwasher even if it is not there. A design approach would be to use an automated assembly fixture to assure that the lockwasher will always be inserted. An alternative design choice would be to make the washer an integral part of the bolt. A third choice would be to eliminate the need for the washer in the product. General corrective actions, in order of precedence, are:

1. Change the process so that the cause of the defect is eliminated.
2. Change the product design to avoid the process.
3. Install some kind of warning system so that the operator will know what is wrong with the process before it produces defective pieces.
4. If these actions are not effective, give the operator special training.

There may be a few hundred failure modes in a FMECA. Not every one requires a design action. In the interest of working effectively and efficiently, the important failures should be determined, by evaluating the risk assessment results, and handled first. The safety-related failures must receive top priority, even though their total rating may not be the highest. In cases where the corrective action is not obvious, a fault tree (Chapter 5) or process analysis maps (this chapter) may help in identifying root causes and actions.

6.2.5 Quality Assurance Plans for Procurement and Process Control in the Design Phase

Some planning is required before going into production. These efforts will allow for a smooth transition from design to production.

6.2.5.1 *Equipment Procurement Plans* Procurement plans should include verification of equipment performance with suppliers. Of course, performance of some large capital equipment cannot be verified until it is installed at the customer's site. In that case, performance at subsystem levels should be verified. Other activities may include testing for component interchangeability, statistical tolerance analysis, establishing machine capabilities, and pilot runs.

6.2.5.2 *Process Control Plans* All the critical points where a major defect can be introduced should be identified before going into manufacturing. Quality function deployment and process FMECA should be applied to generate process control plans, since this is the only way to assure that the right things are controlled. Every control should be traceable to customer satisfaction, which is reflected in the QFD. Plans should include methods for controlling these critical process points, definitions of acceptance standards, and ways of monitoring all defects. The critical points should be highlighted in the manufacturing method sheets.

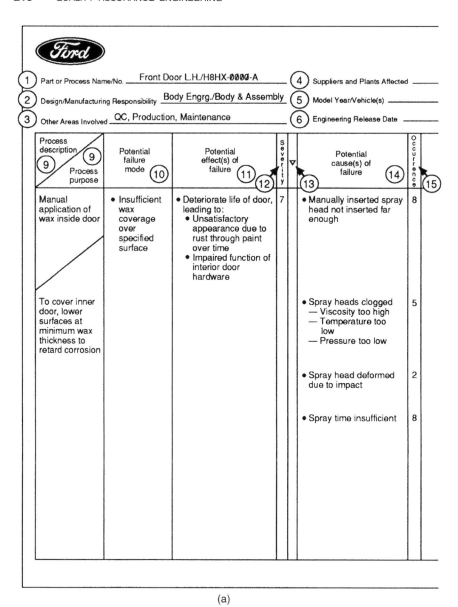

(a)

Figure 6.10 Process FMECA.

6.2.5.3 Component Procurement Quality Plans

Critical dimensions and important material properties should be identified before approving component suppliers. Once defective parts come in, they destroy any hope for a successful pilot run. The supplier and customer must work together because supplier components require

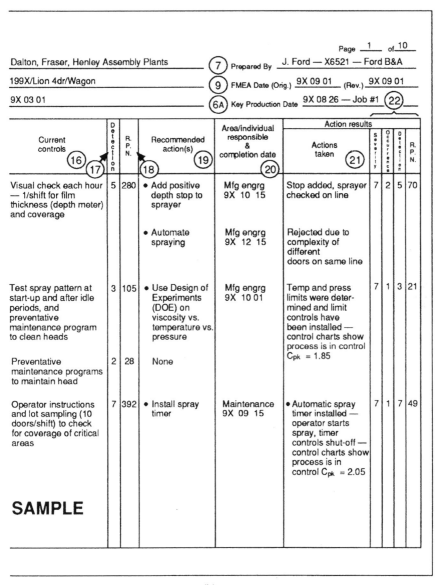

Current controls 16 17	D e t e c t i o n	R. P. N.	Recommended action(s) 18 19	Area/individual responsible & completion date 20	Actions taken 21	S e v e r i t y	O c c u r r e n c e	D e t e c t i o n	R. P. N.
Visual check each hour — 1/shift for film thickness (depth meter) and coverage	5	280	• Add positive depth stop to sprayer	Mfg engrg 9X 10 15	Stop added, sprayer checked on line	7	2	5	70
			• Automate spraying	Mfg engrg 9X 12 15	Rejected due to complexity of different doors on same line				
Test spray pattern at start-up and after idle periods, and preventative maintenance program to clean heads	3	105	• Use Design of Experiments (DOE) on viscosity vs. temperature vs. pressure	Mfg engrg 9X 10 01	Temp and press limits were deter- mined and limit controls have been installed — control charts show process is in control C_{pk} = 1.85	7	1	3	21
Preventative maintenance programs to maintain head	2	28	None						
Operator instructions and lot sampling (10 doors/shift) to check for coverage of critical areas	7	392	• Install spray timer	Maintenance 9X 09 15	• Automatic spray timer installed — operator starts spray, timer controls shut-off — control charts show process is in control C_{pk} = 2.05	7	1	7	49

Page 1 of 10

Dalton, Fraser, Henley Assembly Plants (7) Prepared By J. Ford — X6521 — Ford B&A

199X/Lion 4dr/Wagon (9) FMEA Date (Orig.) 9X 09 01 (Rev.) 9X 09 01

9X 03 01 (6A) Key Production Date 9X 08 26 — Job #1 (22)

SAMPLE

(b)

Figure 6.10 *Continued.*

as much quality as any other component in the product. At least four areas warrant concern and cooperation:

1. Process control implementation throughout the production lines
2. Component qualification

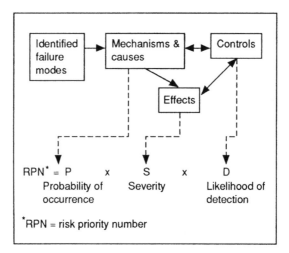

Figure 6.11   Relationship of criticality to FMECA items.

3. Standard and special screening tests (see Chapter 3)
4. Closed-loop failure management

6.3   QUALITY ASSURANCE IN THE MANUFACTURING PHASE

Quality assurance activities in manufacturing are aimed at assuring that processes are always up to standards and that manufacturing protocols for quality assurance are in place. The ownership of each procedure and process must be logically and clearly identified. For instance, if the material is discrepant, it should be the production foreperson who has ownership of the problem—not the quality control engineer. It is the production foreperson's job to ship only good products. The quality control function must go on, even if the quality control engineer is not available.

6.3.1   Evaluation of Pilot Run

If the manufacturing yield is going to be low, the pilot run is the place to correct it. Under no circumstances should the customer relationship be injured by the shipment of shoddy products under pressure of schedules. With a lot of "homework" done in advance, the pilot run yield should be high.

Preparation activities should also concentrate on assuring that process engineers have written process specifications to establish process performance standards. The operating procedures should also be documented to establish human performance standards. The yield depends on both.

Usually the process standards are poorly defined and often are not defined at all. A solder equipment operator is told to produce a good solder joint but is rarely told the best temperature range for the solder. Similarly, rarely does a manufacturing engineer specify when to change bearings, filters, and lubricating oils. They are usually changed after the process starts to go out of control. In over 95% of cases, equipment operators have poor process specification. A typical statement is: "Heat the part to 200 °C." This statement is completely inadequate. It does not tell the operator if 190 °C is acceptable or how closely the variation must be controlled.

Operating procedures are just as important. In the above example, the operator should also be instructed what to do if the temperature is above or below certain values. Other information in the operating procedures should include when to add lubricant, how much lubricant, when to stop the process, how to monitor the process, and how to prevent production of defective units.

Operating procedures also include preventive maintenance procedures. The maintenance procedures should be based on knowledge of the process. They should be designed to prevent manufacture of defective parts. The belts, bearings, and fans should be changed before they have negative effects on quality. In other words, the maintenance personnel must be familiar with statistical process control techniques. In one case, a chemical concentration in a solution was to be controlled closely but the process was always out of control. The reason was that one of the chemicals came in 5-gallon drums and the operator poured all 5 gallons, even though only 1 gallon was required. The operator was never told how to determine the right amount to pour. Without good operating procedures, a process will never be stable, and statistical controls will be meaningless. In any way possible, quality function deployment, discussed in Section 6.2.1, should be used to generate the manufacturing process performance standards and operating procedures (for operation standards). Thus, customer needs are converted to measurable (controllable) engineering characteristics; these engineering characteristics are then deployed to different parts specifications, and manufacturing process performance standards and operating procedures are finally developed from the associated parts specifications. This exercise guarantees that what the manufacturing process is doing really relates to the original customer expectations and anything to be changed will directly affect what the customer wants.

6.3.2 Process Control

It cannot be emphasized enough that someone in the organization should have ownership of the process specifications and operating standards. Every time a defect is produced, the validity of these documents should be questioned. This protocol should be well understood.

Some continuous improvement activities during the manufacturing stage may be validation of interchangeability, experiments to improve relationship of tolerance to product performance, verifying machine capabilities, establishing accuracy and precision of quality assurance measuring devices, improving preventive maintenance, improving operator training, and conducting audits. How well these activities are performed will determine whether the manufacturing process can be controlled.

Process control tools are used to get the best yield out of the existing equipment. Nonetheless, if a new item of equipment is needed, a trade-off analysis should be made. Operating personnel are trained to use process control techniques, which may or may not involve statistical control charts. Many controls can be put in place without control charts. The best strategy is to eliminate the source of problems and avoid costly controls.

Many try to use control charts on a characteristic that is out of specification without proper planning. They take months to find out that the process cannot be controlled without controlling the causes. They waste time and money. Unfortunately, most companies belong in this category. Proper planning requires the use of analysis tools. Fishbone diagrams, design of experiments, and process analysis maps are useful problem-solving tools and prevention tools to improve the process. Lots of people believe the myth that putting statistical process control (SPC) charts in place and doing some fishbone diagrams and probably limited DOE will bring their process back to life without even noticing that they have an inferior process design at the outset.

6.3.2.1 Identifying Causes of Variation
Traditionally, so-called Ishikawa diagrams (fishbone diagrams), such as that in Fig. 6.12, are constructed to identify sources of variations. These variations can come from the "six M's": man, materials, measurements, methods, machines, and mother nature (environmental factors).

Fault-tree analysis is another approach. One of the authors used a simplified fault-tree approach while working with McElrath & Associates, management consultants, at one of the largest steel plants. Since 1981 this tool has been used at several

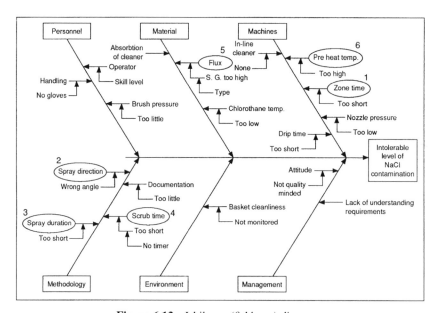

Figure 6.12   Ishikawa (fishbone) diagram.

corporations for statistical process control training with dramatic results. In one case, the cost of a component was reduced by about 50%. The tool offers flexibility and ease of construction.

Fault-tree analysis can be simplified as a process analysis map (PAM), Fig. 6.13. Each level in a PAM identifies the causes of variations at the previous level. Usually three levels are sufficient. If it is necessary to go to more levels later, a sheet of paper can be added to extend the chart.

At each level, the causes attributable to the six M's are identified. Typical causes are:

Man: Machine operator, maintenance technician, supervisor

Machine: Production equipment, tools, instruments, gauges

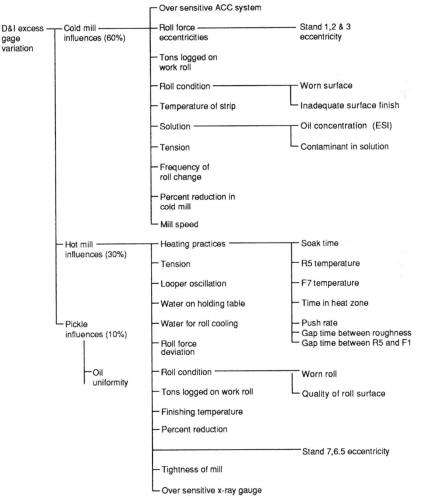

Figure 6.13 Example of a process analysis map (PAM).

Measurement: Production equipment accuracy and measuring instrument accuracy and precision

Methods: Machine control methods, standard operating procedures, maintenance procedures, repair procedures

Materials: Supplier parts, fabrication materials such as epoxies, fluxes, welding rods

Mother Nature: Humidity, temperature, dust, chemicals, contamination, radiation

To manufacture products that are cheaper and of higher quality at the same time, one should follow the famous 20/80 principle, which says that about 20% of causes contribute to 80% of losses. The percent contribution of each cause to the problem is therefore estimated so that process controls are placed at the critical points. The PAM of Fig. 6.13 identifies more than 30 causes for the variation in the thickness of a material.

Controlling all the causes is costly. Some causes can be technically detrimental but may have very little effect on the process. Therefore, one must verify the contribution of the critical causes by design of experiments. A good strategy is to use judgment in selecting three or four causes and experimentally test their influence. Then identify other important variables and check them the same way.

6.3.2.2 Verifying the Influence of Causes

The next step is to run experiments to isolate the critical causes in the PAM. Methods proposed in earlier references and in Reference 15 are among those commonly used. A technique presented below (we learned to appreciate this full factorial technique from Shainin) is reasonably economical and will work most of the time. If one still cannot identify most of the critical causes, then more in-depth and more efficient studies and experiments can be selected from the references. Decide which three or four variables are most critical. Temperature, eccentricity, and roll condition were picked from the PAM in Fig. 6.13. Call them variables A, B, and C. To study the influence of a variable, perform experiments at atleast two levels (high and low) for each variable. To assure the experiment is performed for all the possible combinations, perform $2^3 = 8$ experiments, where 2 represents the number of levels and 3 represents the number of variables.

The experimental design matrix is constructed by using the principles of full factorial experiments as shown in Fig. 6.14. The + and − signs are equivalent to levels 1 and 2 in the Taguchi method and represent the high and low levels of the cause. In this method, half the experiments for the first variable A are assigned at high (+) level and the other half at low (−) level (see column labeled Factor A). The set of four continuous + signs under A is called a run, as is the continuous set of four negative signs. To assign levels for variable B, use an equal number of + and − signs for each run of variable A (see column labeled Factor B). Similarly, for variable C, make half the experiments at high level and half at low level for each run of variable B.

	Factor A	Factor B	Factor C	AB	BC	AC	Yield
1	+	+	+	+	+	+	75
2	+	+	−	+	−	−	29
3	+	−	+	−	−	+	25
4	+	−	−	−	+	−	7
5	−	+	+	−	+	−	48
6	−	+	−	−	−	+	56
7	−	−	+	+	−	−	39
8	−	−	−	+	+	+	4

Figure 6.14 Experimental design matrix. (*Note*: Interaction *ABC* is not shown but can be assessed.)

In these experiments, one needs to know the effects of interactions of two or more variables. No additional experiments are required because the interaction effects can be assessed from the same data. The columns AB, BC, and AC show the influence of these interactions as follows. To construct column AC, multiply the signs of variable A and variable C to determine the sign under variable AC; two positive or two negative signs multiplied yield a + sign, and two *dissimilar* signs multiplied yield a − sign. As shown below, the resulting signs will be used in determining the influence of the interaction. The experimental design in Fig. 6.14 shows that the first experiment will be performed with all three variables at high level. The second experiment will have variable A at high level, variable B at high level, and variable C at low level. There are altogether eight experiments, which include all the possible combinations of A, B, and C.

The last column shows the yield (percent product in the target range) for each experiment. Any other measurement of inherent process quality would be just as appropriate. Looking at the data, one can say that the highest yield is obtained when A, B, and C are at high level, but such conditions may be too expensive to maintain all the time. Also, a hidden variable may be responsible for some of the yield. Scientifically, it is better to establish the influence of the individual variable and maintain control on it all the time. Even better, eliminate the source of variation. In this case, the source of variation, eccentricity, could be eliminated only by a major equipment change, and management elected to use statistical process control instead.

Factor	Avg. yield at high level	Avg. yield at low level	Difference
A	34	36	−2
B	52	18.8	33.2
C	46.8	24	22.8
AB	36.8	34	2.8
BC	33.5	37.3	−3.8
AC	40.0	30.8	9.2

Figure 6.15 Comparison of yields.

To determine the effect of the variable as other factors vary, one may compare the average yields when the variable is at high level and at low level. This analysis is shown in Fig. 6.15. For variable A, the average yield for the first four experiments (A at high level) is 34% and the average yield for the last four experiments (A at low level) is 36%. Since the difference in the yield is only 2%, the effect of variable A does not seem to be significant. Variable B, on the other hand, seems to have a considerable effect. The average yield for B at high level (experiments 1, 2, 5, 6) is 52% versus 18.8% at low level (experiments 3, 4, 7, 8). The influence of interactions is assessed in the same manner. If we wish to know the effect of *AC*, we can compare averages when interaction *AC* is at high level (experiments 1, 3, 6, 8) and when interaction *AC* is at low level (experiments 2, 4, 5, 7). The corresponding yields, 40.0% and 30.8%, show that the interaction seems to influence the yield. If the reader is interested in accuracy, then analysis of variance (ANOVA) and other techniques found in the references may be used.

The data in these experiments show the variables B and C have a significant effect on yield. Often the yield will improve dramatically when just a few variables are controlled. In this case, the statistical process control of only two variables—B and C—produced a complete reversal in yield (Fig. 6.16).

In summary, design of experiments is a powerful tool, and tools such as PAM and fishbone diagrams are just as necessary. The key is to eliminate sources of variation, if possible. Short of that, variability must be reduced first, then the mean of the process must be controlled closest to the target value.

6.3.2.3 Statistical Process Control Ideally, a cause of excess variation should be eliminated by the process engineer in the equipment design. Many times this can be done, but companies normally do not invest enough in this area. For an inherently unreliable process, the control chart may be necessary. The purpose of control charts is to study the variation in the process and monitor its stability. Actually, control charts *do not* control the causes of variation. Control charts *detect* a signal out of control, and other tools are needed to bring the process back

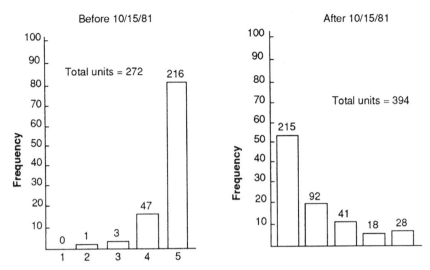

Figure 6.16 Reversal in yield with control of two variables.

into control. Also, once control charts detect a signal out of control, one has to verify whether the causes are sporadic or systematic. A sporadic cause (such as the operator not replacing the coolant in the machine) can be eliminated very quickly. However, the removal of a systematic cause often needs a management decision. For example, a machine simply may not be able to hold the required tolerance because it is not designed for that accuracy. Then a management decision may be made to change the machine.

The generic steps in controlling the process are:

Step 1: Check whether the process is stable by constructing a control chart. If the process average or the standard deviation is not stable over a reasonable length of time, then some uncontrolled causes must be removed before the process can be controlled.

Step 2: Determine the inherent process capability which may be computed at ± 4 standard deviations from the data. To guard against production of nonconforming parts, a customer often specifies a so-called capability index $C_p = 1.33$. C_p is the ratio of engineering tolerance to the traditional 6 standard deviation process spread. For an area less than 100 ppm outside the specification, this ratio should be at least 1.33. (For detailed discussions of noncentered processes, see Section 6.3.3.) If the process capability range is wider than the engineering specifications, then some yield loss is expected, besides extra inspection cost. Therefore, an attempt must be made to remove or control some causes of variation by using information from cause–effect diagrams such as a PAM or fault trees.

Step 3: Keep removing or controlling the causes of variations and verify their effects on control charts until the C_p and ppm goals are met. The ppm goals will determine the maximum allowable size of the standard deviation.

Step 4: Monitor control charts and train operators to recognize the signals for processes going out of control. They should be able to correct the condition before any defective piece is produced. Control charts typically give warning of the process going out of control. Many books on statistical quality control cover such topics [16].

There are two types of basic control charts: (1) control charts for variables, such as X-bar and range charts, and (2) control charts for attributes, such as *p* charts and *c* charts. Modified charts are used for special applications such as precontrol charts, moving average charts, X-bar and *S* charts, and median charts. Most charts are described in Reference 17.

6.3.2.4 *Control Charts for Variables* These charts are used when the variable can be measured, such as a dimension on a shaft. They should be avoided if the component designed can be changed to eliminate the need for control. In a steel part the control changing the chemistry of the steel such that the variation did affect the strength of the steel eliminated charts. The concept of control charts was covered in the Chapter 2.

6.3.3 PPM Control for World-Class Quality

Traditional inspection sampling plans like Mil-Std-105 [18] are no longer appropriate. Sample size in thousands is required when one or more nonconformities are acceptable. For 100 ppm quality (0.01% acceptable quality level), the plan calls for 100% inspection up to a lot size of 1250, and no defects are permitted. If the process is producing more than 100 ppm defects, then practically every lot is going to be rejected. Even if the process produces world-class quality (100 ppm or less), about 12% of the lots are questionable, as shown by the following computation.

Suppose a batch has a quality level of 100 ppm, or 1 defect per 10,000. The probability of a good piece then is 0.9999. The chance that all pieces are within specification is $(0.9999)^{1250} = 0.8825$. In other words, 11.75% of the lots are expected to have 1 or more pieces defective. Experience tells us that if an inspector is checking a sample of 1250 pieces and if the defects are rare, the inspector is likely not to notice the defects. The bad lots are expected to go to the customer. On the other hand, if the bad lots are caught, then the entire production is suspect, resulting in 100% inspection. Regardless of which way the results go, the cost of inspection is too high. A better alternative is to narrow the process variation and monitor the stability of the mean at the specified target. There is a trade-off between cost of inspection versus cost of engineering. Investing in engineering is generally cheaper if the root causes of variation are controlled (preferably eliminated).

Global competition, requiring higher quality at lower costs, has forced international companies to do the right things at the right time. As a result, the informal standard of nonconforming levels of less than 100 ppm has emerged. As mentioned earlier, world-class quality level is a moving target set by the best producer in the world. For a complex product and emerging technologies, it may not be possible to achieve this standard. Such products may have a ppm as high as 15,000. The moral is that whoever comes into the global market first with low price and high quality sets the standard for others.

The criteria for producing less than 100 ppm nonconformance can be explained with the help of Fig. 6.17. Since a process is rarely exactly at the center of the engineering specification, analyze the worst side, which will produce more nonconformance. (If the process is centered, either side will do.)

In this case, side AB is the worst case. Table A in the Appendix shows that for $Z = 4\sigma$, the nonconformance rate is 30 ppm, which meets the requirements of a maximum of 100 ppm. Therefore, if the distance AB is at least equal to 4 standard deviations and if the process is stable, then the process will always produce parts within the 100 ppm level. This criterion is explained in a slightly different way in industry [16]. It is written as

$$C_{pk} = \frac{Z}{3} = 1.33 \quad \text{or} \quad Z \geq 4.0$$

where C_{pk} is the capability index and Z is given by the expression

$$Z = \frac{|\text{specification} - \text{process average}|}{\text{standard deviation}}$$

if the process average is somewhere within the engineering specifications (see Chapter 2).

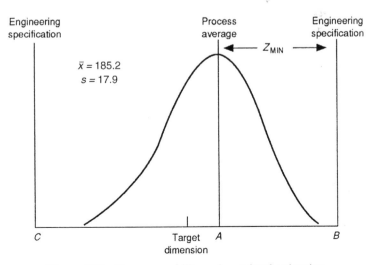

Figure 6.17 Process capability and ppm level estimation.

Example 6.3   A process is required to maintain a value within 150 and 225 lb/in.^{2}. (a) Calculate the nonconforming ppm level produced by this process. (b) What should be the standard deviation for less than 100 ppm nonconformance?

Solution   (a) The nonconformance for side *AB* of Fig. 6.17 is

$$Z_{AB} = \frac{|225 - 185.2|}{17.9} = 2.22$$

From Table A in the Appendix, the percent nonconformance is 1.32. The nonconformance for side *AC* is

$$Z_{AC} = \frac{|185.2 - 150|}{17.9} = 1.97$$

From Table A, the fraction nonconformance is 2.44. The total nonconformance is

$$Z_{CB} = 1.32 + 2.44 = 3.76\% = 37,600 \text{ ppm}$$

The worst side is *AB*. The criterion is $Z_1 \geq 4.0$. That is,

$$\frac{|225 - 185.2|}{\text{Standard deviation}} \geq 4.0$$

From this equation, the standard deviation should be less than $(225-185.2)/4$ or 9.95

It has been mentioned that, to avoid inspection, a process must be stable. If the average keeps moving, controlling the standard deviation may not help. The secret of controlling the average is to make sure the maintenance and operating procedures are sound and well understood. It is usually the preventive maintenance people who really have control over the equipment.

Complete control on a process requires monitoring seven parameters: product upper specification limit (USL), product lower specification limit (LSL), product target *T*, process mean *M* (central tendency), process standard deviation (variation), process skewness (a measure of symmetry), and process kurtosis (a measure of peakedness), in most applications. However, C_{pk} discussed earlier considers only four of the seven parameters. Thus, the C_{pk} value is not unique; two processes can have the same C_{pk} value but different distributions. For example, consider two processes with $T = 15$, USL $= 20$, and LSL $= 10$. Process A has ($\mu_A = 15$, $\sigma_A = 1.25$ and process B has $\mu_B = 18$, $\sigma_B = 0.50$. Both processes have $C_{pk} = 1.33$ but are totally different processes.

6.3.4   Working with Suppliers

One hundred percent inspection is costly for the buyer as well as for the supplier. They must work as a team to reduce the cost of inspection and increase the yield. This will lower the costs of components. The buyer must give a clear indication of what quality characteristics are critical. Simply asking suppliers to use statistical process control will raise costs. A very large automobile company sent such a directive to its suppliers. One of the suppliers started to put SPC on hundreds of dimensions, even though on 80% of them it did not make any difference whether they were in control or not. The result was that too much money was wasted on controlling dimensions that did not matter and not enough effort was put on critical dimensions. Some suppliers were controlling the dimensions but not their causes. They could never keep the processes stable and always fell back to 100% inspection. One of the best tools is to have proper communication for tracking continuous improvements in process capabilities. The savings resulting from improved yield and less inspection should be shared by the buyer and the supplier.

Eventually the ppm goals will be met by the supplier, and the buyer no longer needs to monitor. This is the most effective tool for a just-in-time system where the supplier ships parts straight to the production floor of the buyer. Since no ware-house inventory is required, the savings are enormous.

Caution: Buyers of components should have the same standard of excellence for their own final assemblies. Frequently, a buyer will demand perfection from the sup-plier but its own final product is shoddy.

6.3.5 PPM Assessment

Quality has become a prime factor in procurement. Buyers want to know the ppm level of components. Several organizations are involved in writing standards for ppm levels. One such standard is the ANSI/ EIA-555 Lot Acceptance Procedure for Verifying Compliance with the Specified Quality Levels in PPM (1989), devel-oped by the Electronic Industries Association (2000 I Street, Washington, DC 20006) and published in conjunction with the American National Standards Institute (1430 Broadway, New York, NY 10018). It classifies nonconforming devices and components into five categories.

PPM-1: Functional nonconformances that are inoperative

PPM-2: Electrical nonconformances that fail to meet specified electrical parameters

PPM-3: Visual/mechanical nonconformances that fail to meet specified visual/ mechanical characteristics

PPM-4: Hermetic nonconformances that fail to meet hermetic requirements of a product

PPM-5: All nonconformances including PPM-2, PPM-3, PPM-4, and all other specification nonconformances, excluding administrative requirements

This standard contains two methods to estimate the ppm level. Method A is the worst-case estimate and is covered below. Method B is based on lot acceptance rate and can be found in the standard. The worst-case estimate is given by

$$\text{ppm in any class} = \frac{0.7 + \text{total nonconforming}}{\text{total number inspected}} \times 10^6$$

The number 0.7 represents a minimum number of nonconforming devices expected with roughly 50% confidence when a sample of n items is taken from an infinitely large population.

Example 6.4 Assess the ppm level when 3000 components are inspected and no nonconforming devices are found.

Solution

$$\text{ppm} = \frac{0.7 + 0}{3000} \times 10^6$$

$$= 233$$

Example 6.5 A manufacturer wants to develop a sampling plan that will allow less than 100 ppm nonconforming devices. Compute the sample size required when (a) no nonconforming devices are acceptable and (b) one nonconforming device is acceptable.

Solution (a) For acceptance number equal to 0,

$$\text{ppm} = 100 = \frac{0.7 + 0}{N} \times 10^6$$

Therefore,

$$N = \frac{0.7}{100} \times 10^6 = 7000$$

(b) For acceptance number equal to 1,

$$\text{ppm} = 100 = \frac{0.7 + 1}{N} \times 10^6$$

Therefore,

$$N = \frac{1.7}{100} \times 10^6 = 17,000$$

6.4 QUALITY ASSURANCE IN THE TEST PHASE

In a large job-shop system such as a nuclear power generator, the test stage has a very critical influence on the schedule. Every precaution should be taken to design and manufacture the system right so that the system will pass the acceptance test in the least amount of time. The same concept applies to mass-produced products such as washing machines, coffee makers, and automobiles, where the test cycle is relatively short but the financial risk of rejecting the entire production is very large. In determining the adequacy of the test, a closed-loop system between field failures and the test program is necessary.

6.4.1 Qualification Versus Production Testing

Qualification tests are those applied to prototypes and initial production. These tests are made rigorous to prove the product will work during its intended life. The tests are usually destructive. Production tests are much less rigorous. Their usual purpose is to make sure the product meets functional requirements.

Make sure the safety-critical features are validated for complete design qualification at least once or twice a year depending on the risk tolerance. A company shutdown completely because a relay's gasket surface was not as smooth as it used to be. Leak tests in production were too mild to detect fine leaks. An engineering life test on the seal would have discovered it easily.

The biggest mistake quality control departments make is that they rarely perform complete qualification tests once the device is in production. One of the authors was once asked to investigate the cause of "wet bikes" going into the sea and never coming back. At least 11 persons died when they took these motorcycle-like watercraft into the sea for recreational purposes. The vehicles are designed to be driven on water, but unfortunately a relay absorbed salt water and short-circuited the electrical system. During early design, the relays were leak-tested under the toughest environmental conditions with highly accurate leak-testing equipment. However, the production units were not subjected to such tests as thermal shock and accelerated aging, with the result that marginal units managed to go to the customers. If the company had done qualification tests periodically, these defective relays would have been caught in the factory. The company is out of business now.

The best guide for ensuring a good system is to qualify several production units at a predetermined interval. The frequency should depend on how good the process controls are. This is the best insurance against production errors.

6.4.2 Industry Standards

A word to the wise. Don't use industry standards as the *only* standard. These usually set a minimal level of performance. Every product usually requires unique tests, which can be determined from analyses such as failure-mode, effects and criticality analysis, fault-tree analysis, and preliminary hazard analysis.

6.5 QUALITY ASSURANCE IN THE USE PHASE

For an existing product, quality assurance should continue in the use phase. Instead, most companies concentrate on reducing costs in the manufacturing phase. Meanwhile, many customers are turning away because of substandard characteristics unrelated to manufacturing. For example, car buyers may not like many small buttons and complicated instructions on their radios or they may prefer a fuel injection system instead of a carburetor system. Therefore, quality assurance must survey the real concerns of customers. This information should be fed back to product and process engineering to make improvements, to update the QFD results, and to ensure that similar mistakes are not made again. That's where the secret of quality control lies!

6.6   TOPICS FOR STUDENT PROJECTS AND THESES

1. Critique the role of the manufacturing engineer in controlling the manufacturing processes. Develop a model for manufacturing engineering for assuring high quality at lower costs.

2. Most companies are unable to control quality because they fail to control operating and maintenance procedures. Develop a quality control model highlighting the role of these procedures.

3. Most practitioners complain that their management's commitment to quality has been ceremonial rather than real. Develop a model for top management involvement in quality assurance.

REFERENCES

1. *Total Quality Management: A Guide for Implementation*, DoD 6000.51-6 (draft), U.S. Department of Defense, Washington, DC, March 23, 1989.
2. H. E. Vannoy and J. A. Davis, Test Development Using the QFD Approach, Technical Paper 89080, SAE, Warrandale, PA, 1989.
3. J. R. Mauser and D. Clausing, The House of Quality, *Harvard Business Review*, May–June 1988.
4. R. N. Kackar, Taguchi's Quality Philosophy: Analysis and Commentary, *Quality Progress*, 1986.
5. D. Shainin, Better than Taguchi Orthogonal Tables, *ASQC Quality Congress Transaction*, 1986.
6. G. Taguchi, *System of Experimental Design*, Vols. 1 and 2, D. Clausing (Ed.), Unipub-Kraus International Publications, White Plains, NY, 1987.
7. C. R. Hicks, *Fundamental Concepts in the Design of Experiments*, McGraw-Hill, New York, 1973.
8. K. R. Bhote, *World Class Quality*, American Management Association, New York, 1988.
9. G. Taguchi, Development of Quality Engineering, *ASI Journal*, vol. 1, no. 1, 1988.
10. C. J. Wang, Taguchi's Robust Design Methodology in Quality Engineering, unpublished training booklet, Chrysler Corp., 1989.
11. M. Phadke, *Quality Engineering Using Robust Design*, Prentice-Hall, Englewood Cliffs, NJ, 1989.
12. R. V. Leon, A. C. Shoemaker, and R. N. Kackar, Performance Measures Independent of Adjustment, *Technometrics*, vol. 29, no. 3, 1987.

13. G. Taguchi and M. S. Phadke, Quality Engineering Through Design Optimization, *Conference Record, IEEE GLOBECOM '84 Meeting*, IEEE Communications Society, Atlanta, November 1984.

14. Y. Wu, Reliability Study Using Taguchi Methods, *ASI Journal*, vol. 2, no. 1, 1989.

15. G. E. P. Box, W. P. Hunter, and J. S. Hunter, *Statistics for Experimenters: An Introduction to Design, Data Analysis, and Model Building*, John Wiley & Sons, Hoboken, NJ, 1978.

16. Continuing Process Control and Process Capability Improvement, Ford Motor Co., Dearborn, MI, 1985.

17. J. M. Juran, and F. M. Gryna (Eds.), *Quality Control Handbook*, McGraw-Hill, New York, 1988.

18. *Sampling Procedures and Tables for Inspection by Attributes*, Mil-Std-105E, Naval Publications and Forms Center, Philadelphia, 1989. This standard was cancelled. The current preferred standard is Mil-Std-1916.

FURTHER READING

Anderson, V. L., and R. A. McLean, *Design of Experiments*, Marcel Dekker, New York, 1974.

Besterfield, D. H., *Quality Control: A Practical Approach*, Prentice-Hall, Englewood Cliffs, NJ, 1979.

Continuing Process Control and Process Capability Improvement, Ford Motor Co., Dearborn, MI, 1985.

Control Chart Methods of Analyzing Data, ASQC Standard B2/ANSI 21.2, American National Standards Institute, New York, 19XX.

Deming, W. E., Out of the Crisis, Massachusetts Institute of Technology, Center for Advanced Engineering Study, Cambridge, MA, 1986.

Diamond, W. J., *Practical Experiment Designs*, Van Nostrand Reinhold, New York, 1981.

Feigenbaum, A. V., *Total Quality Control*, 3rd ed., McGraw-Hill, New York, 1983.

Grant, E. L., and R. S. Leavenworth, *Statistical Quality Control*, 5th ed., McGraw-Hill, New York, 19XX.

Guide for Quality Control Charts, ASQC Standard B1/ANSI 21.2, American National Standards Institute, New York, 19XX.

Ishikawa, K., *Guide to Quality Control*, rev. ed., Asian Productivity Organization, 1976.

Juran, J. M., *Juran on Planning for Qualify*, ASQC Quality Press, Milwaukee, 1988.

Juran, J. M., and F.M. Gyrna, Jr., *Quality Planning and Analysis*, McGraw-Hill, New York, 1970.

Kearns, D. T., Chasing a Moving Target, *Quality Progress*, October 1989, pp. 29–31.

Ostle, B., and L. C. Malone, *Statistics in Research*, 4th ed., Iowa State University Press, Ames, 1988.

Ott, E. R., *An Introduction to Statistical Methods and Data Analysis*, Duxbury Press, North Scituate, MA, 1977.

Process Qualify Management and Improvement Guidelines, Code No. 500-049, AT&T, Indianapolis, 1988.

Sampling Procedures and Tables of Inspection for Variables for Percent Defective, Mil-Std-414.

Scherkenbach, W. W., *The Deming Route to Quality and Productivity, Roadmaps and Roadblocks*, Ceepress Books, Washington, DC, 1986.

Schilling, E., *Acceptance Sampling in Quality Control*, Marcel Dekker, New York, 1982.

Shoenberger, R., *Japanese Manufacturing Techniques*, The Free Press, New York, 1982.

Taguchi, G., *Introduction to Quality Engineering*, Asian Productivity Organization, Unipub-Kraus International Publications, White Plains, NY, 1986.

CHAPTER 7

LOGISTICS SUPPORT ENGINEERING AND SYSTEM SAFETY CONSIDERATIONS

7.1 LOGISTICS SUPPORT PRINCIPLES

A company will not be in business long if it does not support a product throughout its life. This support includes giving attention to the customer's need to keep the product operating. *Logistics engineering*, the branch of engineering that meets this customer need, bridges the gap between design engineering and product support functions. Design engineering decisions in the concept stage help to reduce logistics costs such as those of distribution, transportation, maintenance, spares, and documentation [1].

To reduce support costs, engineers design systems for reliability and maintainability, but their success has been questionable. Lerner maintains the following [2]:

> In most widespread use is the reliability and maintainability audit. Here an independent group of auditors analyzes the designs of engineers and estimates such variables as mean times between repairs and cost. If these do not meet specifications, the engineers are told to modify their designs. This is where the process breaks down. Engineers do not as a rule know which part of the design needs improvement, and the design process is so far along that most often they wind up accepting less reliability and maintainability.

Logistics engineering involves all the support functions needed to preserve and maintain the integrity of a system once it has been shipped to the customer. Thus, it is an engineering effort that minimizes costs and malfunctions during such

Assurance Technologies Principles and Practices: A Product, Process, and System Safety Perspective, Second Edition, by Dev G. Raheja and Michael Allocco
Copyright © 2006 John Wiley & Sons, Inc.

activities as storage, handling, transporting, spares provisioning, repair, preventive maintenance, and training.

From a different angle, Coogan [3] explains:

> For simplicity's sake, we could say that every system is made of two major subsystems: the operating subsystem and the logistics subsystem. The operating subsystem is the set of functions that perform the actual mission of the system. The logistics subsystem is the set of functions that are necessary to sustain the operating system.

The fact that logistic analysis is an engineering function means that it should be done in the engineering phase. When logistics engineering is done properly, the design of the entire system will be changed for the better.

7.2 LOGISTICS ENGINEERING DURING THE DESIGN PHASE

Blanchard suggests the following measures of logistics effectiveness [4]:

Reliability factors (failure rates, component relationships)

Maintainability factors (downtime lengths, frequencies, costs)

Supply support factors (probability of spares availability, inventory considerations, mission completion considerations)

Test and support equipment factors

Organizational factors (direct and indirect labor time, personnel attrition, delays)

Facility factors (turnaround time, facility utilization, energy utilization)

Transportation and handling factors

Software factors (quality, reliability)

Availability factors (inherent, achieved, and operational)

Economic factors (inflation, cost growth, budgets)

Effectiveness factors (capacity, accuracy, and factors related to safety, reliability, maintainability and dependability, and life-cycle costs)

This list shows that logistics support analysis requires inputs from all assurance technologies to assess their effect on support costs. Many trade-offs are required to optimize these costs. Support costs can be further reduced by investing in software such as that for fault isolation and automated testing.

7.2.1 Logistics Specifications for Existing Products

Writing a good specification is the key to logistics engineering. It is here one can reduce life-cycle costs dramatically without a major investment in design changes. For a new product, the logistics engineering analysis must be performed before the final specification is issued. Some inputs needed are FMECA, FTA, maintainability analysis, MTBF, estimates of repair times, estimates of preventive maintenance workload, information related to shipping, storage, and facilities, and training needs. This information is vital to improving the design. For example, if it is found that highly skilled electronic technicians are required for troubleshooting, it may be possible to redesign the equipment for technician transparency. That is, the troubleshooting can be simplified by a standard modular design so that the lowest-level technician can remove the failed module and replace it with a new one.

For a product that already has a history, Lerner [2] presents a technique used on the Lockheed C-130 tactical airlifter. The landing gear shown in Fig. 7.1 generated excessive unscheduled maintenance worker-hours. Analysis of data showed that key repairs were removal and replacement of the wheel-and-tire assembly and maintenance of the mechanisms that lower and retract the wheels. The technique is described in this case history from Reference 2:

Repairing the wheel/tire assembly was complex, needing several types of equipment such as jacks, torque wrenches, and thread protectors operated by level-five specialists. The repair exposes hub bearings to dirt and grit.

A main reason for frequent tire failures was heat transfer from the brakes through the wheel to the tire. The main reason for retraction mechanism failure was friction from tight clearances. Finally, both systems had many similar components that were not interchangeable or standardized.

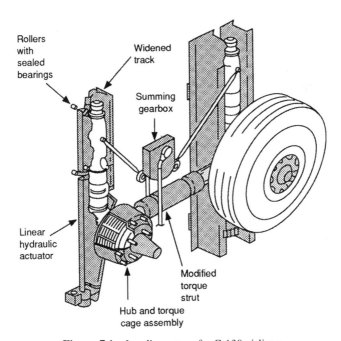

Figure 7.1   Landing gear of a C-130 airliner.

From these key deficiencies, the engineers generated SDTRs (support-ability design-to requirements):

- Avoid close tolerances to raise reliability and diminish the skill required of maintenance personnel.
- Make parts interchangeable and reduce the total by 50%.
- Eliminate separate jacks with a self-jacking assembly on the landing gear.
- Do not allow wheel/tire removal to expose the hub or require specialized tools.
- Design wheel to keep heat transfer from the brakes to a minimum.

After the design engineers saw these concrete requirements, they came up with a much more supportable design. Rollers spring-loaded against their tracks replaced a close tolerance show-and-track guide mechanism. Jacking modules appeared in the lower part of the shock strut. Three components replaced the landing gear wheel: the hub, the torque cage with the brakes, and the wheel rim/tire assembly. The rim assembly is not in direct contact with the brake assembly, reducing heat flow, and perforations in the brake assembly admit cooling air. The modular design meant that each part could be repaired separately and with no special tools. Also, parts were standardized throughout.

Results were spectacular. Retrofits nearly tripled mean time between failures, reduced spare parts by 60%, shortened ground time for wheel/tire replacement fourfold, halved repair personnel, and lowered skill levels. Although the modular design added 600 lb to gear weight, longer lasting carbon brakes saved 400 lb, so performance was unaffected.

7.2.2 Logistics Specifications for New Products

Consider the chemical plant in Fig. 5.3 of Chapter 5. Even from design sketches like this, many logistics concerns can be identified. Some such concerns will be:

How can material handling equipment be placed underground? What material handling devices are required?

Is a crane needed?

Can all repairs be made underground?

Can all preventive maintenance be performed underground?

Are there any safety concerns during repairs?

What spare parts will be needed?

Can corrosion cause damage?

What kind of quality assurance help is needed?

What information is needed in preventive-maintenance and repair manuals?

What kind of training personnel will be needed?

What kind of maintenance personnel will be needed?

Are all provisions considered for initial installation of the equipment?

What items must be operational at all times (minimum equipment list)?

Hundreds of questions such as these can be anticipated in early design. But, if special equipment is needed, it is better to design it at this stage. If some repairs would require enormously expensive spares, it is better to remove that failure mode by design or provide a backup at this stage. In fact, if very large and bulky material handling equipment is required, an underground design may be inefficient. Perhaps an on-the-ground facility with containment would be better. With advanced analysis and design changes, support costs can be reduced dramatically.

7.2.3 Design Reviews

Logistics engineers often get involved in design reviews too late. They should participate in conceptual design reviews, identify all the support costs and concerns, and recommend design changes. A good goal for logistics engineering is suggested by the U.S. Air Force as "double R/half M," which means double the investment in reliability cuts maintenance costs in half. The increased investment in reliability is small compared to the giant reduction in maintenance costs. Principles of design reviews are covered in Chapters 3 and 6.

7.2.4 Logistics Support Analysis

Logistics support analysis is a composite of systematic actions taken to identify, define, analyze, quantify, and process logistics support requirements. It evolves as the product development takes place. Maintenance analysis is a major element of this analysis. For military systems, Mil-Std-1388 [5] describes many tasks such as:

1. System-level logistics support analyses
   a. System design and operational concept influence
   b. Identification of supportability, cost, and readiness drivers
   c. Support concept development
   d. Trade-offs
2. Element-level logistics support analysis
   a. Support system optimization
   b. Resource requirements identification
   c. Task and skill analysis
   d. Early fielding analysis (to assess impact of introduction of the new product on existing systems)
   e. Postproduction support analysis
3. Supportability assessment and verification
   a. Supportability assessment plans and criteria
   b. Support concept verification

    c. Verification of resource requirements

    d. Identification and correction of logistics problems

7.2.5 FMECA for Logistics Support Analysis

Maintenance, of course, is one of the major elements of logistics engineering. It is an essential tool for design analysis and maintenance planning. FMECA provides an analytical baseline for a variety of other purposes—for example, for reliability analysis (Chapter 3) and for manufacturing process analysis (Chapter 6). The procedure described below is for maintenance analysis.

Figure 7.2 shows a FMECA model worksheet for maintenance planning [6]. Multiple use of FMECA in the same program requires careful planning to prevent duplication of effort and to properly interface with reliability, safety, and maintainability engineers. A separate worksheet is completed for each replaceable or repairable unit, called a *logistics support analysis candidate* (LSAC). A separate worksheet is also required for each structurally significant item (SSI) and functionally significant item (FSI) as defined in Section 7.5.1. Brief descriptions of the blocks in the worksheet are given below.

- *System/Subsystem Nomenclature*: Refers to the end product such as an aircraft or a specific subsystem.
- *Part Number/FSCM/Model Number*: Identifier for the item. FSCM refers to Federal Supply Code for Manufacturers.
- *Mission*: The specific task the item is expected to perform.
- *Indenture Level of Item*: The level of the item in a functional breakdown order.
- *Zone Identification*: A geographical code unique to military logistics. See Appendix B of Mil-Std-2080A(AS) [6].
- *Number per System*: The number of identical items in the system.
- *Reference Drawing*: Drawing number of the item being analyzed.
- *LSACN*: Logistic support analysis control number assigned to the item.
- *Item Description*: Description of the item in terms of its function and major components or assemblies.
- *Compensating Provisions*: Redundancies and protective features in relation to functions and functional failures.
- *Item Reliability Data*: Information such as inherent and operational mean time between failures and failure rates; maintainability data such as mean time between maintenance actions (MTBMA) and mean time to repair should also be included. Other useful information is the burn-in and screening requirements and safe life for limited-life items.
- *Minimum Equipment List*: Can the end item be operated if this item has failed or is inoperative for maintenance?

Revision no. _____

System/Subsystem Nomenclature	Part number/FSCM/model number	Mission	Prepared by:	Date
			Reviewed by:	Date
			Approved by:	Date

— Item data —

| Indenture level of item | Zone identification | Number per system | Reference drawing |

| LSACN | Item description | Compensating provisions (for each function/subfunction) |

Item reliability data
MTBF (inherent) _____ MTBF (operational) _____
MTBMA _____ MTTR _____ Burn-in time _____

Minimum equipment list. (Can this aircraft or support equipment be dispatched or operated with this item inoperative? If so, list any limitations.)

Classification of item
_____ Significant
_____ Nonsignificant

Functions/subfunctions		Functional failures (failure mode)		Failure causes (engineering failure mode)		Hidden function	Mission phase	Functional failure effects (local, next higher level, and end effects)	Functional failure detection method	Severity class	Wearout life (for each failure cause)	R & M data (for each failure cause) and Remarks
No.	(for each function)	Ltr.	(for each function)	No.	(for each functional failure)							

Sheet _____ of _____

Figure 7.2 FMECA worksheet for maintenance analysis.

243

- *Classification of Item*: Is it significant or nonsignificant? A significant item is one whose functional failures would have safety or major economic or operational consequences. Others are nonsignificant. (Any higher-level item containing a significant item is also a significant item.)
- *Functions/Subfunctions*: List of primary as well as secondary functions, which may also have an effect on safety. For example, a primary function of a fuel pump is to pump fuel at a specified rate. A secondary function may be to contain fuel within specified boundaries.
- *Failure Modes*: A list of possible functional failures. For example, a pump at too low a flow rate may pump too much fuel, or it may not pump at all.
- *Failure Causes*: A list of all possible causes for each failure mode. For example, the causes of a fuel pump malfunction may be a drive-gear failure, a shaft failure, or a broken oil line.
- *Hidden Function*: List the functional failures that will not be evident to the operating personnel while they perform their normal duties.
- *Mission Phase*: The part of the mission in which failure would be critical. The rubber seal on a solid booster rocket may fail disastrously during lift-off, but with no effect in orbit.
- *Functional Failure Detection Method*: List of detection provisions that minimize the effects of failures.
- *Severity Class*: (I) Catastrophic, (II) critical, (III) marginal (IV) minor [7].
- *Wear-out Life*: List of life expectancies of the item for each failure cause.
- *R&M Data and Remarks*: List based on lower-level reliability and maintainability data, of inherent and operational MTBF, MTBMA, MTTR, and burn-in/screening time (if applicable) for each failure cause. The nature of the maintenance required should be included where applicable.

7.2.6 Time-Line Analysis

Critical actions are usually multiple events that are either time constrained or have safety ramifications. They require a strict protocol for their safe accomplishment. For these activities, time-line analysis provides a means to identify and analyze the sequential requirements of the maintenance and the operating tasks. The analysis determines whether the actions can be performed on a phased or parallel basis and ensures that they are in correct order of protocol. This analysis can be applied to the following:

- Tasks that require more than one person.
- Tasks that involve safety consequences or must by nature be accomplished on a phased basis. Examples are fueling and oxygen servicing.
- Tasks that require safety equipment to be operational. For example, a flare unit that is designed to burn leaked gases from a chemical plant should not be down for maintenance during operating hours.

7.2.7 Level-of-Repair Analysis

Level-of-repair analysis (LORA) should be done as soon as the preliminary design is approved. The purpose of this analysis is to determine the least life-cycle costs for various alternatives involving repairing, discarding, and replacement. The analysis includes items that seem uneconomical, such as already-established safety practices and protocols.

A major portion of level-of-repair analysis deals with the level at which the repair is economical. Depending on the life-cycle costs, it may be cheaper to repair at the organization, the intermediate, or the depot level. (For life-cycle cost analysis, see Chapter 4.) Mil-Std-1390 [8] contains information on level-of-repair programs.

7.2.8 Logistics Support Analysis Documentation

In military logistics, a great quantity of records are required because of complex systems and variable personnel skills. In a commercial system, one must make a value judgment. For military systems, Mil-Std-1388-2A [5], containing about 600 pages, spells out most of the documentation required. It suggests the following records:

Record A: Maintenance requirements

Record B: Item R&M characteristics

Record Bl: Failure-modes and effects analysis

Record B2: Criticality and maintainability analysis

Record C: Operation and maintenance task summary

Record D: Operational and maintenance task analysis

Record Dl: Personnel and support requirements

Record E: Support equipment and training material description and justification

Record El: Unit under test and automatic test program

Record F: Facility description and justification

Record G: Skill evaluation and justification

Record H: Support items identification

Record HI: Application-related support items identification

Record J: Transportability engineering characteristics

Mil-Std-1388-2A, like most military standards, requires tailoring. It contains data logs so that the process can be automated. The process is iterative through all phases of the equipment life cycle and makes use of past histories.

7.3 LOGISTICS ENGINEERING DURING THE MANUFACTURING PHASE

The principles of logistics engineering apply equally well to manufacturing operations. Japan's famous kanban, or just-in-time, system would not be possible without the use of these principles. With logistics-engineered transportation and distribution, a steel plant needs to keep only 1 hour's worth of inventory (to be sure, an efficient system of quality control for materials and services must be in place). BMW in its plants in Germany introduced logistics engineering in 1976; automobile production has increased 60% during 1976–1986, as compared to only 20% for the average German manufacturer. According to Pretzch [9], logistics at BMW is seen as a cybernetic function structured according to control-loop principles.

Computer-integrated manufacturing in many U.S. automobile plants is also a good example of logistics principles applied to manufacturing.

7.4 LOGISTICS ENGINEERING DURING THE TEST PHASE

Without verification tests, logistics engineering problems are bound to grow by leaps and bounds. These tests can be done on mock-ups and partial systems. Sometimes verification can be done by detailed paper analysis and simulation. Generic guidelines are given below; additional tests may be indicated by analysis.

7.4.1 Tests for R&M Characteristics

Reliability and maintainability (R&M) characteristics tests are performed to verify MTBM, MTBF, MTTR, and other specified features. R&M-related logistics tests can be merged with the standard reliability and maintainability tests.

7.4.2 Tests of Operating Procedures

Operating procedure tests are conducted to see if the facilities, training, and procedures are adequate as planned. Most of these tests involve audits. All the test goals can be developed from logistics support analysis records.

7.4.3 Tests for Emergency Preparedness

When a major disaster strikes—a power blackout, train accident, or a major failure at an oil refinery—the maintenance crew must restore the system to operable condition under intense pressure. In such situations, technicians are likely to make mistakes, resulting in even longer downtime or another accident. For this reason, in large systems it is customary to perform emergency drills to expose deficiencies in maintainability and support requirements. The system may need a major redesign effort. For example, it often happens that rescue equipment is available, but the facility is not designed to give access to it.

7.5 LOGISTICS ENGINEERING IN THE USE PHASE

Logistics engineering in the use phase involves making sure that materials, supplies, spares, training, maintenance needs, and similar requirements are continually evaluated and improved. The most important function is generally maintenance. If old or degraded components are replaced, overall reliability shows little degradation with age. This section covers reliability-centered maintenance and measuring its effectiveness.

7.5.1 Reliability-Centered Maintenance

Reliability-centered maintenance (RCM) refers to a preventive maintenance program designed to preserve the inherent reliability of the equipment. The program emphasizes maintenance tasks selected on the basis of the reliability characteristics of the equipment and a logical analysis of the consequences of failures.

7.5.1.1 RCM Analysis Planning   Reliability-centered maintenance planning must be done during the design phase—not in the use phase. Since the RCM analysis is done on the basis of consequences of failure, *FMECA is absolutely necessary.* RCM may require certain features to be an integral part of the design. For example, if a noise level inside a casting is to be monitored, provisions must be made so that monitoring devices can be installed.

7.5.1.2 RCM Process   The basic steps in RCM analysis are the following:

1. Develop significant items that are critical to safety and major functions. An item may be functionally significant (loss of function would have a significant consequence on the equipment) or structurally significant (failure would result in a major reduction in residual strength of the structure). It is important to distinguish the types of significant items because different decision logic is applied to each (Figs. 7.3 and 7.4) [10].
2. Perform failure-mode, effects, and criticality analysis.
3. Analyze failure consequences.
4. Assign an appropriate maintenance task corresponding to one of the strategies in Section 7.5.1.3.

7.5.1.3 RCM Strategies   There are at least four basic RCM strategies:

1. *On-Condition Monitoring*: This calls for removal or repair when the condition of the part warrants it. Observation is directed at specific failure modes, which give identifiable physical evidence of potential failure. (For example, if the treads on an automobile tire are less than 1/16 inch thick, it may be prudent to replace the tire.) Each component is inspected at periodic intervals. It remains in service until it reaches the unacceptable condition. This strategy allows all components to realize their individual full life. Condition can also be monitored if an operator can distinguish certain sounds or

248

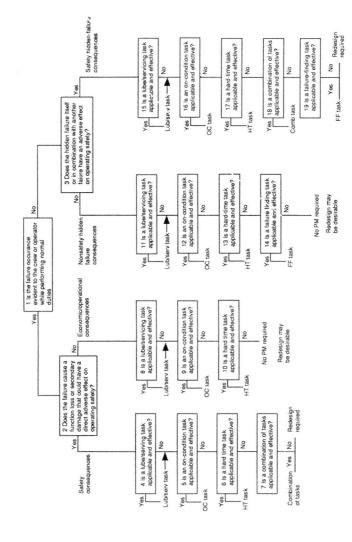

Figure 7.3 RCM decision diagram for functionally significant items.

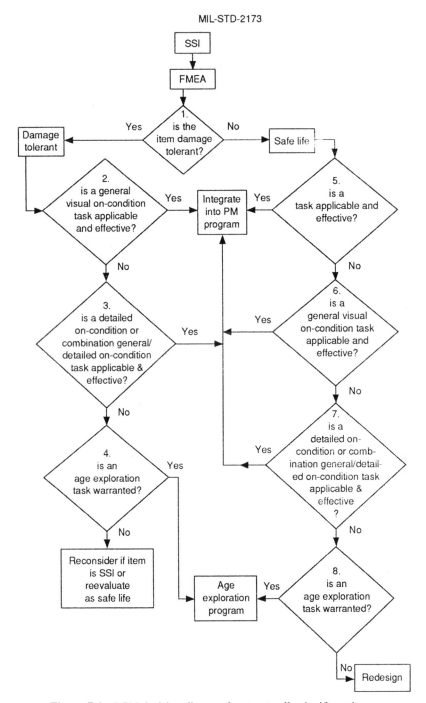

Figure 7.4   RCM decision diagram for structurally significant items.

smells as a warning of failure: for example, electrical insulation may give off a smell if overheated; or an automobile brake squeaks before complete wear-out.

2. *Internal Condition Monitoring*: This strategy measures internal parameters that change before a failure. For example, electronic circuits show higher junction temperatures, which can be monitored remotely with X-ray guns or thermal scans. An electrical utility computer can monitor pressure and temperature transducers in transformers in the electrical distribution network. Acoustic emissions can be monitored to detect internal flaws in the structure of an aircraft.

3. *Hard-Time Tasks*: For many items, the probability of failure becomes significantly greater after a certain operating age. This is usually because of a wear-out mechanism. The overall failure rate of such items can be reduced by imposing a definite ("hard") time limit on them. This prevents the use of an item at an age when the probability of failure is beyond reasonable limits.

    Hard-Time Tasks (scheduled removal and replacement of items) are established on economic life limits when operating experience indicates that the scheduled discard or rework task is adequate on purely economic grounds. A safe-life limit for hard-time tasks is imposed only when safety is involved and no on-condition observation is practical. An item is removed *before* the specified maximum age. On aircraft structures, the safe life is usually set at one-half the minimum life, which is defined as life at which the probability of failure becomes larger than a specified value.

    To determine the age–reliability relationship, a technique called *age exploration* is used. It is a process involving controlled testing and analysis of operating data. For safety-critical items, statistical analysis of probability of failure is done. In addition, the probability of unintentional events is analyzed.

4. *Group or Block Replacement*: This task is like a hard-time task based on the economic life of an entire group of identical items—sometimes a group of items. A simple example is the replacement of light bulbs on a busy street; a power company changed all the light bulbs during a scheduled visit rather than dispatch a maintenance crew more than 50 times a year to change one bulb at a time. Similarly, a nuclear power plant may decide to replace a block of bearings at the same time. Some companies replace several different items of roughly the same age during scheduled shutdowns.

7.5.2 Measuring the Effectiveness of Logistics Engineering

Chapter 4 describes how to assess effectiveness of maintainability engineering. A similar approach should be taken for logistics engineering. The investment in logistics engineering is productive as long as the total of all the logistics costs, when avoided costs are taken into account, is a decreasing function of time and customer satisfaction goals are met.

7.6 LOGISTICS SUPPORT ENGINEERING AND SYSTEM SAFETY

Logistics engineering addresses the operational life cycle considerations needed to assure the integrity of the product, process, or system after it has been implemented into the field. Obviously, there are system safety matters that should be discussed involving the succession of the system through its life cycle, progressing to the end of expected life, and possibly beyond, toward eventual disposal or recovery. Furthermore, there are risks to address during disposal, recovery, and possible reuse of recyclable materials, which may contain hazardous materials, chemicals, gases, fibers, and long radiological exposure.

7.6.1 Product, General, and Professional Liability

Product, general, and professional liability claims defense can be enhanced by the application of logistics and system safety engineering. There are safety-related risks that must be identified, eliminated, or controlled throughout the remaining life cycle. These logistics and system safety efforts should continue to the actual end of the life cycle and potentially past the expected end state. Because of enhancements in engineering and technology, systems may last longer than expected and consequently have associated risks.

Regardless of how many times an original system developer/manufacturer may have changed hands, merged, or resized, there is a liability chain to consider if the product, process, or system is still in operation. Any upgrades that have been made (within the industry) to enhance safety are the responsibility of the current entity (owner or corporation). Advancements related to system safety improvements to existing products, processes, or systems raise the safety bar, so to speak. Consequently, if a competitor introduces a new safety device in a specific product, other manufacturers must apply similar safety devices. As a recent example, consider the automotive industry with the introduction of side impact air bags, running lights, and backup alarms. Fortunately, in modern technological societies, the general public appears to be more aware of safety concerns and as a result has become more risk averse. Consequently, the risks associated with the possible extended life of products, processes, or systems must be eliminated or controlled to an acceptable level.

7.6.2 Analysis of Changing Risks

Hopefully, during initial design, an extensive system hazard analysis should have been conducted, which identified life-cycle risks and the related controls. However, it is never too late to conduct a hazard analysis. Unfortunately, there are many cases when such an analysis was not conducted or was inadequately conducted. A further complication involves the refinement, upgrading, revision, and enhancement of existing systems. Any change to a system no matter how mundane, such as changing the size of a fastener, may introduce additional risks or may change the risk profile (the risks and controls identified within the system hazard analysis). In actuality, the system hazard analysis should be a living analysis and as changes are planned the risks are to be evaluated.

7.6.3   Life-Cycle Logistics and System Safety

As discussed, concurrent engineering involves system-engineering practices like maintainability, system safety, quality, logistics, human factors, software performance, and system effectiveness. The disciplines should work together in the development of system and subsystem specifications that address all requirements throughout the life cycle.

The safety-related risks that may be coassociated with logistics engineering can be extensive and depend on specific outputs of hazard analyses for products, processes, or systems. However, some key considerations are addressed below, starting with production and deployment, and proceeding through retirement and disposal.

7.6.3.1   Production and Deployment Considerations   At the production and deployment phase, requirements may be inadequate if all the system disciplines have not worked together. The system hazard analysis should have addressed all known life-cycle risks and consequently safety requirements developed. The logistics/safety-related requirements may come into play at the start of production and deployment. The following questions come to mind:

Is the product facility adequate to support safe production flow?

Is the facility protected from physical damage, fire, explosion, flood, earthquake, and vehicle, environmental, and aircraft damage?

Can latent hazards be inadvertently introduced into the product, process, or system?

Can personnel inadvertently introduce latent hazards due to handling, error, injury, contamination, misapplication, or damage?

Is the facility large enough? Is there access to and from the facility?

Is there adequate storage for the finished product, process, or system?

Are there single-point events or common causes that can introduce disruption of production or business?

Are there key personnel who can inadvertently adversely affect the product, process, or system?

Can latent hazards be inadvertently introduced into the flow or supply chain?

Is the finished product, process, or system adequately protected, packaged, stored, transported, assembled, installed, and tested?

Are there appropriate literature, instructions, and training for customers?

Are there appropriate safe operating procedures, contingency procedures, cautions, or warnings?

Have the end user or customer needs been met, from a logistics and system safety view?

Have the functional requirements and constraints met the needs of the customers, from a logistics and system safety view?

Do the physical design variables meet functional requirements, from a logistics and system safety view?

Are there contingencies should the flow or supply chain be interrupted?

Are there real-time safety-related risks associated with the product, process, or system?

7.6.3.2 *Process Runs*

Without the understanding of safety-related risk mitigations, there may be instances where latent damage and hazards can be introduced inadvertently during process design and process operations. For example, the product, process, or system can be adversely affected and the following questions should be considered:

Is the process itself hazardous?

Can fires, explosions, or toxic outgassing occur?

Can contamination be inadvertently introduced?

Are process controls adequate to assure stability in the run?

Do the process variables control design parameters, from a logistics and system safety view?

Is process monitoring adequate?

Can the environment adversely affect the process?

Is the science associated with the process proven?

What are the possible deviations that can occur in the process?

Are supportive analyses and calculations appropriate or correct?

Are the failures/malfunctions associated with automation analyzed and mitigated?

Are there errors in software, algorithms, logic, coding, or specifications, which can cause inadvertent operation or shutdown, or adversely affect the end state?

Is process testing adequate?

Has process safety analysis been conducted?

Are safety requirements met?

Are mock-ups or simulations appropriate? Do they relate to actual process operations?

7.6.3.3 *Production Inspection*

As manufacturing progresses, there should be points of inspection along the line to assure that system engineering standards (involving maintainability, system safety, quality, logistics, human factors, software performance, and system effectiveness) are being met. Inspection is usually conducted by quality engineering and generally the following questions concerning inspection criteria should be addressed.

Are the inspection criteria adequate to identify safety-related risks or mitigations?

Are inspectors trained for safety?

Is inspector training adequate?

Are the inspection procedures adequate and documented?

Are records appropriate—do they identify hazards or inadequate controls?

Is there excessive documentation?

Do inspectors have adequate tools or equipment?

Will inspectors be exposed to safety-related risks?

Can inspectors inadvertently introduce latent hazards into the product, process, or system?

Is data collection appropriate to support analysis?

Do inspectors conduct real-time analysis?

Is there independent evaluation, or audits, by quality?

Are instrumentation and controls designed to support maintenance and inspection?

Are inspectors exposed to physical hazards, robots, or automated devices?

Are mitigations designed to protect inspectors from hazards?

Can interlocks, guards, or safety devices be bypassed?

Are there adequate cautions, warnings, or alerts?

Are inspectors trained to respond to contingencies?

Are there specific controls to assure that the product, process, or system is protected?

7.6.3.4 *Quality Control and Data Analysis* Quality engineering applies methods and techniques to assure that the total system conforms to specifications. The overall objective is to develop a set of requirements that, if met, results in a product, process, or system that is fit for its intended use.

Unfortunately, most of the system requirements are not given the attention and the knowledge needed. Therefore, it is prudent to train the inspectors in reporting hazards that were unknown until this time. Logistics provides the support functions to preserve and maintain the integrity of the product, process, or system once delivered to the end user. It is argued that either objective (fit for intended use, and preserving and maintaining the integrity of the product, process, or system) will not be met unless the safety-related risks are eliminated or controlled.

The following are additional questions involving common quality, logistics, and system safety considerations including data analysis.

Are there criteria to determine if the overall life-cycle process is in control?

Are failures, interruptions, deviations, malfunctions, anomalies, incidents, or accidents occurring?

Are there criteria to proactively mitigate failures, interruptions, deviations, malfunctions, anomalies, incidents, or accidents before occurrence?

Do data collection and subsequent action enable mitigating decisions?

Is the right data collected with emphasis on criticality and severity?

Are statistical methods appropriately applied?

Is analytical data used logically (inductively or deductively) to show correlations and relationships between an initiator through to the outcome?

Is data manually or automatically collected?

Is data collection and analysis error mitigated?

Are the criteria for acquiring regulating data defined?

Is there a procedure for acceptance or rejection of data?

7.6.3.5 Storage, Transportation, and Handling After manufacturing, the product, process, or system is stored and transported. The product, process, or system may go through a middle entity that may sell, warehouse, modify, ship, and distribute. Logistics provides the support through this supply chain. Consequently, without appropriate logistics and system safety requirements, the product, process, or system may be exposed to damage and additional latent hazards can be introduced. Here are some questions to consider:

Can the product, process, or system be damaged during storage?

Are there perishable items or materials that may degrade over time in storage?

Have environmental risks been evaluated and mitigated?

Has storage shelf life been determined and incorporated in removal procedures when kept beyond this period?

Are there special storage requirements?

Are contingency procedures established in the event of an incident or accident?

Are there monitoring and inspection requirements while in storage?

Is the product, process, or system adequately packaged for protection?

Are there appropriate cautions, warnings, or alerts indicated?

Do the warning signs last as long as the system?

Do the warning signs catch attention easily?

Is there an appropriate set of instructions, literature, and other written materials?

Are there any special storage requirements or environmental considerations?

Are there hazardous/dangerous handling or storage requirements?

Are there maintenance/upkeep requirements for storage?

Are there requirements concerning incompatibility with other stored products, processes, or systems?

Are there automated detectors/monitors that require service?

Are security risks mitigated? (Intruders can introduce latent hazards during storage and transportation, for example, reprogram devices, introduce contamination, or install dangerous devices.)

Is the product, process, or system adequately protected against shock or vibration?

Are there special requirements for shipping: air transport, sea transport, railroad, or highway? Are there special routes, flight plans, railways, or roadways to be considered?

7.6.3.6 *Construction, Installation, Assembly, Testing, and Initial Operation* Accidents or incidents can occur during construction, installation, assembly, testing, and initial operations. The hazard analysis should include installation procedures. Latent hazards can also be inadvertently introduced during this time line. Bear in mind the following questions:

Is there a detailed field or site plan that defines the construction, installation, assembly, testing, and initial operating requirements?

Is there a specific hazard analysis that addresses risks and mitigations?

Are there appropriate safety programs and plans?

Have pre-site surveys been conducted?

Is the site suitable for the product, process, or system?

Are there utilities, power, water, and communication assists?

Is the site accessible for intended functions, maintenance, and operation?

Are the environmental risks assessed and mitigated?

Are there suitable emergency resources that are accessible in case of injury, fire, flood, or earthquake?

Are there effective amelioration arrangements such as doctors, emergency crew, and the vicinity of a trauma center?

Will the site be adequately isolated—not cause exposure to the general public nor be exposed to other external hazards?

Are there appropriate evacuation and contingency plans, with equipment and resources?

Are there adequate security controls, physical and informational?

Is there an integrated test plan—functional, operational, and architectural?

Is there a replacement program and resources in the event of an incident or accident?

Are there backups for one-of-a-kind key people, designs, items, parts, algorithms, and codes?

7.6.3.7 *Operations, Maintenance, and Upkeep* Generally, most safety analyses address hazards and related risks that consider operations and maintenance. It is apparent that latent hazards and accidents can occur during this phase. Inappropriate decisions made concerning the system can result in initiating and contributory hazards and accidents. In order to be proactive, operational, maintenance, and

logistical planning should commence as soon as system maturity allows. Here are additional questions to consider:

Meeting the existing codes, requirements, regulations, or standards does not assure acceptable risks. Are inappropriate assumptions made concerning conformance to existing safety codes, requirements, regulations, or standards?

Are operations adequately defined, with appropriate safety requirements?

Is there a reliability-centered maintenance (RCM) plan (as covered in this chapter) with appropriate safety requirements?

It there an appropriate operational and support hazard analysis?

Autonomous systems may have to be accessed for programming, teaching, adjustment, and calibration. Have these risks been assessed?

Are physical devices (energy forms) stabilized, naturalized, and brought to a zero energy state before access?

Can safety devices fail or malfunction during access?

Are there appropriate ergonomic considerations to allow access?

Have a vicinity or zonal analysis been conducted, evaluating physical and environmental risks?

Are there adequate support tools, methods, devices, parts, and replacement units?

Is existing technology outdated and not supportable?

Is there a need for key people in order to maintain and upkeep the system?

Are there emergency and contingency backups, in the event of incidents or accidents?

Has a formalized accident and incident investigation been conducted?

Have monitoring and adjustment been done to assure stability or equilibrium of the system?

Has planning, associated with the extension of the system's useful life, been done?

Has reliability monitoring been conducted to determine wearout, and is there a plan to replace components before they fail?

Has a calculation been done for determining replacement spares for components and parts, related to system performance and mean supply delay time (SDT)?

Has consideration been given to standardization and interchangeability?

Will there be preventive maintenance in support of system safety?

Will there be fault monitoring, detection, and isolation to assure stability?

Has consideration been given to human factors associated with maintenance and operations, skill levels, definition and analysis of procedures, training needs, tools, devices, and instructions?

Have the risks associated with the use of advanced maintenance designs been assessed and controlled? What about autonomous self-diagnostics, self-repair, artificial intelligence, and fuzzy logic?

Has consideration been given to the risks associated with the use of software during maintenance and testing?

Are warnings, cautions, alerts, or notices appropriate?

Are there potentials for miscommunication between humans or between humans and machines? Are there potential language barriers? Is there excessive complexity?

Are safety devices or controls identified?

Has field monitoring been conducted to support system safety through customer feedback, field survey, interview, or inspection?

Are documentation and records appropriate?

Is there a voluntary recall program, with quick response in the event of hazard and risk manifestation?

Is there an appropriate system modification, with quick response in the event of hazard and risk manifestation?

Are there appropriate safety plans, meetings, and participants, in support of continuous system safety? Are there on-going activities, analyses, and refinements?

7.6.3.8 Retirement and Disposal

7.6.3.8 Retirement and Disposal Unfortunately, many products, processes, and systems contain materials that can be dangerous under specific conditions; consequently, there are many risks to address. The field of hazardous waste control is extensive (and not the subject of this book). There may be so many risks to evaluate that it would be appropriate to conduct a specific supportive hazard analysis to address the risks of decommissioning and disposal of the product, process, or system. There are many factors to consider involving acute or chronic exposures: the toxic nature of the solid, gas, liquid, mist, or fume; the flammability of the gas, liquid, mist, or fume; the corrosiveness of the solid, gas, liquid, mist, or fume; the radioactive capabilities; carcinogenic effects; and persistence and synergistic effects. Consider the complexities involving the use of new materials, such as composites, plastics, chemicals, metals, electrical and electronic subsystems, and microelectronics, which can contain, for example, iron, lead, cadmium, copper, beryllium, cobalt, aluminum, and gallium. In evaluating the decommission and disposal phase, the following questions should be addressed:

Has a specific decommission and disposal plan been developed, specifically for the product, process, or system?

Does the plan include the provision to identify acute or chronic hazards within the design?

Do the acute or chronic hazards present real-time or ongoing risks?

Does the plan include a listing of dangerous or hazardous solids, gases, liquids, mists, or fumes?

Are there controls associated with dismantling, disassembly, removal, and transportation to an appropriate site?

What are the considerations for recycling of materials or by-products?

What are the current codes, standards, or requirements associated with hazardous material disposal?

Are the changes in the law being monitored and tracked?

Are there state-of-the-art changes in controlling associated risks?

Are there means to provide appropriate licensing, documentation, and records?

Are there contingency plans in the event of an incident or accident?

Are the transportation risks assessed—provisions for security, route planning, and stability of the cargo?

Is there a need for special equipment, processes, sequences, and precautions to dismantle the product, process, or system?

Are there special handling requirements?

Does the site support future decommissioning requirements?

What are the site environmental risks?

What are the techniques, methods, processes, or procedures for final disposal of hazardous solids, gases, liquids, mists, or fumes? Are there adequate risk controls?

7.7 TOPICS FOR STUDENT PROJECTS AND THESES

1. Write a generic design review procedure for a design for logistics support.
2. Perform a logistics support analysis for a commercial product.
3. Perform a logistics support analysis for a mass-production manufacturing process.
4. Perform a logistics support analysis for a job-shop manufacturing plant.
5. Suggest an approach for analyzing the number of spares required for a system.
6. Criticize the reliability-centered maintenance procedure. Suggest improvements.
7. Select a product, process, or system and develop a logistics engineering plan that considers all life-cycle phases.
8. Expand a logistics support analysis to address system safety.
9. Explain how logistics enhances system safety.
10. Select a product, process, or system and develop a system hazard analysis; define the mitigations appropriate to logistics support.

REFERENCES

1. W. Finkelstein, The New Concept and Emerging View of Logistics. In: *Logistics in Manufacturing*, John Mortimer (Ed.), IPS Publications, Kempston, Bedford, UK, 1988, pp. 201–205.
2. E. J. Lerner, Designing for Supportability, Aerospace America, June 1989.

3. C. O. Coogan, A Systems Approach to Logistics Engineering, *Logistic Spectrum*, Winter 1985, p. 22.

4. B. Blanchard, *Logistics Engineering and Management*, Prentice-Hall, Englewood Cliffs, NJ, 1974.

5. *Logistic Support Analysis*, Naval Forms and Publications Center, Philadelphia, PA 1987, Mil-Std-1388

6. *Maintenance Engineering, Planning, and Analysis for Aeronautical Systems, Subsystems, Equipment, and Support Equipment*, Mil-Std-2080A(AS).

7. *Procedures for Performing a Failure Mode, Effects, and Criticality Analysis*, Mil-Std-1629.

8. Level of Repair Analysis, Naval Publications and Forms Center, Philadelphia, PA, Mil-Std-1390.

9. Hanns-Ulrich Pretzch, BMW Logistics: A Step in the Future. In: *Logistics in Manufacturing*, John Mortimer (Ed.), IFS Publications, Kempston, Bedford, UK, 1988, pp. 145–149.

10. *Reliability Centered Maintenance Requirements for Naval Aircraft, Weapon Systems and Support Equipment*, Mil-Std-2173(AS).

FURTHER READING

Blanchard, B. S., and E. E. Lowery, *Maintainability: Principles and Practices*, McGraw-Hill, New York, 1969.

Definition of Effectiveness, Terms for Reliability, Maintainability, Human Factors and Safety, Mil-Std-721, Naval Publications and Forms Center, Philadelphia.

DoD Requirements for a Logistic Support Analysis Record, Mil-Std-1388-2A, Naval Publications and Forms Center, Philadelphia.

CHAPTER 8

HUMAN FACTORS ENGINEERING AND SYSTEM SAFETY CONSIDERATIONS

8.1 HUMAN ENGINEERING PRINCIPLES

Human errors can cause a product to fail or can result in an accident. The purpose of this chapter is to introduce the concept of *design for immunity* to human errors. A product design, which is immune to errors, contributes toward a *robust design*. For example, disposable cigarette lighters come in two styles: with and without a cap. A lighter without a cap is not a robust design because the flame is supposed to go out in a few seconds but sometimes does not. The user is responsible for making sure the flame is extinguished. Some people have burned their chests and faces because of such a flame. Lighters with caps do not pose this hazard. They cut off the oxygen supply, extinguishing the flame.

There are many attempts to blame human error as the so-called cause of an accident. Humans will make errors; consequently, systems should be designed to tolerate human error especially when addressing catastrophic risks. Minimally, at least three independent events should occur prior to harm resulting.

An investment in human factors design reduces life-cycle costs. The cost reduction is in terms of saved human lives, reduced training and skill requirements, lower liabilities, and fewer disasters.

Other examples of human errors and design for immunity are the following: In older automobiles, a driver can travel at speeds far above the safe limit. In many new cars, there is a limit on the top speed. In a software project, a new programmer

Assurance Technologies Principles and Practices: A Product, Process, and System Safety Perspective, Second Edition, by Dev G. Raheja and Michael Allocco
Copyright © 2006 John Wiley & Sons, Inc.

wrote a wrong command and caused a major disaster. The software was changed to limit the ability of the system to go beyond certain boundaries.

These examples show that mishaps can be anticipated and prevented during early design, when design changes are much cheaper. The changes are made on paper before the actual product is manufactured.

There are many products on the market which are not immune to human errors. The latest models of automobiles have radios with many small buttons and labels. In trying to read and use them, a driver may be totally distracted from the road. Indeed, several accidents have already occurred because a radio was being tuned. The operation of showers in hotels and institutions is not standardized. In some places the knob is rotated counterclockwise to increase hot water, while in other places the reverse is true. Many older people in nursing homes have received severe burns from not knowing which direction to turn the knob. But confusion would be eliminated if all designers followed the guidelines already available in Mil-Hdbk-1472 [1]. Other mishaps can be prevented by performing appropriate analyses, as shown in this chapter.

8.2   HUMAN FACTORS IN THE DESIGN PHASE

During detail design an engineer should take at least these actions (described in more detail later):

1. Use checklists and standards in developing specifications.
2. Perform design reviews for human interfaces in normal and emergency conditions.
3. Use lessons learned.
4. Review the results of hazard analyses (Chapter 5), which include hazards generated by human error.

8.2.1   Use of Checklists and Standards in Specifications

Checklists and standards contain proven design principles. Mil-Hdbk-1472, for example, contains checklists and design guidelines that give information on display location; illumination levels; standardization of switches, knobs, buttons, and signs; workplace dimensions; access dimensions for maintenance; interpretation of warning lights; alarms and their dimensions; and much more.

This generic list of good human engineering principles can be followed:

1. Do not standardize if interchangeability will cause an accident. (A repair technician in a hospital connected an oxygen hose to a carbon monoxide source, and vice versa. The fittings were identical. They should have been different for each source.)

2. Standardize according to typical human behavior. (Usually a person turns the handle clockwise to lower a window of an automobile. But some cars are designed for counterclockwise rotation on the driver side. During an emergency, a driver might think that the window is stuck.)

3. The color blue should always be used to indicate cold and red to indicate hot conditions. (In some hotels, blue knobs have been installed for hot water. The person who installed the knobs may not have paid attention to the color code.)

4. Repetitive operations should always be automated. (They are monotonous for humans.)

5. The location of safety devices should be standardized. The emergency horn actuator in some cars is in the center of the steering wheel; in others, it is on the rim of the steering wheel; and in still others it is on a stick on the side. (A person renting a car on a business trip may not know where to push the horn in an emergency.)

6. Controls in electronic devices should be designed so that the quantity (voltage, temperature, current, etc.) is increased in the clockwise direction.

7. The controls in hydraulic and pneumatic systems should rotate counterclockwise to increase the quantity (e.g., water, gas).

8. The color green should indicate a "go" signal, red a malfunction, yellow caution, and blinking red an emergency.

9. A design should not depend only on a lamp to give a warning, because the lamp itself may have burned out. An alternative is to keep the lamp lit all the time and have it blink in an emergency. If the lamp is not lit, someone will check for the reason. Another option is to provide a backup audio alarm.

10. For remote controls, the worst situation should be considered. (Rear electric windows in automobiles have been rolled up by the drivers only to crush a child's fingers. Some new designs incorporate fiber-optic sensors for sensing fingers on the glass edge.)

11. Small, round parts, such as those in toys, can be swallowed by children. Parts should have square or rectangular shapes so that they cannot be swallowed easily.

12. Critical warnings should be large, clear, and easily noticeable. (Do not depend on the user to get this information from the operator's manual.) The warning sign should be permanently printed or secured. It should not fall off.

13. Backup devices should be clearly marked. (An operator at a nuclear plant caused an accident by trying to control an unenergized secondary pump instead of the primary pump. The control panel did not indicate which one was active.)

14. If a human can input the wrong information, such as a wrong number on a computer panel or a keyboard, fail-safe features should be incorporated.

15. Do not design for color interpretation if any operator is color blind. Mil-Hdbk-1472 [1] recommends color coding in various designs. Incidentally, most color coding involves red and green, where most color blindness exists.

According to McConnell [2]: "Most color weak individuals show deficiency either in their response to red or green or to both these hues. About 5% of the people in the world are totally blind to one or more hues on the color circle." McConnell further points out that the main types of color blindness involve a red–green deficiency or blue–yellow deficiency. A person suffering from red–green deficiency will see the world in blue and yellow. Green grass will appear to be blue and a fire engine will appear yellow to such a person. The much smaller proportion of people with blue–yellow deficiency will see the world in red and green. Distinguishing black and white is not a problem with any type of color blindness.

16. Instruments should avoid glare and blurring on indicators. Some unfortunately allow moisture to penetrate and their dials cannot be read.

17. Knob settings should be labeled for clarity and durability. (Labels fall off, lettering fades, and glare interferes with legibility.)

18. The sequence of control use should proceed in numerical order. In one case, a system failed because the push buttons were numbered 1, 2, 3, 4, 5, 7, 6. The numbers 6 and 7 were not in sequence because of an engineering change. The operator assumed they were in sequence without reading them.

19. Positive and negative battery terminals should have different sizes to prevent incorrect cable attachment.

20. For safety-critical complex systems, the connector colors should match wire colors to prevent miswiring. (At least 14 fire control warnings on a Boeing 747 airplane were found miswired in 1989. Color coding helped prevent future problems.)

21. Numerals on instrument dials should be placed so that they are not covered by the pointer.

22. Instruments in control rooms should be mounted so that they are clearly visible to the operator. (Some controls in the Three Mile Island nuclear power plant were found behind the operator.)

23. Frequently used displays should be directly in front of the operator and controls should be on the side, within easy reach.

24. Warnings should be bright enough or should have sufficient contrast that they are legible under all illuminating conditions.

25. A computer should give positive feedback to the operator regarding acceptance or rejection of data. Otherwise, the operator may start pressing the wrong buttons in desperation.

26. An operator should not be required to remove many parts to inspect items such as filters, fan belts, and corroded parts.

27. Provision should be made to prevent accidental actuation of controls. Critical control buttons may be designed so that they do not protrude from the control panel, or guards may be provided.

28. Sharp edges should be avoided on parts to prevent cuts and lacerations.

29. Designs that require operator's to be alert to dangers should be avoided.

30. Enough clearance is needed on consoles and operating equipment to avoid injuries to an operator's body parts.

31. Enough access space is needed for humans as well as tools and equipment for repairs.

32. Software that is capable of distinguishing between valid and unsafe commands should be required.

33. Humming noise in equipment should be avoided; it may cause drowsiness or irritation.

34. If there are too many digital controls, operators cannot monitor them. Replace the critical controls with analog dials.

35. Avoid the need for verbal communications in a noisy environment. For example, when the pilot and first officer perform startup checks in an airliner, it is difficult to distinguish if the first officer calls out "ON" or "OFF."

8.2.2 Design Reviews for Human Interfaces

When human interface analysis (HIA) deals with a specific machine, it is called *human–machine interface analysis*. Human interface analysis may involve a repair or maintenance procedure or even a person making software program changes.

The best technique is operating and support hazard analysis. O&SHA consists of writing the operating procedure and analyzing what could go wrong with each step in the procedure. It is a technique for preventing accidents but can be used for preventing any failure attributable to human error. For example, if the attention of an operator monitoring an instrument dial strays, an out-of-limits indication may go unnoticed. If this failure mode can cause an accident, then some other means of monitoring should be designed. An example of an O&SHA format is shown in Fig. 8.1. The procedure is the same as the maintenance engineering safety analysis in Chapter 5.

For example, suppose a failure mode is: "Technician forgets to install seal on a bolt on an aircraft engine while performing preventive maintenance." The design solution is to make the seal an integral part of the bolt. The technician will never have to worry about forgetting to put the seal on.

A specialized tool called *link analysis* is used in the aircraft industry. An analyst observes which controls in the aircraft cockpit are used more frequently. Design improvements are made by placing the most frequently used controls in favorable positions.

8.2.3 Using Lessons Learned

A new design should not be infested with the same problems as previous ones. The information for design improvements can come from failure histories, data banks,

System/subsystem title _____ Activity _____	Operating hazard analysis	Revision _____ Date _____ Prepared by _____				
Task number	Task description	Hazard	Hazard Effect	Hazard category	Potential accident prevention measures	Proc. no.	Resolution

Figure 8.1 Example of operating and support hazard analysis.

dealers, distributors, users, and competitors. One of the best sources of information is the technique for human error rate prediction.

8.2.4 Review of Hazard Analyses

In Section 8.2.2, the operating and support hazard analysis was suggested for the human–machine interface. Actually, there are several other interfaces such as the human–software interface and the human–environment interface. The most important analysis for these interfaces is preliminary hazard analysis, which covers many human-induced hazards. (Chapter 5 covers PHA and several other analyses.) A human factors specialist should be involved in coming up with any design changes resulting from the analysis [3]. Other hazard analyses to be reviewed for human errors are maintenance engineering safety analysis (MESA) and any additional analysis with human involvement. Gibble and Moriarty [4] take a life-cycle approach in the technique called *human factor/safety functional analysis*. They postulate that a human error can be initiated in almost every life-cycle phase, whether it be the design, production, maintenance, or operational phase. For example, Fig. 8.2 shows that short circuiting an electrical contact in an airplane is a hazard that can be introduced in any of the above phases. During design, the failure modes can be designed in by the designer. A production operator can allow a defect to cause a short circuit. Similarly, the pilot of the aircraft can override

Hazard (SSHA)	Design	Production	Maintenance	Operational
Contact C-4 shorts and causes spoilers to deploy inadvertently.	Contacts have several failure modes designed in by engineers.	Faulty processes allow defect to cause short circuit.	Maintenance worker error during installation.	Pilot overrides automatic setting.
Hazard	Design	Production	Maintenance	Operational
Ordnance fuse out-of-line; rotor allows initiation in out-of-line position.	Not enough tolerance built into rotor system by designer.	Error in production, or misassembly	Maintenance check does not detect problem.	Pilot uses wrong fuse setting.

Figure 8.2 Illustration of human factor/safety function analysis.

an automatic setting and cause the same accident that a short circuit would have initiated.

When reviewing for human errors, the designer has to assume the worst case. No matter how careful an operator may be, there is a good chance an error will be made. The system should be designed to overcome such actions. For example, the operator may be required to use both hands to operate a machine; this assures that the operator cannot put a hand in an unsafe portion of the machine. Similarly, radio controls can be placed on the steering wheel of an automobile (Fig. 8.3). This allows drivers

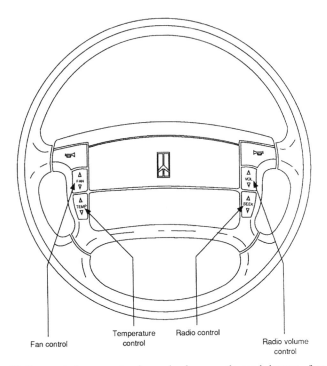

Fan control   Temperature   Radio control   Radio volume
                     control   control

Figure 8.3 Radio control on a steering wheel—an enhanced human factors feature. (Courtesy of Delco Electronics.)

to use the radio, without taking their eyes off the road. In addition, the steering wheel includes heater and fan controls.

8.3   HUMAN FACTORS IN THE MANUFACTURING PHASE

Errors by humans in manufacturing can introduce latent defects, which may not be detected by inspection and test procedures. These defects affect safety and reliability of products. They may affect the safety of the humans themselves. This section covers such concerns and precautions.

8.3.1   Types of Manufacturing Errors and Controls

Errors cited below are frequently encountered. Some suggestions for preventing them are provided.

8.3.1.1   *Errors of Illusion*   Inspectors often see only what they want to see. If they have been seeing 100-ohm resistors on circuit boards, a 1000-ohm resistor can get by without their knowing it. If they are used to seeing a seal on an aircraft engine, a missing seal often will not be noticed. Such errors may be prevented by mechanizing inspection. In the case of a missing seal, however, mechanized inspection may not be viable. Then some kind of feedback can be designed into the system. For example, a pressure gauge in the cockpit can give a warning of low pressure. Inspection may not be necessary. Those who are not convinced about the errors the subconscious mind can make should try the experiment in Fig. 8.4 [5]. Read the statements in the figure at your own pace and observe anything unusual or incorrect.

Over 50% of the people who examine the figure cannot find anything unusual in their first try! The reason: their minds are made up on what they want to see. (They do not want to be confused with facts, as the joke goes. Each of the triangles contains a redundant word—for example, "Paris in *the* spring."

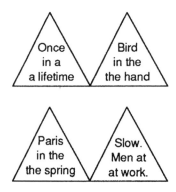

Figure 8.4   An experiment dealing with the subconscious mind. (From Ref. 5.)

A reader who thinks this is a freak experiment should try the following [6]. Let two or more people read the phrase below and count the number of times the letter F appears. Very rarely will they come up with the same count.

THE FIRST FINE FISHING DAY OF THE YEAR WE FINALLY FLEW
TO ALASKA FOR FIVE DAYS OF REAL FISHING.

8.3.1.2 Errors of Vision Inspectors not only see what they expect to see, their eyes retain the image of the previous piece for roughly 1/16 second. This is the reason for saying that 100% inspection is rarely 100% effective. Such repetitive inspections should be done by robots.

8.3.1.3 Errors of Insufficient Knowledge This is very common. For example, a production foreman was told to change a chemical solution for cleaning steel parts before starting every shift. He was never told that the parts (for smoke detector alarms) must be very clean to prevent rust. One day, in the interest of getting higher production, he did not change the solution. The parts rusted, making the smoke detectors inoperative. The company had to recall the alarms. Unfortunately, only 6% of the consumers returned the alarms. The company, according to U.S. law, was liable for the alarms still in the market. Needless to say, the company is bankrupt.

It is important, therefore, that the procedures clearly reflect the risks. According to assurance technology principles, the solution should have been changed by an automated device. Even better, the chemical process should have been avoided altogether by using stainless steel parts.

8.3.1.4 Errors of Engineering Oversight Production operators and inspectors are not the only ones who make mistakes. Many mistakes are attributable to oversight on the part of the engineers who design the process equipment. In one case, a soldering temperature was actually 40 °F below the temperature shown on the dial. The equipment produced many cold solder joints. The oversight in this case was that the equipment designer did not give any guide for calibration. Nobody, therefore, calibrated the equipment.

8.3.2 Preventing Inspection Errors

Inspection is done by production operators and inspectors. In either case, certain techniques can be used.

Compartmentalize the Work: Divide the inspection into smaller tasks rather than one large task. If there are 300 components on a circuit board, ask the operators to check all resistors first, then the transistors, and so on. If the inspection is done according to the sequence in the parts test, it will become tedious. Another way of compartmentalizing is to divide the circuit board assembly in several sections and have all boards inspected for the first section, then for the second section, and so on until all sections are inspected.

Comparison Inspection: In this method, a known good assembly is placed as a standard model and all the production is compared visually with it. If the product is complex, an automated means such as robot vision or artificial intelligence technology may be used.

Visual Aids and Tables Instead of Paragraphs: Humans like to have information that is easily accessible. Inspection instructions several text pages long are hard to remember. Humans have proved to be much more effective when they use photographs, visual aids, and tabular columns for instructions instead of text in paragraphs.

Job Sampling: Harris and Chaney [6] recommend a job sampling technique. In this technique, a known number of errors are planted (as in the error seeding technique for software) and inspection reliability is measured by the percent errors caught. This tool should not be used as a performance appraisal tool but as a training tool.

In addition to above practices, the usual motivating factors have a big role in inspection effectiveness. The supervisory environment can encourage as well as discourage errors. Sound workplace design and application of ergonomic principles will also help.

8.4 HUMAN FACTORS IN THE TEST PHASE

Tests for human factors are complex because they are done system-wide. They include tests to make sure there is fail-safe fallback mode for errors in manufacturing, in service, and in actual use. They cannot be done easily without the persons who are going to use, repair, and maintain the system. At least four tests should be done, as outlined in this section.

8.4.1 Tests for Stereotype Behavior

Humans act in a certain manner unless trained otherwise. The left-handed person does tasks in a certain way and so does the right-handed person. Their actions must be studied to prevent mishaps. For example, a typical way to reduce electrical load on a power system is to turn a control counterclockwise. If the control is not designed this way, then the load will increase instead of decreasing. In the case of toys, as mentioned earlier, children will tend to swallow small parts. That is why a reputable toy maker makes sure the small parts are not separable from the toy or they are not round.

The best way to test for stereotype conditions is to observe the use of the system by humans on an experimental basis. A large toy maker observes children using and misusing toys through one-way mirrors. The parents are given toys and asked to report the findings. A recent test on automobiles revealed that drivers who rent cars for business purposes cannot always easily remove the ignition key from a late-model car. The car has an unusual mechanism of removal. A driver can become very frustrated if the key cannot be withdrawn; the driver should not have to struggle for minutes to figure out how to extract the key. A driver will try to remove a key in a certain natural manner; the driver should be able to remove it in that manner or be able to guess the correct procedure quickly. If 90% can guess and 10% cannot, the design is unacceptable.

8.4.2 Tests for Emergency Preparedness

Many disasters have been caused because, in emergencies, people have no time to think. They react to instincts and make many mistakes. One of the reasons for a blackout in New York was that a technician increased the load on the power system instead of decreasing it. In another power company, an operator was making desperate adjustments in the control room on a backup pump instead of the primary pump. Many persons in the Bhopal, India, chemical accident died because they did not know the reason for the emergency alarm. Thinking it signaled an ordinary accident, they ran toward the plant instead of away from it.

To minimize the effects of a mishap, every organization with potential for an accident should conduct emergency preparedness tests to assure that people will do what they are expected to do. These tests are called *emergency drills*. Everyone—including the people living in the vicinity—is instructed to perform certain tasks. Periodically, a mock emergency is created to see how people will react in controlling the accident. Observers then interview the participants and make improvements in the system design and procedures.

Typical findings are that people cannot turn fire extinguishers on, sprinklers do not work, people turn controls in the wrong direction, handicapped people cannot escape, evacuation takes much longer than expected, and people forget to open windows in case of toxic smoke. The most frustrating finding is that no one seems to be in charge. The purpose of emergency preparedness tests is to discover all the roadblocks before a real disaster takes place.

In emergency preparedness tests, the following errors should be looked for [7]; the information will be useful for preventing future accidents.

Errors of Substitution: An operator habitually selects a wrong control or device: for example, turning on windshield wipers instead of lights in a car, turning on hot water instead of cold water in a hotel, pressing the car's accelerator instead of the brake pedal.

Errors of Selection: An operator carelessly or inadvertently selects a wrong control or setting. This error is similar to the error of substitution but is not a result of habit patterns. For example, a pilot hits a wrong switch on the cockpit controls; a computer user selects wrong commands without reading the instructions.

Errors of Reading: An operator reads wrong information. For example, an automobile driver reads the AM dial instead of FM dial on the radio; a nuclear plant inspector reads the 10-to-100 scale assuming it is the 1-to-10 scale and therefore does not detect a radiation leak.

Errors of Irritation: An operator is liable to make mistakes in uncomfortable weather or when surrounded by too many confusing controls or alarms. Inability to read dials and settings also causes irritation.

Errors of Warning: When caution, warning, and danger signs are not properly posted, an operator can introduce an error. Do not rely on operators to remember all the instructions.

Errors of Alertness: An operator may not be alert because of lack of oxygen or presence of toxic gases, for example.

Errors of Lack of Understanding: An improperly trained operator may be afraid to ask questions. During emergencies this person may make too many mistakes.

Errors of Haste: An operator is not able or not trained to perform work in an allocated time. In such cases wrong controls may be initiated.

Errors of Sequencing: An operator may use shortcuts and bypass certain operations or perform them in the wrong sequence, not knowing it may cause an accident. A DC-10 jet with more than 300 passengers crashed on takeoff at Chicago because the maintenance technicians took a shortcut in mounting one of the engines.

Errors of Overconfidence: It is sometimes said that a senior repair technician is more likely to be involved in an accident than a trainee. The trainee will follow the procedures while a senior technician will make independent decisions. An experienced Occupational Safety and Health Act (OSHA) inspector electrocuted himself when he checked a high-voltage power installation without wearing insulated gloves.

Errors of Reversal: An operator may operate a device in the reverse direction because there is no clear indication on the control. For example, an operator may flip a switch up instead of down, or may rotate a knob counterclockwise instead of clockwise.

Errors of Unintentional Activation: An operator may wear oversize clothes or metal rings, which may accidentally activate controls. Other examples are a driver's coat sleeve may get tangled on the speed selector and shift it to an undesired setting, or long hair may get tangled in an escalator at a subway station.

Errors of Oversight and Omission: An operator may simply forget a task. For example, a pilot may not call out certain checks before takeoff. A commercial airliner crashed after takeoff because a second officer did not call out for setting the wing flaps in the proper position.

Errors of Mental Overload: An operator may not be able to attend to critical tasks because of too many tasks, fatigue, or personal problems. For example, too many alarms at a nuclear power plant may prevent the operator from responding to the right ones; a business traveler renting an automobile late at night may not take time to know the locations of the windshield wiper control and defroster.

Errors of Physical Limitations: An operator may respond incorrectly or not respond at all because of inability to reach or handle the work. There are limitations imposed by height, strength, left-handedness or right-handedness, and so forth.

Errors of Casual Behavior: An operator may not take the task seriously. When the driver of an automated mass-transit train left to have a chat with someone, the train took off without him. This threw passengers into panic.

8.4.3 Tests for Amelioration

Tests for amelioration assess the ability to provide immediate medical and rescue services to injured people. The tests may be designed to around the following:

1. How long it takes to travel to a first aid location
2. How well the medical services are performed
3. How long it takes to provide emergency help to those close to death
4. How long it takes to get an ambulance
5. How long it takes to reach the hospital
6. Emergency devices available in the vicinity and in the ambulance

8.4.4 Tests for the Human–Machine Interface

Tests of the human–machine interface can be part of emergency preparedness tests. They are done mainly to assure the effectiveness of the interface. For example, control room equipment may be tested to determine whether:

- A control can be accidentally pressed and arm a weapon or activate a process at the wrong time.
- Specialized tools are secured close to the equipment so they are always accessible.
- Functions are testable individually so that the operator can identify a problem easily and quickly.
- A unit is easy to disassemble and reassemble for repair.
- Functions of knobs, switches, and lights are standardized. Sometimes two or more suppliers design different portions of a system without coordination and total confusion results; in one place the on position of a switch is up and in other places the on position is down.
- Glare on instruments hinders observation.

These and many other ideas can be obtained from standards, checklists, and personal experience.

8.5 HUMAN FACTORS IN THE USE PHASE

During normal use, many incidents happen that can result in major failures, mishaps, accidents, or near misses. They should not be taken lightly. There should be formal reporting of such incidents. Many hospitals, for example, use a data collection system based on incident/accident reports. The reports are summarized and corrective action is taken to prevent recurrence of the events. There must be a tracking system for such incidences to assure timely preventive action and for validation.

Other errors in use are those related to maintenance and repair. Errors induced by maintenance personnel may degrade reliability and introduce unsafe conditions. Techniques such as maintenance engineering safety analysis (Chapter 5) may be used to prevent these errors.

In the use of equipment, fatigue and monotony often play an important role. In the use of aircraft, human performance clearly can result in accidents.

8.6 ADDITIONAL CONSIDERATIONS INVOLVING HUMAN FACTORS AND SYSTEM SAFETY

In taking into account the human element of a system, an objective of human factors and system safety is to help enhance system design to assure acceptable risk. The human is a principal element in all systems considering the fact that humans design and operate systems and humans are imperfect and errors are expected. When humans, for example, are stressed, distracted, or confused, the error potential can increase and accidents can occur. It is appropriate then to enhance designs to accommodate the human element by designing systems with error tolerance, thereby decreasing the risk of harm as a result of human error. Designers can make errors in developing systems and as a matter of fact, errors can be made throughout the system life cycle. Systems then are to be designed to allow for such errors to occur with predetermined fallback modes. Systems should be designed to accommodate the human rather than attempt to change the human. Or modify the design, to accommodate for poorly designed systems.

8.6.1 Human Variability

Humans interface with the system throughout the life cycle. They create designs, develop programs, contrive prototypes, conduct calculations, write specifications, design processes and procedures, and physically assemble and construct systems. Humans are the end customers ... the users. Systems are to be designed to accommodate not only human behaviors but also any physical interaction. Designers are to consider the end user population, people's variability—size, shape, dimensions, backgrounds, physical states, ages, health, stereotypes, and conventions. System safety can be adversely affected at any particular time. System safety and human factors engineers have to evaluate any potential for harm or hazards that can be the result of inappropriate human interaction. Consequently, there are many methods and techniques to assess the human–machine interface. There are also many complexities to evaluate.

8.6.2 Human Engineering Complexities

Human factors engineering is the application of the human biological and psychological sciences in conjunction with the engineering sciences. This is to achieve the best fit between the human and machine. The objective also addresses effectiveness and system safety. The principal disciplines involved are anthropometrics, physiology, and bioengineering.

Anthropometrics information describes the dimensions of the human body, through the use of bony landmarks to which heights, breadths, depths, distances, circumferences, and curvatures are measured. For engineers, the relationship of these dimensions to skeletal "link joint" systems is of importance, so that the human body can be placed in various positions relative to positions and equipment [8].

8.6.3 The Human Machine

The human interfaces with the machine through physical links in a form of close physical coupling. Consider a pilot sitting in a cockpit, or a truck driver sitting in a cab of a tractor, or a heavy equipment engineer sitting in the cab of a bulldozer. Consequently, in considering the physical aspects of the interface, biomechanics becomes applicable. Biomechanics explains characteristics of the human body as a biological system in mechanical terms. Biomechanics have been developed from research that relates physics (mechanics) and mathematics and from anatomy, physiology, and anthropometrics. In a sense the human is analyzed like a mechanical device, for physical stress and strain, physics of movement or failure, kinetics, torque, tinsel stress, elongation, and compression.

8.6.4 Human Behavior

The human remains the most complex machine especially when the behavioral characteristics are being considered. The behavioral science aspect of human engineering involves cognitive psychology, and the evaluation of mental psychology, for example, the study of sensation, perception, memory, thinking, and motor skill. Additional complexities involve the understanding of organizational and social psychology, sociology, and psychometrics [9, pp. 2–3].

8.6.5 Human Motivation

One of the most important attributes of safety is to motivate people toward safety-related objectives. Motivation is the impulsion to achieve particular wants, which lead to the satisfaction of certain needs. Generally, the participants associated with safety must be motivated; managers, engineers, analysts, users, and other stakeholders must be motivated toward accident prevention. However, depending on one's point of view, different levels of motivation will be apparently based on our behaviors and attitudes about risk [10, pp. 86–90]. Motivation is perhaps one of the most complex issues in the field of human behavior—people have many needs, all continually competing and influencing behavior. Motives determine management choices. They often result in wrong choice.

8.6.6 Motivation and Safety Culture

Generally, depending on an individual's duty within an organization, it may be hard to relate to a common objective. An individual should be made to understand how he/she contributes to a particular objective—accident prevention. People should be able to associate and relate to the particular safety-related risk—to connect the dots so to speak and relate a particular duty, task, function, or activity and be capable to make an association to a particular risk. When this connection or association is made, safety motivation can be enhanced. A positive safety culture equates to common safety motivation throughout an organization.

8.6.7 Human Error

Human error can be considered the source of accidents especially when errors are made during the creative efforts in engineering and science, during the development or revision of a product, process, or system. What is human error, since it is so important? Human error is an act occurring through ignorance, deficiency, or accident. Human error is the departing from or failure to achieve what should be done. Errors can be predictable and random. Errors can also be categorized as primary or contributory. Primary errors are those committed by personnel immediately and directly involved with the accident. Contributory errors result from actions on the part of personnel whose duties preceded and affected the situation during which the result is harm. Errors can also be considered as the difference between a computed, observed, or measured value or condition and the true, specified, or theoretically correct value or condition.

Errors can be deviations from accepted norms, deviations from procedures, errors in assumptions, or decision errors associated with personal perceptions of risk. The decision maker can be biased due to preconceived notions about risk. Decision makers can be risk takers, risk avoiders, or somewhere in the middle of the continuum. Many errors that have adversely affected system safety can be traced back to an inappropriate decision involving the creation and later design of the system. External stressors such as the politics of engineering or management, scheduling stress, or cost limitations—all of these pressures can have an effect on decisions, which can consequently result in decision errors.

8.7 REAL TIME AND LATENT ERRORS

Errors are also considered unsafe acts that can be initiators or contributors within accidents. These hazards can be the result of active real-time unsafe acts or latent unsafe conditions. Active contributors can be associated with the real-time performance of individuals within the system. Latent unsafe acts could be the result of acts of omission or commission, by management, designers, and planners. Unlike active contributors, latent initiators and contributors may not be readily apparent; they may lie dormant within the system for a long period until the exact moment of initiation. Consider an error in an assumption, calculation, or line of code in software.

8.8 ANALYSES IN SUPPORT OF HUMAN FACTORS AND SYSTEM SAFETY

Almost any system analysis technique can be applied to address the human element and system safety, considering that mostly anything that influences the system can ultimately affect the human aspects and safety. There are in estimation about 550 methods and techniques in use. Discussed below are a number of techniques that may be useful in addressing system safety in the context of this book.

8.8.1 Human Interface Analysis

Within any system, the functions to be performed by personnel, hardware, and software should be evaluated. There are many instances where the human can adversely affect the system throughout its life cycle—full scale engineering, production, deployment, operations and maintenance, system retirement, and disposal. Consequently, at any time when the human interfaces with the system, either the human or system can be adversely affected. Human interface analysis is an all-encompassing term that actually involves the application of many other techniques, for example, link analysis, ergonomics evaluation, sequence analysis, task analysis, job safety analysis, simulation, prototyping, workload assessment, and error analysis. The overall objective of human interface analysis is to optimize the human with the other integrated system elements of hardware, software, and environment. From a safety perspective, the interface hazards have been identified, eliminated, or controlled to an acceptable level. Interface hazards include unsafe acts such as errors, deviations from procedures, miscalculations, miscommunication, oversights, and omissions. These hazards may also include unsafe conditions that are the result of misjudgment, decision error, and inappropriate human action, which results in latent conditions.

8.8.2 Link Analysis

Link analysis is a means of evaluating transmission of information by type (visual, auditory, tactile), rate, load, and adequacy. It indicates the operational relationship between two units, whether they are two people or a person and a piece of equipment. Analysts are concerned with positions, arrangements, and frequency of interchanges, but not with time in some cases. A link is any connection between two elements; an element is a person, control, display, piece of equipment, or station. Link analysis attempts to reduce the lengths of the most important or most frequent links at the expense of those of less importance or frequency. Links are rated according to their use–importance relationships. Importance is an indicator of its criticality or serverity.

Link Analysis Procedure   Link analysis can be conducted with other techniques such as mock-up, bread boarding, prototyping, or simulation. The analysis enables the evaluation of the physical layouts of areas of concern where humans may interface, for example, workstations, cockpits, vehicle cabs, monitoring stations, or control systems.

A link analysis involves the following

- A drawing, schematic, mock-up, or prototype is prepared, indicating the locations of the elements to be studied. Where distances are important, the details should be to scale.
- Links are drawn between the various elements between which communication will exist.
- The frequency of each link is determined. This frequency may be obtained by counting the number of times that each task element in a procedure is to be accomplished and the number of times that the procedure is to be carried out.
- The importance of each link is established according to its criticality or severity. Criticality or severity can be based on importance of the task to be accomplished, level of difficulty, need for speed and control by performer, frequency with which each task must be accomplished, total time involved for its accomplishment during a complete series of tasks, and the safety effect.
- A value is assigned to each link by multiplying frequency or total time by importance.
- Elements are then arranged so that the links with the highest use–importance values are the shortest in length. Consolidation of related elements may reduce motions that could be tiring, be interrupted or interfered with, or lead to errors.

Operations and training can be enhanced by the use of a more appropriately designed link interface. From a safety point of view, the importance of each link can be established according to elements of risk: severity, likelihood, exposure. Analogies can also be made between links and hazards or hazard controls. Error analysis can also be conducted in conjunction with link analysis. Consider a link that is made out of order, or out of sequence, or an inappropriate action occurs during a link communication or action either by a human, control device, or display.

Link analysis can also be conducted and presented with many other supportive methods; for example, multidimensional drawings, schematics, mock-ups, or prototypes are used. Real-world-like simulators or demonstrators are also used. Various spreadsheets, worksheets, association charts, and flow diagrams [11, pp. 117–123] are also used.

8.8.3 Critical Incident Technique (CIT)

The critical incident technique [12, pp. 301–324] is a means by which previously experienced accidents or incidents information can be determined by interviewing persons involved. It is based on collecting information on hazards, near misses, and unsafe conditions and practices from experienced personnel. It can be used beneficially to investigate human–machine relationships in past or existing systems and use the information learned during the development of new systems, or for the modification and improvement of those already in existence.

CIT is a method of identifying unsafe acts or unsafe conditions that contribute to potential scenarios or actual accidents within a given population by means of a stratified random sample of participant-observers who are selected from within the population. Observers are selected from specific departments or operational areas so that a representative sample of the operational system existing within different contributory categories can be acquired.

CIT Procedure   The trained interviewer questions a number of persons who have performed particular tasks within certain environments and asks them to recall and describe unsafe acts or unsafe conditions that they have made or observed or that have come to their attention in connection with particular operations. Participants describe as many specific "critical incidents" as they may recall, regardless of whether or not harm occurred. When the potential accident contributors are identified, a decision is made concerning a risk priority for allocating resources for eliminating or controlling the risks.

Reapplication of CIT is conducted using a new stratified random sample to detect new problems and to measure the effectiveness of the controls applied. Should there be any future changes to the system, CIT should be reapplied to determine any changes that may affect system risk.

8.8.4 Behavior Sampling

Behavior sampling [13, pp. 283–298] is based on the statistical principle of random sampling. By observing a portion of the total, a prediction can be made concerning the makeup of the whole. Behavior sampling can be defined as the evaluation of an activity performed in an area using a statistical measurement technique based on a series of instantaneous random observations or samples.

Behavior sampling is based on the principle that the observations taken will conform to the uniqueness of the normal distribution or Gaussian distribution. The characteristics of a normal distribution are that the curve is symmetrical around the mean, which is also equal to the median and mode. The larger the number of samples or the number of observations made, the closer the plot of them will approach the normal curve, and the more confidence one can have that the sample readings are representative of the population.

In applying behavior sampling, it is necessary to determine the percent of time a person is behaving safely and the percent of time the person is behaving unsafely.

The analyst can observe the person throughout the total shift or observe the person at certain times. A record is made on whether the person is working safely or unsafely. The total number of observations is noted and safe and unsafe behavior observations are prorated.

Behavior Sampling Procedure   In general, the following steps are conducted to accomplish behavior sampling.

- Trained observers and analysts are to conduct hazard analysis and sampling.
- Conduct hazard analysis to identify unsafe acts, unsafe conditions, inappropriate and unsafe behavior.
- Prepare a list of unsafe acts that are appropriate to the operation or system.
- Select trial observation periods of time by a random process.
- Subjects should not be biased.
- Make the trial observations and instantaneously decide whether the observed behavior is safe or unsafe.
- Observe people within a selected group from a single observation point or conduct a walk-through.
- Determine the number of observations required for desired accuracy and level of confidence.
- Select random time periods when the observations are to be made.
- Conduct the actual study by making the number of random observations required and recording the observed behavior according to the safe or unsafe classifications.
- Compute the observational statistics.
- Repeat above steps to acquire appropriate sample.
- Construct a behavior statistical control chart and calculate the mean percent of time each person is involved is unsafe acts or the mean percent unsafe behavior for the entire group.
- Compute the upper and lower control limits.
- Determine if stability exists in terms of percent unsafe behavior for the test period.
- If the situation is not stable, introduce hazard controls for the population studied and repeat the process until stability in unsafe behavior in achieved.
- After attaining stability, introduce enhancement in system safety and repeat the behavior sampling on a scheduled basis.
- Reconstruct the behavior statistical control chart to evaluate the period of enhancement.
- Continue to modify system safety until unsafe behavior has been reduced to a minimum or seek to redesign the system.
- Monitor the system by repeating the behavior sampling procedure if necessary.

8.8.5 Procedure Analysis

A procedure is a set of instructions for sequenced actions to accomplish a task such as conducting an operation, maintenance, repair, assembly, test, calibration, transportation, handling, emplacement, or removal. This analysis is a review of the actions that must be performed, generally in relation to tasks; the equipment that must be operated or maintained; and the environment in which personnel must exist. The analysis is sometimes designated by the activity to be analyzed. Some examples are test safety analysis, operation safety analysis, and maintenance safety analysis.

In conducting a formal review of actions to be performed, hazards can be identified as a result of considering inappropriate actions, actions that are out of sequence, inadvertent actions, or actions that do not occur when needed. Any procedure that can potentially affect a system should be evaluated from a system safety view.

Procedure analysis can be conducted with other supportive techniques, for example, functional flowcharts, decision trees, time-line analysis, fault trees, and operational-sequence diagrams.

Procedure Analysis Method   Generally, the method of conducting this analysis involves the following.

- Define a set of instructions that describe a sequenced arrangement of actions that have a common objective (specific procedure).
- Define the interfaces and interactions of the actions for the particular procedure.
- Prepare an operational-sequence diagram, flowchart, or schematic, detailing interfaces and interactions.
- Evaluate each action within the sequence by identifying hazards as a result of inappropriate action, out-of-sequence action, delayed action, or omitted action.
- Define system risk in the context of identified hazards.
- Redesign procedure action to eliminate or control risk.
- Reevaluate procedure as a result of redesign.

8.8.6 Life Support/Life Safety Analysis

Commonly, these analyses involve the evaluation of risks related to human occupancy of a particular environment or facility. The objective is to ensure that the environment in which a person exists will not be harmful. The complexity of the analysis will vary based on the exposures and intricacy of the environment. Consider some environments with hazardous exposures that do not support human life: outer space, under water, high altitudes, temperature and pressure extremes, toxic environments, hazardous occupancies, confined spaces, physical and health hazards, and special operations. Typically, hazard analysis, compliance to criteria, and checklists are used within these analyses. Consideration is given to determining if the environment will sustain life under potentially inhabitable situations and conditions: ingress, egress, life support, damage containment, oxygen deprivation, pressure changes, fire and explosion prevention or containment, radiation, toxic fumes, dusts, mists, gases, biohazards, and animal containment and control.

Life Support/Life Safety Analysis Methods Analysis methods vary based on the particular exposure or operational context. Generally, the procedure includes the following.

- Conduct an analysis of the environment that the human must inhabit.
- Determine the physiological and psychological effects on the human.
- Identify initiator or contributory hazards related to potential physiological and psychological effects.
- Take steps to design out the hazards and provide engineering controls, safety devices, and safety procedures.

8.8.7 Job Safety Analysis

Job safety analysis is one of the original safety analysis techniques developed in the 1930s and 1940s. It is a procedure that identifies hazards associated with each step of a job. It is a simple but effective approach commonly used at the supervisory level. A job is decomposed into individual steps and each step is analyzed for particular hazards. The job safety analysis is an excellent training tool and it can be posted by a particular workstation for reference. Generally, a line supervisor and operator will have important input during the analysis process. It is also appropriate to have participation of the process designer. Ideally, it is more important to design the process to preclude hazards, possibly eliminating the need for job safety analysis.

Job Safety Analysis Procedure

- Break down the job or operation into elementary steps.
- List them in their proper order.
- Examine each step to determine hazards.
- Develop hazard controls to mitigate the risk.
- Reevaluate the analysis should any changes occur.

8.8.8 Human Reliability

There are many methods of quantitatively predicting and evaluating the performance of a human within a system; this overall process is called human reliability (HR). HR can be applied to any activity, set of procedures, or tasks, and results of the performance can be used to determine success or task accomplishment or error. HR is a method to provide a quantitative way of describing the performance of the human element in the system and to integrate this quantitatively into the system reliability objective.

Human reliability analysis (HRA) [13, pp. 259–264] can be useful in diagnosing those factors (stressors that affect humans) in the system that can lead to less than adequate human performance. It is possible to isolate the error rate estimated for a particular task and to determine where errors are possible. Once sources of error are identified, steps can be taken to correct them. HR can be used to enhance human performance, which can be accomplished by applying design controls that have been identified as an output of analysis. Comparisons can be made between design alternatives in terms of human and system performance. Each alternative is analyzed to determine its probability of successful task accomplishment. The alternative with the highest success probability would be selected.

Human Reliability Analysis (HRA) Method   Usually the following steps outline the method of conducting HRA.

- Select the analysis team and provide appropriate training.
- The team should be familiar or knowledgeable about the system, tasks, procedures, and operations to be evaluated.
- Consider the range of human actions, interactions, links, and interfaces.
- Construct an initial system model (event, logic tree, fault tree, flowchart).
- Identify specific human actions that are appropriate or inappropriate.
- Develop descriptions of human interactions, links, interfaces, and associated information to complete the model.
- Identify failure modes or hazards that can affect humans, errors of omission or commission and performance shaping factors.
- Evaluate the impact on system performance as a result of inappropriate human actions.
- Estimate error probabilities for various human actions and interactions and determine sensitivities and uncertainty ranges.
- Review, validate, and verify analysis results.
- Document the HRA.

8.8.9 Technique for Error Rate Prediction (THERP)

THERP [14] is well known within the nuclear industry. The method depends on task analysis to determine the error situations. Potential system or subsystem failures or hazards are defined, after which all the human operations involved in the failure or hazard and their relationship to system tasks are modeled in the form of a human event tree. Error rates for both correct and incorrect performance of each branch of the event tree are estimated.

THERP Approach   Generally, the following steps are conducted in THERP.

- Describe the system, system functions, and human performance characteristics.
- Describe the tasks, procedures, and steps to be taken and performed by the human.
- Identify all error potentials and hazards.

- Estimate the likelihood of each potential error in each task, procedure, and step, and the likelihood that the error will be undetected.
- Estimate the severity of the undetected or uncorrected error.
- Develop controls to eliminate or control the risk of error.
- Monitor the system and conduct reevaluation if needed.

Performance Shaping Factors   Since probability is applied in the mathematics, it is necessary to determine all possible errors, to determine error rates, to determine the degree of dependence or independence among tasks and errors, and to determine the factors that affect error likelihood. These factors are referred to as performance shaping factors (PSFs). Depending on the PSF operative in the error situation and the degree of dependence among tasks, the correct error value will shift up or down in the range of values provided.

Examples of PSFs are listed.

Temperature, humidity, air quality

Noise and vibration

Work hours

Availability of resources

Actions by co-workers

Actions by supervisors

Task speed

Task load

Sensory deprivation

Distractions

G-forces

Vibration

Personality

Motivation

Knowledge

Task complexity

Workload

Human interface

Ergonomics

Fatigue

Hunger

8.8.10   A Technique for Human Event Analysis (ATHEANA)

Over recent years, the U.S. Nuclear Regulatory Commission (NRC) has sponsored the development of a new method for performing human reliability analysis (HRA).

The analysis addresses errors of omission (EOOs) and also errors of commission (EOCs). Generally, the analysis characterizes and quantifies errors in order to integrate data into probabilistic risk assessment (PRA) models. Consideration is given to potential human failure events (HFEs), unsafe actions (UAs), and their error-forcing contexts (EFCs). HFEs, UAs, and EFCs are critical elements of the ATHEANA method and are defined as follows:

HFE: A basic event that is modeled in the logic models of a PRA (event and fault trees) and that represents a failure of a function, a system, or a component that is the result of one or more unsafe actions. A HFE reflects the PRA systems modeling perspective.

UA: An action inappropriately taken, or not taken when needed, by plant personnel that results in a degraded plant safety condition.

EFC: The situation that arises when particular combinations of performance shaping factors (PSFs) and plant conditions create an environment in which unsafe actions are more likely to occur.

ATHEANA was designed to identify the types of events that could lead to serious outcomes that have not been previously identified. The approach is derived from a characterization of serious accidents that have occurred in the nuclear and other industries in the past.

8.8.11 Human Error Criticality Analysis (HECA)

The HECA [15; 17, pp. 315–320] method is a task analysis based on operational procedures and human error probability for each human operational step, which assesses error effects on the system. The results of analysis show the interrelationship existing between critical human tasks, critical human error modes, and human reliability information of tasks. The results of analysis are corrective actions for improvement in system reliability and safety.

HECA Approach   Normally, the following activities are conducted to accomplish HECA.

- Define tasks and procedures associated with the operation under evaluation.
- Perform task analysis.
- Construct an event tree.
- Estimate human error modes, human error probability, probability of hardware failure, and error-effect probability.
- Calculate the human error probability of human tasks via an event tree.
- Calculate human reliability of task and human reliability of operation.
- Calculate the criticality index of error modes and criticality index of tasks and complete HECA worksheet.

- Develop and analyze a criticality matrix.
- List the critical human tasks, critical human error modes, and reliability information, and provide the event tree.

8.8.12   Workload Assessment

This procedure enables the evaluation of workload [12, pp. 135–138] on operators, where maintained situational awareness is required for system safety. Consider safety-critical operations (e.g., transportation, medical procedures, munition handling, and nuclear power). It is expected that when a human must conduct a safety-critical operation the associated workload should not have an adverse effect that may result in loss of situational awareness, distraction, fatigue, reduced alertness, confusion, or fixation. This procedure provides for the evaluation of task loading or the ability of personnel to conduct assigned tasks, in the allotted or available time. Many supportive methods and techniques can be used to accomplish workload assessment: task analysis, time-line analysis, simulation, survey questions, observation, and interviews. The objective is to define an optimal workload that does not have an adverse effect on system safety. Workload is to be evaluated from a physical metabolic, biomechanical, and ergonomic perspective as well as from a psychological perspective.

8.9   TOPICS FOR STUDENT PROJECTS AND THESES

1. Develop a human factors analysis approach for product reliability.
2. Develop a human factors analysis approach for quality control inspection.
3. Develop a human factors analysis approach for maintainability.
4. Develop a human factors analysis approach for manufacturing operations.
5. Develop a human factors analysis approach for software system safety.
6. Tailor a human factors analysis approach for a consumer product.
7. Tailor a human factors analysis approach for a nuclear power plant or a toxic chemical plant.
8. Develop a model for estimating human reliability for an airline maintenance technician.
9. Develop a model for estimating human reliability for a toxic chemical plant or an oil refinery.
10. Develop guidelines for including human engineering requirements in a product design specification.
11. Conduct an experiment using the job sampling technique in Reference 6 and state your conclusions and recommendations.
12. Develop a technique for hazard analysis not covered in this chapter.
13. In Section 8.4.2, several types of human errors are pointed out. Can you suggest other types of errors? Develop a design checklist to avoid such errors during emergencies.

14. Select a complex human intensive system and apply three human factors analyses, report your results, and explain how the selected techniques enhance your efforts and system safety.

15. Conduct research of a specific complex human intensive system and acquire or estimate quantitative data associated with error rate prediction. Select a quantitative analysis technique and discuss your findings.

16. Contrast and discuss the differences between quantitative and qualitative human factors analysis methods and select a particular method to address a specific human-interface problem.

17. Select a complex human intensive system and apply a system hazard analysis. In support of this activity, apply two human factors analyses and report your results.

18. Select a complex system associated with a highly hazardous environment and conduct appropriate analyses to support system safety.

REFERENCES

1. *Human Engineering Design Criteria for Military Systems, Equipments and Facilities*, Mil-Hdbk-1472. Naval Publications and Forms Center, Philadelphia.

2. J. V. McConnell, *Understanding Human Behavior*, Holt Rinehart Winston, New York, 1980.

3. B. S. Dhillon, *Human Reliability with Human Factors*, Pergamon Press, New York, 1986.

4. J. W. Gibble and B. H. Moriarty, Human Factors in Accident Causation, lecture notes at University of Southern California, 1981.

5. J. M. Juran and F. M. Gryna, *Quality Planning and Analysis*, McGraw-Hill, New York, 1980.

6. D. H. Harris and F. B. Chaney, *Human Factors in Quality Assurance*, John Wiley & Sons, Hoboken, NJ, 1969.

7. A. B. Leslie, Jr. The Human Factor: Implications for Engineers and Managers, *Professional Safety*, November 1989, pp. 16–18.

8. K. H. E. Kroemer, H. J. Kroemer, and K. E. Kroemer-Elbert, *Engineering Physiology— Bases of Human Factors/Ergonomics*, 2nd ed., Van Nostrand Reinhold, New York, 1990, p. 1.

9. T. B. Sheridan, *Humans and Automation: System Design and Research Issues*, John Wiley & Sons, Hoboken, NJ, 2002.

10. W. W. Lowrance, *Of Acceptable Risk—Science and the Determination of Safety*, William Kaufmann, San Francisco, 1943 and 1976.

11. A. Chapanis, *Human Factors in Systems Engineering*, John Wiley & Son, Hoboken, NJ, 1996.

12. W. E. Tarrants, *The Measurement of Safety Performance*, Garland STPM Press, New York, 1980.

13. M. Modarres, *What Every Engineer Should Know About Reliability and Risk Analysis*, Marcel Dekker, New York, 1993.

14. R. A. Bari and A. Mosleh, Probabilistic Safety Assessment and Management, PSAM 4. In: *Proceedings of the 4th International Conference on Probabilistic Safety Assessments and Management*, Volume 1, Springer-Verlag, London, 1998.

15. J. A. Forester, K. Kiper, and A. Ramey-Smith, Application of a New Technique for Human Event Analysis (ATHEANA) at a Pressurized Water Reactor. In: *Proceedings of the 4th International Conference on Probabilistic Safety Assessments and Management*, Springer-Verlag, London, 1998.

16. F. J. Hyu and Y. H. Huang, *Human Error Criticality Analysis of Igniter Assembly Behavior.*

FURTHER READING

Bass, L., *Products Liability: Design and Manufacturing Defects*, Shepard's/McGraw-Hill, New York, 1986.

Hammer, W., *Product Safety Management and Engineering*, Prentice-Hall, Englewood Cliffs, NJ, 1980.

Johnson, W. G., *MORT Safety Assurance Systems*, Marcel Dekker, New York, 1980.

Joint Army–Navy–Air Force Entering Committee, *Human Engineering Guide to Equipment Design*, John Wiley & Sons, Hoboken, NJ, 1972.

Juran, J. M., *Quality Control Handbook*, McGraw-Hill, New York, 1987.

Kolb, J., and B. S. Boss, *Product Safety and Liability*, McGraw-Hill, New York, 1980.

Lupton, T. (Ed.), *Proceedings of First International Conference on Human Factors in Manufacturing*, IPS Publications, Kempston, Bedford, UK, 1984.

Sanders, M. S., and E. J. McCormick, *Human Factors in Engineering and Design*, McGraw-Hill, New York, 1987.

Swain, A, and H. Guttman, *Handbook of Human Reliability Analysis with Emphasis on Nuclear Power Plant Applications*, NUREG/CR 1278, Nuclear Regulatory Commission, Washington, DC, 1983.

CHAPTER 9

SOFTWARE PERFORMANCE ASSURANCE

9.1 SOFTWARE PERFORMANCE PRINCIPLES

The trend of software recalls is going up, especially for medical devices and automobiles. Among the recalled devices was a programmable pacemaker that malfunctioned when a patient walked by a store's antitheft device. Another was an X-ray machine in a hospital that produced 80 times the recommended radiation dose and killed three patients. The supplier insisted that the X-ray machine software met customer expectations because it did everything the hospital asked for in the specification. The hospital agreed that the software met the requirements, but maintained that it did not meet implied expectations. Although the hospital did not explicitly ask for a limitation on X-rays, it expected the equipment to perform in a safe manner. A court ruled in the hospital's favor and held the equipment manufacturer liable.

In 2005, some automotive recalls were as follows:

A software bug on a new concept car disabled gauges and warning lights.

The engine control module (ECM) software in certain 2003–2005 hybrids was improperly programmed, making the engine run slightly lean. Eventually, this could cause the malfunction indicator lamp (MIL) on the instrument panel to come on and lead to failure of an important part of the vehicle's emission system, the catalytic converter. Also, the ECM could misinterpret normal oil pressure rise during the first engine start after an oil change. In another

Assurance Technologies Principles and Practices: A Product, Process, and System Safety Perspective, Second Edition, by Dev G. Raheja and Michael Allocco
Copyright © 2006 John Wiley & Sons, Inc.

hybrid car, the wrong version of the software downloaded, resulting in stalling at higher speeds.

In another recall, if unfavorable electrical tolerances coincided in the electronic logic unit, it was possible that the fuel level and range of remaining fuel (in vehicles with on-board computer) would be displayed incorrectly. This means that the display indicates a higher fuel level to the driver than is actually available and therefore, in some circumstances, the vehicle may run out of fuel, causing a loss of power and thus the potential for an accident.

A software bug in the electronic management unit of a European car's fuel pump could make the engine stall if the fuel tank was below one-third full.

These examples show that compliance with engineering specifications may satisfy contractual requirements but may not fulfill legal requirements. Therefore, software specifications require thorough review to include "what the customer could not articulate" and truly reflect the latent expectations of the user.

Assurance technologies principles apply to software just as they do to hardware, although some tools and techniques are unique to each. Tools for hardware such as FMECA, FTA, and hazard analysis can also be applied to software. Confusion about how to assure software performance comes from the lack of standard definitions of software reliability, maintainability, and other such requirements. This chapter suggests definitions that can be considered valid because of their acceptance in important contracts [1, 2].

9.1.1 Software Quality

Statements of software quality such as "the software shall be accurate and precise" or "the software shall conform to the specification" are too broad. Contracts should define software quality in terms of factors such as the following:

Correctness: The extent to which a program satisfies its specifications. A general-purpose method for proving software correctness should include an easily applied and validated way of specifying assertions concerning the correct operation of the software and a way of indicating variance from correct operation (errors).

Interoperability: A measure of the ease with which one subsystem can be coupled with another. Verification procedures and testing should be as thorough for interfaces with related systems as for the main system. It is important to make sure that the users of all systems realize the implications of any interface, especially where these affect system status or timing.

Flexibility: A measure of the ease with which a program can perform (or be modified to perform) functions beyond the scope of its original requirements.

Efficiency: A measure of the use of high-performance algorithms and conservative use of resources to minimize the cost of operations.

Validity: The ability of a program to provide performance, functions, and interfaces that are sufficient for effective application in the intended user environment. The distinction between this and the definition of *correctness* should be noted. Whereas correctness pertains to specifications, validity pertains to the application as well as to the specifications.

Generality: The ability of the computer program to perform its intended functions over a wide range of usage modes and inputs, even when a range is not directly specified as a requirement.

9.1.2 Software Reliability

Software reliability is the probability that software will perform its assigned function under specified conditions for a given period of time (the time here usually refers to the period during which all the user conditions are expected to occur). Reliability is not a function of degradation, since the software components do not degrade with time. This definition is a generic one. The following specific items may be included as part of reliability requirements [2]:

Operational Requirements

- Predominantly control
- Predominantly computational
- Predominantly input/output
- Predominantly real time
- Predominantly interactive

Environmental Requirements

- Hardware interfaces
- Software interfaces
- Human interfaces
- Degree of human interaction
- Variability of hardware
- Training level of operators
- Variability of input data
- Variability of outputs

Complexity Considerations

- Number of entries and exits
- Number of control variables
- Use of single-function modules
- Number of modules

- Maximum module size
- Hierarchical control between modules
- Logical coupling between modules
- Data coupling between modules

Duty Cycle

- Constant mission usage
- Periodic mission usage
- Infrequent mission usage

Nonoperational Usage

- Training exercises
- Periodic self-test
- Built-in diagnostics
- Self-maintenance

Qualitative Characteristics

- Correctness
- Validity
- Efficiency
- Portability
- Resilience
- Reusability
- Fault tolerance
- Clarity
- Testability
- Readability
- Interoperability

Many of the qualitative characteristics mentioned above are defined in Section 9.1.1. Others are defined as follows:

- *Resilience*: Sometimes referred to as robustness, resilience is the measure of a computer program's ability to perform in a reasonable manner, despite violations of usage and input conventions.
- *Reusability*: A measure of the ease with which a computer program can be used in a different application from the one for which it was developed.
- *Fault Tolerance*: The ability of a computer program to perform correctly, despite the presence of error conditions.

- *Clarity*: The ability of a computer program to be easily understood. Clarity is a measure not only of the computer program itself but also of its supporting documentation.
- *Readability*: A measure of how well a skilled programmer, not the creator, can understand a program and correlate it to the original and new requirements.

Software reliability may also defined in terms of "the frequency and criticality of program failure, where failure is an unacceptable effect or behavior under permissible operating conditions." Software reliability can be represented by the rate at which errors are uncovered and corrected. The following factors should be considered in measuring reliability:

- Does the program check for potentially undefined arithmetic operations such as division by zero?
- Can software perform the function when not required?
- Can software give unreasonable or harmful output?
- Are loop terminations and multiple-index parameter ranges tested before use?
- Are subscript ranges tested before use?
- Are error recovery and restart procedures tested?
- Are input data validated?
- Are output data validated for reasonableness?
- Are test results validated?
- Does the program make use of standard library routines rather than individually developed code to perform common functions.

9.1.3 Software System Safety

A simplistic definition is that a software system is safe if the software will not cause the hardware to create an unsafe condition and will be fail-safe if the hardware causes the software to malfunction. The Tri-Service Software System Safety Working Group (Army–Navy–Air Force) defines software system safety as "the optimization of system safety in the design, development, use, and maintenance of software and its integration with safety-critical systems in an operational environment." Associated definitions are:

Safety-Critical Computer Software Components or Units: Those components or units whose errors could result in a hazard.

Safety-Critical Functions: Functions in hardware, software, or both whose errors could result in a hazard.

Safety-Critical Path: A path through the software leading to the generation of a safety-critical function.

9.1.4   Software Maintainability

Software maintainability is the probability that a program will be restored to working condition in a given period of time when it is being changed, modified, or enhanced.

Software maintainability may be defined qualitatively as the ease with which software can be understood, corrected, adapted, tested, and enhanced. Quantitatively, maintainability can be assessed indirectly by measuring the following maintainability attributes: problem recognition time, administrative delay time, maintenance-tool collection time, active correction (or modification) time, local testing time, global testing time, maintenance review time, and total recovery time. It is a catchall term used to summarize all the features of a program that allow it to be easily altered or expanded. It includes these factors:

Modifiability: A measure that takes into account the extent to which likely candidates for change are isolated from the rest of the computer program. (For example, isolation of input and output routines that are hardware- or human-dependent would increase a program's modifiability.)

Portability: A characteristic of software that allows it to be used in a computer environment different from the one for which it was originally designed. Use of standard high-level languages is one of the ways to increase portability. It can also be defined as the ability of computer software to run with different kinds of hardware and software. Portability has become an important aspect of maintainability. Newer and superior hardware is being introduced so rapidly that companies are willing to pay extra for this feature.

Availability: The percentage of a specified time period that a system is in an operable state. It can be calculated by dividing the system's uptime by total scheduled time. It can be reported weekly, monthly, or annually. Although availability is sometimes used as a measure of maintainability exclusively, actually it is a measure of reliability and maintainability together.

Testability: A characteristic of a program that allows its functional requirements to be logically separated for step-by-step testing.

Modularity: A measure of the number of independent parts in a software system. A high modularity helps to localize design modifications.

Prognostability: The ability to self-correct or warn well in advance of the failure.

9.1.5   Software Logistics Engineering

Although the term *software logistics engineering* has seldom appeared in the literature, it is an important concept nonetheless. It may be defined as that branch of engineering that deals with software support functions such as maintenance, trainability, preparation of technical manuals, security, and preservation of software. This includes on-line help through call centers—automated wireless communication to fleet managers responsible for maintenance and diagnostics tools.

9.1.6   Some Important Definitions

Since the software field is still groping for standardization in many areas, it will be helpful to define some terms not already covered. They are compiled largely from References 3, 4, and 5.

Computer Software Component (CSC): A distinct part of a computer software configuration item (CSCI). CSCs may be further decomposed into other CSCs and computer software units.

Computer Software Configuration Item (CSCI): Software that is designated by the procuring agency for configuration management.

Computer Software Unit (CSU): An element specified in the design of a computer software component that is separately testable.

Coupling: A measure of the strength of module interconnection; modules with strong interconnections are highly coupled and those with weak interconnections are loosely coupled.

Design Walk-Through: An informal design evaluation by an independent reviewer, usually another programmer familiar with the project.

Do-While: A structured control flow construct that repeats a sequence of statements while a specific initial condition is true. Note that the sequence of statements is not executed if the specific condition is initially false.

Firmware: Software that resides in a nonvolatile medium that is read-only in nature and is completely write-protected when functioning in its operational environment.

Hazard: An inherent characteristic of a thing or situation which has the potential of causing a mishap. Whenever a hazard is present, the possibility, regardless of degree, of a mishap occurring exists.

Hazardous Operation/Condition: An operation (activity) or condition (state) that introduces a hazard to an existing situation without adequate control of that hazard, or removes or reduces the effectiveness of existing controls over existing hazards, thereby increasing mishap probability or potential mishap severity, or both.

Higher-Order Language: English-like user-oriented languages oriented toward problem solving rather than detailing the work of the machine.

IF-THEN-ELSE: A control structure that implements conditional instruction execution.

Independent Verification and Validation (IV&V): An independent test and evaluation process that ensures the computer program satisfactorily performs, in the mission environment, the functions for which it was designed. Verification is the iterative process of ensuring that, during each phase of development, the software satisfies and implements only those requirements that were approved at the end of the previous phase. Validation is the test and evaluation process to ensure that the software meets all system and software performance requirements.

Integration Test: The process of testing several unit-tested modules as a whole, to assure compliance with the design specification.

Lines of Code: The actual number of instructions at the source code level of a programming language that collectively perform a software function.

Mishap: An unexpected, unplanned, or undesired event or series of events that has harmful consequences. Harmful consequences are injury; death; occupational illness; damage to or destruction of equipment, facilities, or property; or pollution of the environment.

Morphology: Shape of a software structure, measured in terms of depth, width, fan-out, and fan-in.

Operating System: Software that controls the execution of computer programs and that may provide scheduling, debugging, input–output control, accounting compilation, storage assignment, data management, and related services.

Preliminary Design Review (*PDR*): A formal, technical, and management review of the software development effort that concentrates on the top-level structural design and its traceability to the requirements.

Product Baseline: Configuration items that have design frozen at established program milestones (system design review, preliminary design review, critical design review) and are ultimately subjected to formal testing and configuration audits prior to delivery.

Requirements Review: A formal review of the software requirements specification to determine whether the software requirements specification (SRS) is acceptable to both developer and requester and adheres to the system specification and the software plan.

Safety-Critical Computer Software Component (*SCCSC*): Computer software component (unit) whose inadvertent response to stimuli, failure to respond when required, or response out-of-sequence or in unplanned combination with others can result in a critical or catastrophic mishap, as defined in Mil-Std-882B.

Software: All instructions, logic, and data, regardless of the medium on which they are stored, that are processed or produced by automatic machines and that are used to control or program those machines. Software also includes firmware and documentation associated with all of the above.

Software Configuration: An instantaneous snapshot of the physical representation of software at some point in time. This representation takes two forms: (1) nonexecutable material generated to document or complement the computer program and (2) executable material processed directly by a computer.

Software Requirements Specification (*SRS*): A document that concentrates on four aspects of a software project: information flow and interfaces, functional requirements, design requirements and constraints, and testing criteria to establish quality assurance.

Stub: A dummy procedure used to test a super-ordinate procedure.

Support Software: All software used to aid the development, testing, and support of applications, systems, test, and maintenance software.

System Specification (*SS*): The document that defines overall system requirements without a detailed regard for the implementation approach. The document specifies functional characteristics and performance objectives of the system, interface characteristics, environment, overall design concepts, reliability criteria, design constraints, and predefined subsystems.

9.2 SOFTWARE PERFORMANCE IN THE DESIGN PHASE

Studies show that about 60% of software errors are specification or logic design errors; the remainder are coding- and service-related errors. With such a high magnitude of errors from the design phase, at least 40% of the software budget should be allocated to the design phase. Many service-related errors also can be prevented by proactive designs.

9.2.1 Software Quality Assurance in Design

Since the emphasis is on accuracy and precision, the entire software development process must be monitored. The most widely used procedure is identified in DoD-Std-2167A [3] (Fig. 9.1). It is intended to include reliability and safety features, but these items are not detailed. The procedure shows that the quality assurance function starts in the system definition phase. At this stage, the system performance—that of hardware and software together—is analyzed. Then the software

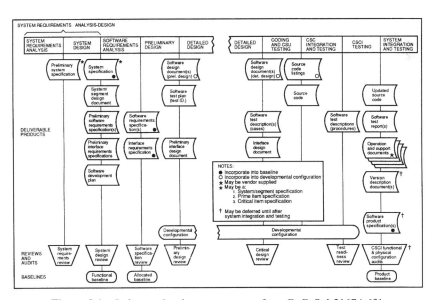

Figure 9.1 Software development process from DoD-Std-2167A [3].

system specification is developed along with an interface specification. At this time, the functional baseline is established. Design reviews are held at each stage of the development and a baseline is established after each major development, all the way to acceptance testing.

9.2.2   Software Reliability in Design

While errors in hardware often can be seen, software errors are invisible except to the programmer. Tiny bugs can have disastrous effects. They have killed sailors, devastated space programs, and threatened to collapse stock markets. These bugs breed fast in complex software, where thousands of lines of code are trusted to hundreds of programmers who work only on a module or two and have limited information on interfacing modules. The vast majority of systems are deeply flawed from the viewpoint of reliability.

Software has no random failures and has infinite life, once it works. There is no wear-out. The traditional hardware testing strategies therefore may not work. Often the right test combinations do not occur until much later during use. Most software engineers agree that practically 100% of software defects are human errors, mainly in three areas: specification, coding, and modification. We have depended on the intuition of the programmer, who does not necessarily perform design reviews, and logic designers, who do not perform fault-tree analyses. "Such language-specific smart editors promise at most freedom from syntactic errors and simple semantic errors," according to Fred Brooks [6], Kenan Professor of Computer Science at the University of North Carolina.

In view of the complexity of defining a failure, one has to consider two kinds of definitions. One is explicit and is in the contract. The second definition is implied. In safety-related incidents, the courts tend to accept the hidden implied definition. The following definitions of *failure* and *fault* [7] may be considered explicit.

A failure is:

1. The termination of the ability of a functional unit to perform its required function.
2. The inability of a system or system component to perform a required function within specified limits. A failure may be produced when a fault is encountered.
3. A departure of program operation from program requirements.
4. The event, or inoperable state, in which any item does not, or would not, perform as previously specified.

A fault is:

1. An accidental condition that causes a functional unit to fail to perform its required function.
2. A manifestation of an error in software. A fault, if encountered, may cause a failure. It is synonymous with *bug*.

3. An immediate cause of failure (e.g., maladjustment).

4. A condition, action, or element of software that, when encountered during testing or operation of the system or components, results in a failure.

9.2.2.1 *Software Design Techniques for Reliability* Derating, safety margins, and fault tolerance are design provisions for hardware reliability. Somewhat similar provisions exist for software reliability. A study reported in Reference 2 suggests these approaches:

1. *Control Complexity*: This is the most important factor to be controlled in software. Higher complexity tends to make computer programs incomprehensible. It affects other factors such as readability and modularity. Software must be reviewed, discussed, and tested by programmers to assure the complex relations can be understood by the user. Complex software is prone to errors and makes it difficult to avoid faults. On the other hand, it allows many functions to be performed with less memory per function. Usually the reliability per function is increased. Such trade-offs must be carefully evaluated.

2. *Perform Top–Down Functional Decomposition*: This structural breakdown of computer software configuration items helps keep the code traceable and readable. The generic hierarchy is system, subsystem, program, module, routine, unit, and instruction. The following hierarchy applies to government contracts:

   a. Computer software configuration item (software aggregate for configuration control).

   b. Computer software component (functionally or logically distinct part of a CSCI).

   c. Unit (lowest-level logical entity that completely describes a nondivisible function to allow independent implementation and testing).

   d. Module (lowest physical entity that may be assembled or compiled alone). *Note*: If the module is a logical entity, it will appear higher in the hierarchy as a *unit*.

   e. Instruction (single executable statement of code for a single action).

   f. Operation (action to be performed by the instruction).

   g. Operand (addresses in the computer memory where data reside).

3. *Modularize*: The segmentation of computer programs into single-purpose, single-entry, single-exit modules is the most popular technique discussed in the literature.

4. *Design Hierarchically*: Most authors acknowledge the significant reduction in complexity that can be achieved by enforcing a hierarchical structure of module segment calling and controlling relationships.

5. *Follow Structured Approaches*: The advantages of structured techniques are well known. They tend to simplify complex software. Top–down or chain structures are examples.

9.2.2.2 *Preventing Specification and Design Errors*   Software teams can use design reviews. Software fault-tree analysis (SFTA) is very productive, but its use is not widespread. The industry has traditionally relied on so-called IV&V— independent verification and validation. A tool known as *soft-tree analysis*, which applies to hardware–software interfaces, has been tried successfully. But the problem is, as Patrick Hoff of the Naval Surface Weapons Center puts it: "These fools can require intimate workings of the software. It is at this point that many or even most engineers throw in the towel. In spite of multidisciplinary engineering backgrounds, most are not prepared for this type of analysis."

We used failure mode, effects, and criticality analysis with the help of programmers and influenced many changes in specifications. FMECA, in fact, is very powerful if used or performed at several layers such as at the functional level, logic design level, or code level [8]. Boeing has developed software for software sneak circuit analysis [9]. All these analytical tools and many more often reveal weaknesses in specifications, particularly with regard to fallback modes, testability, fault masking, fault tolerance, clock synchronization, and maintainability. Section 9.2.6 contains an example of one such tool (software system failure-mode and effects analysis) applied to software reliability.

9.2.3   Software Maintainability in Design

Many programmers have learned through hard experience that, unlike hardware, software often deteriorates in reliability when a failure is fixed. In one automotive software design, it used to take about 6 weeks to make an engineering change. Now it takes almost 6 months to make a similar change. The reason is that it is safety-critical software with over one million permutations and combinations, and it is impractical to test all of them. Another reason is that a programmer repairing a module often is unaware of all the interface effects and ends up creating more serious errors. Good configuration control may provide some relief, but writing good documentation and using interface traceability matrices are even more effective.

The major tools for maintainability are modularity, top–down structured programming, and sequential programming as opposed to spaghetti coding. In top–down structure, the branches of the software are not allowed to be connected sideways to other branches. To search for an error, one has to look below or above only for a fault. Ensuring modularity and testability and using fault-isolation techniques are also powerful aids to maintainability.

Artificial intelligence tools can help too. For example, a tool developed at the Navy Materiel Command in Dover, New Jersey, called *software fingerprinting*, sets limits on complexity in the modules and establishes a profile ("fingerprint") of the baseline. The program identifies critical paths and attempts to limit their number to less than 10. Since no two profiles are supposed to be alike, each profile becomes a fingerprint. The set of fingerprint parameters can consist of source code, object code, module function, program flow mapping, and similar items. Changes in any of these items will result in a fingerprint change. If an unexpected change takes place in any module, its fingerprint becomes different, and it can be identified instantly. Among many other artificial intelligence techniques is the use of expert systems that can detect multiple failure modes and determine the effect of changes on the system.

Those who are concerned about controlling software errors must allocate time and financial resources to analytical tasks. It is cheaper to prevent the problem than to spend 100 times more during troubleshooting. The following ranking done by Chenoweth and Schulmeyer [10] on a Rome Air Development Center (RADC) study gives some insight into sources of errors. The data show the importance of interface design. The user interface and routine-to-routine interface errors add up to 13.32%.

Source of Error	Portion (%)
Logic	21.29
Input/output	14.74
Data handling	14.49
Computational	8.34
Preset database	7.83
Documentation	6.25
User interface	7.70
Routine-to-routine interface	5.62

Even though software maintainability is a qualitative attribute, General Electric [5] defines some quantitative measures:

1. Problem recognition time
2. Administrative delay time
3. Problem diagnostic time
4. Maintenance tool collection time
5. Specification change time
6. Active correction time
7. Local testing time
8. Regression/global testing time
9. Change review time

9.2.4 Software System Safety in Design

The definition of software system safety (Section 9.1.3) implies that the prevention of software hazards must be done during the entire life cycle, with software and hardware analyzed jointly.

In order to proceed with hazard analysis, a *software hazard* needs to be defined. By itself the software does not cause any harm. On the other hand, by itself it is of no use. It only becomes useful—as well as dangerous—when it is integrated with hardware. Therefore, a software hazard is one that can introduce error or malfunction resulting in an unsafe condition in the hardware.

The U.S. Air Force Inspection and Safety Center [11] classifies software hazards into four broad categories:

1. An inadvertent or unauthorized event that can lead to an unexpected or unwanted event.
2. An out-of-sequence event; a planned event occurs but not when desired.
3. Failure of a planned event to occur.
4. The magnitude or direction of event is wrong. This is normally caused by an algorithm error.

Some possible hazards in software are:

- Routines that disable interrupts.
- Conditions that call for an exact match of floating-point numbers (this can result in an infinite loop).
- Conditions that result in division by zero. Many times, a very small number may be truncated to zero by the software.
- Precision and scaling of data resulting in a hazardous condition.

9.2.4.1 *Software Safety Risk Assessment*
In hardware, risk is defined as a product of severity and frequency for each potential accident. In software, once the bug is out, it is not expected to appear again. Therefore, risk may be assessed mainly in terms of severity ratings. The Tri-Service Software System Safety Working Group has recommended the following categories:

I. Software exercises autonomous control over potentially hazardous hardware systems, subsystems, or components without the possibility of intervention to preclude the occurrence of a hazard. Failure of software or failure to prevent an event leads directly to an occurrence.

IIa. Software exercises control over potentially hazardous hardware systems, subsystems, or components, allowing time for intervention by independent safety systems to mitigate the hazard. However, the systems by themselves are not considered adequate.

IIb. Software item displays information requiring immediate operator action to mitigate a hazard. Software failures will allow or prevent the hazard's occurrence.

IIIa. Software item issues commands over potentially hazardous hardware systems, subsystems, or components requiring human action to complete the control function. There are several redundant, independent safety measures for each hazardous event.

IIIb. Software generates information of a safety-critical nature that is used to make safety-critical decisions. There are several redundant, independent safety measures for each hazardous event.

IV. Software does not control safety-critical hardware systems, subsystems, or components and does not provide safety-critical information.

9.2.4.2 *Software Safety Tools* The U.S. Air Force *Software System Safety Handbook* [11] suggests the following safety analysis tools:

1. *Preliminary and Follow-on Software Hazard Analysis*: This is done at the concept stage. It is a brainstorming technique to identify portions of the software such as modules or routines that can create hazards. Analysis is done on system specifications, subsystem specifications, interface specifications, functional flow diagrams, lessons-learned data, program structures, and information related to testing, coding, storage, modification, and use.

2. *Software Fault-Tree (Soft-Tree) Analysis*: In this technique, the hardware and software trees are constructed with standard symbols (Chapter 5). The two trees are then linked at their interfaces so that the entire system can be analyzed. A slightly different way of accomplishing this is to construct a hardware failure tree in which the software interfaces are identified. A software fault tree is then added to the interfacing points.

3. *Use of Cut Sets*: This technique identifies single-point and multiple-cause failures and is shown for hardware in Chapter 5. In the case of the software system, the technique is applied to soft trees. Computer programs for cut set analysis are commercially available.

4. *Common Cause Analysis*: Examination of cut sets can reveal some causes that can result in the system failing in spite of many redundancies. A cut set is a path or a set of components that, if it fails, can cut off the system function. These are also referred to as *common causes* for the downstream failures. In the functional block diagram in Fig. 9.2, cut set 1 is the common cause of failure of components B through H to operate. Similarly, the failure of component D is the common cause of failure of redundant components E, F, G, and H to operate. Chapter 5 shows a cut set algorithm based on fault trees. For example, if a power supply is common to two data registers, a failure in it will disable both data registers. Similarly, a voltage transient in the power line can cause a malfunction in several portions of a system. Computer programs are available for this analysis.

5. *Software Sneak Circuit Analysis*: See description of this technique in Chapter 5 for hardware.

6. *Use of Petri Nets*: This is a mathematical model invented by Carl Petri in 1961. One of the advantages is that the user can describe the model graphically. The system is modeled in terms of conditions and events and traces the conditions as events take place in time. In a way, it is a simulation technique and requires complex software. Readers interested in this technique may refer to a paper by Nancy Leveson [12].

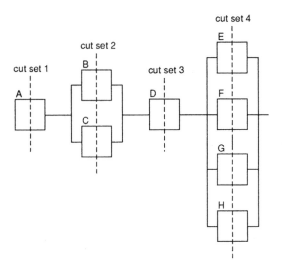

Figure 9.2   Cut sets identified for a block diagram.

7. *Use of Mil-Std-882*: By far, the most widely used guidelines are in this standard. It requires analysis and design improvements in several stages. Hardware hazard analysis procedures are covered in Chapter 5 and are equally applicable to software.

Steven Mattern [13] conducted a survey of the users of software safety analysis tools (the users are few), with the help of the Tri-Service Software System Safety Working Group. The results are given in Table 9.1.

TABLE 9.1   Survey Results

Software Hazard Analysis Tool	Users	Software Hazard Analysis Tool	Users
Fault-tree analysis	8	Hierarchy tools	1
Software preliminary hazard analysis	4	Compare and certification tool	1
Traceability matrices	3	System cross-check matrices	1
Failure modes and effects analysis	3	Top–down review of code	1
Requirements modeling/analysis	2	Software matrices	1
Source code analysis	2	Thread analysis	1
Test coverage analysis	2	Petri net analysis	1
Cross-reference tools	2	Software hazard list	1
Code and module walk-throughs	2	BIT/FIT plan	1
Sneak circuit analysis	2	Nuclear safety cross-check analysis	1
Emulation	2	Mathematical proofs	1
Subsystem hazard analysis	2	Software fault-hazard analysis	1
Failure-mode analysis	1	Mil-Std 882B series 300 tasks	1
Prototyping	1	Topological network trees	1
Design and code inspections	1	Critical function flows	1
Checklist of common software errors	1	Blackmagic	1
Data flow techniques	1		

The listing of hazard analysis tools in Table 9.1 shows that fault-tree analysis ranks highest in number of users. The consensus is that for most effective use, fault-tree analysis should be applied to both hardware and software as components of one system. Figure 9.3 shows a fault tree for a hardware–software–human system. All three elements are analyzed in the same tree.

For a safe and fault-tolerant design, the criterion is that the failure of the system should not be the result of a single-point failure. Therefore, the top level should always have an AND gate. In the system of Fig. 9.3, AND gates were designed into the product at the second level because the AND gate at the top was not viable. When this criterion is used, many faults at lower level can be tolerated. For one design, about 4 months of analysis was eliminated at the lower levels because of the protection. Designs using this criterion have been safe and efficient and result in lower life-cycle costs.

9.2.5 Software Logistics Engineering

Support provisions, including trained programmers and software engineers to change, modify, or enhance the software, are part of software logistics engineering. So, too, is documentation. User manuals should be started at the same time as

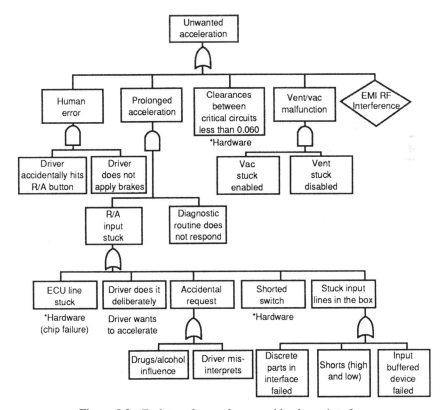

Figure 9.3 Fault tree for a software and hardware interface.

coding. They should not be written at the last moment and must be tested as part of system tests.

Another aspect of logistics engineering is to plan and execute data collection. A closed-loop feedback system should be used during the entire software development process to assure high reliability and maintainability. Data analysis should continue after the software is operating. It will provide valuable information for the next project.

9.2.6 Software System Failure-Mode and Effects Analysis

Software system failure-mode and effects analysis (SSFMEA) applies to software systems where software controls the hardware. Most technical products belong in this category, such as aircraft, power plants, refineries, and complex systems. In SSFMEA, the focus is on identifying system weaknesses through software functional flowcharts so that software specifications can be made complete, clear, comprehensive, and unambiguous. The goals are (1) to determine whether the software is fault-tolerant with respect to erroneous inputs, no inputs, valid but unreasonable inputs, and unreasonable outputs; and (2) to identify missing requirements in the system specification or software specifications.

Usually, hardware improvements are also required for the software to be effective such as the need for redundant sensors, redundant processors, and an independent means of verification. A software engineer should never assume that the hardware will work perfectly; it does not. The biggest concern is the interface reliability of the hardware where loose or bad connections can cause undiagnosable software failures. Many times there are trade-offs to be made between hardware and software improvements. SSFMEA emphasizes software, although there is no harm done if hardware is improved as a by-product. The reason for emphasizing software is that software errors usually result in enormous downtime and the causes are not obvious to the user. Hardware errors, in contrast, are often so obvious that even nontechnical people can easily notice them; if a tire on an automobile is flat, anyone can detect the cause of failure.

9.2.6.1 *The Objective*
The main objective of SSFMEA is to enhance the role of software in assuring the safety and reliability of the system. In doing so, the hardware system is also enhanced because the software will not be able to do its job without the help of fault-tolerant hardware. The second objective is to reduce software life-cycle costs. SSFMEA analysis is done as soon as the specification is written and is continually updated as the design work progresses, well before coding. This allows many engineering changes to be done in a short time and at the lowest cost.

9.2.6.2 *The Methodology*
SSFMEA is simple. A functional flowchart from the software specification is developed. Each block in the flowchart is treated as a component of the software system. Then the potential functional failures and hazards in the system are analyzed during execution of each software function. An attempt is made to control the hazard and critical failures through software

design changes or hardware control, whichever offers a viable solution and lower life-cycle costs.

9.2.6.3 Demonstration of Methodology

Consider a simple example of SSFMEA. The flowchart in Fig. 9.4 shows four software components in an aircraft. For the *Check engine speed* component, the system failure modes and causes are:

1. Software receives erroneous speed reading. *Causes*: A stuck-at condition in the microprocessor; sensor inaccurate.
2. Software receives no inputs on speed. *Causes*: Sensor failed; microprocessor failed.
3. Software is unable to check the speed. *Causes*: Sensor response cycle may be longer than the software waiting time.

Once the causes of faults and errors are identified, the design improvements become more obvious. The following design precedence should be applied:

• Remove cause of the hazard from the design.
• Design for fail-safe consequences and to prevent major damage.
• Design for fault tolerance.
• Provide self-checking software every time a critical operation is performed.
• Implement a prognostic design to alert the user before any mishap takes place.

In this example, the cause of the hazard can be removed by having the software wait until the instrument responds—event driven rather than time driven. Make sure you reanalyze the new implementation. One may argue that the software waiting

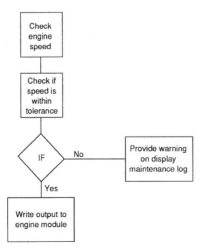

Figure 9.4 Four components of a software item.

time was based on the sensor response time in the first place, so why should we worry? We still need to worry because the response time can change, as the instrument gets older. Sometimes additional circuits are connected to the sensor, which can change the response time. The upshot is that, if the sensor does not respond in a reasonable time, then the failure of the sensor should be reported—instead of a false alarm.

Figure 9.5 shows a partial flowchart of an aircraft early warning system, and Fig. 9.6 shows a SSFMEA format for the software components. As noted, the SSFMEA will identify missing requirements, as is evident in this example. The first software element in Fig. 9.6, *Check exhaust temperature*, is associated with

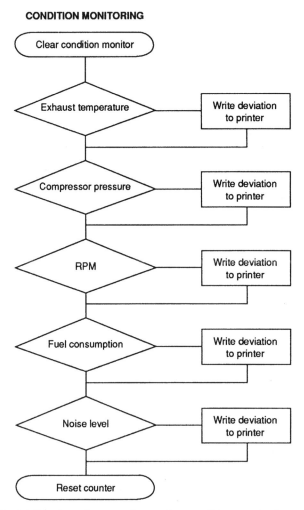

Figure 9.5   Partial flowchart for an early-warning condition monitoring system in a jet engine.

Software element	System failure mode	Cause	Effects	Criticality	Recommended software/hardware changes
1. Check exhaust temperature (within tolerance)	a) Erroneous input	Sensor malfunction	software reports large deviation	IIC	a) Software should switch to redundant sensor, when large deviation is reported. Add redundant sensor.
	b) No input	Sensor failed	same as (a)	IIC	b) Same as (a)
	c) Computational error	Voltage transient Temporary fault	same as (a)	IIB	c) Use time redundancy
	d) Temperature reading very high	Exhaust temperature	engine fails	IB	d) Software to alert pilot to shut down engine and report on maintenance log.
2. Check compressor pressure	a) Pressure too low	Clogged or dirty filter, micropro-cessor error	possible engine failure	IC	a) Software to alert pilot to shut down engine and report on maintenance log.
	b) Pressure too high	Compressor malfunction	possible engine failure	IC	b) Same.
	c) Compressor OK but erroneous pressure reported	Algorithm error, sensor malfunction	false alarm	IIA	c) Include independent check on compressor pressure.

Figure 9.6 Software system FMEA.

the failure mode *Erroneous input* because of the malfunction of the sensor. The current design requires that the deviation should be reported on the printer. From a SSFMEA point of view, reporting a hazard may not be good enough; the jet engine may fail in midair before it receives maintenance. The proper action is to switch over to a redundant sensor. This is the missing requirement that should be added to the system specification. Similarly, the third failure mode, *Computational error*, caused by voltage transients, should be *forgiven* by the software design. A common technique in such cases is the use of *time redundancy*, which calls for running the same algorithm at three different times; at least two of three outputs must match. Thus, if an error is temporary, it will not match the remaining two outputs.

9.2.6.4 *Software–Hardware Interface Control*

Often hardware failure will prevent software from functioning. A typical software engineer's response might be: "Well, if the hardware has failed, it is not a software problem." This argument is not accepted in the courts. The legal system does not differentiate between hardware and software. It is concerned with the system. It is the responsibility of the manufacturer to *make* the system perform in spite of the hardware failure. Consider software controlling a fan through a microprocessor to cool an engine. If the microprocessor has failed, the software is helpless. The question is: Can the software still control the fan? The answer is yes, if the software and hardware engineers want to work together. There are many ways the software can be kept in charge. One way is to provide an external switch through which the software can turn on the fan.

9.2.6.5 *Other Uses of SSFMEA*

So far we have discussed only SSFMEA's role in design productivity. However, SSFMEA can be used for many other purposes. One important use is in developing test cases. Many companies spend thousands of dollars in developing test cases and are still unable to do a thorough job because it takes a monumental amount of time and money to check all paths. SSFMEA points out important failure modes, and at least the test cases related to these failures ought to be developed. Other tools such as soft trees can identify additional test cases.

SSFMEA can be used for software maintainability and availability analysis. Since it predicts failures, the associated downtimes can be estimated. If the downtimes are not acceptable, the software can be redesigned to include fault tolerance, modularity, and early diagnostic features.

9.2.6.6 Implementing a SSFMEA Program

Implementing hardware FMEA has taken years in some cases and has still been done wrong (this is discussed in Chapter 3). Because it has been practiced wrong for many years, people assume that the wrong way is the right way. Most often, people violate the first rule of FMEA—that the analysis be performed by a team. Often, it is done by one person who is really not familiar with the internal working of the design and therefore not qualified. But in a team, the same individual can be very effective by being curious and asking a lot of questions based on personal experience. An effective SSFMEA is one performed by a team of hardware, software, service, and system engineers with guidance from a system safety or reliability engineer.

A major misuse of FMEA is in predicting failure modes and failure rates. These are not the primary goals of FMEA. The real goal—to make design improvements—is mostly overlooked. Some perform FMEA so late that no one is allowed to make major engineering changes. The investment in FMEA for these situations is not justified. The only way to make it dramatically effective is to do it very early in design, before final specification is released. And since only management can make it happen, the process must be institutionalized.

9.2.7 Software Performance Specification

The features discussed so far should be an integral part of a software specification. Unfortunately, companies rarely spend enough time on developing a clear and complete specification, and, as a result, many designers and programmers spend more time on testing and fixing the problems. That is why the software is often well behind schedule. Figure 9.7 presents guidelines used by a large company for writing performance specifications. Additional items can be added as appropriate.

9.2.8 Software Design Review Checklist

In any design review, a checklist is a good place to start. Every company should develop its own checklists for software quality, reliability, safety, and maintainability. We found a checklist (Fig. 9.8) from a major aerospace company to be one of the best. It was developed mainly for software system safety but can be applied to reliability. The list is slightly modified to include some additional features.

SYSTEM SPECIFICATION
INTRODUCTION

Software is often an integral part of a larger configuration that includes hardware, facilities, and personnel. This configuration, called a "system," must be carefully specified prior to commencement of software planning and development.

The SYSTEM SPECIFICATION indicates the overall functional, operational, and performance characteristics of a system. In it both hardware and software elements are described and the requirements for each are specified.

The software organization is normally not responsible for the generation of the SYSTEM SPECIFICATION. However, because the system document serves as a baseline for all software documents, software personnel should participate in its development.

A recommended format for the SYSTEM SPECIFICATION is presented in this Appendix.

1.0 SCOPE

This section provides an overview of the system and defines scope for performance, design, development and test requirements. In addition, system objectives and characteristics are discussed.

2.0 APPLICABLE DOCUMENTS

This section lists all documents which apply (in whole or in part) to the specification.

3.0 REQUIREMENTS

This section contains a descriptive and quantitative definition of all system requirements. The following topics are considered:

1. The performance and design requirements for the system.

2. The requirements related to operating, maintaining, and supporting the system, to the extent these requirements define or constrain design of the system.

3. The design constraints and standards necessary to assure compatibility of system items.

4. The definition of the principal interfaces between the system being specified and other systems with which it must be compatible.

5. The functions of the system, and the principal interfaces between and within each function.

6. The allocation of functional performance and the definition of specific design constraints.

7. The identification and use of existing equipment, computer programs or operating procedures.

Unless purely descriptive by nature, requirements are stated in quantitative physical terms with tolerances that can be verified by subsequent tests, demonstrations, or inspection. Requirements stated in this section are the basis for the tests specified in Section 4 of the specification.

3.1 System Definition

This paragraph lists the elements to be developed or included as part of the system. Elements include hardware, software, and operating procedures. A graphical representation (block diagram) of these items should also, be shown if appropriate.

Figure 9.7 Guidelines for writing system specifications. (From Ref. 5.)

3.1.1 Functional Description

This paragraph contains a brief description of the system including a list of all functions. The intended use of the system is described to the extent necessary for an overview understanding.

3.1.2 System Diagrams

This paragraph incorporates the system level functional schematic diagrams and includes the top-level functional control and data flow of the system. The level of detail that is required must identify all system functions and their relationship with system elements.

3.1.3 Interface Definition

This paragraph describes the functional and physical interface: 1) between the system and other systems with which it must be compatible; and 2) between all functions within the system.

3.1.4 Existing Resources and Procedures

This paragraph lists existing elements which the system is to be designed to incorporate. This list identifies the elements (e.g., hardware, software) by reference to nomenclature, specification number and/or part or model number. If the list is extensive, it may be included in an appendix which can be referenced in this paragraph.

3.1.5 Operation and Organization

This paragraph includes an overview description of the intended operation of the system. Also, the anticipated deployment of the system both geographically and organizationally (e.g., the number and location of installations) is described.

3.2 Characteristics

System characteristics are described in detail in the paragraphs that follow.

3.2.1 Performance Characteristics

This paragraph contains system performance characteristics and provides sufficient guidance for technical development. It describes what the system should accomplish and specifies both upper and lower performance limits. The following considerations are included:

1. Quantitative criteria covering endurance capabilities of the equipment that are required to meet the user needs under stipulated environmental and other conditions. Minimum total life expectancy is stated.

2. Quantitative expected performance for each operational mode (e.g., average rates, peak rates, quantity of inputs).

3. Programming language requirements (for software), if language is dictated by other system elements.

4. Other essential aspects of system performance which cannot be more appropriately located under another heading.

3.2.2 Physical Characteristics

This paragraph considers the following topics (as required):

1. Weight limits, electrical and equipment heat generation specifications.

2. Dimensional limitations, space, operator station layout, and access for maintenance.

3. Requirements for transport and storage, such as tie down, pallets, packaging, and containers.

4. Durability factors to indicate degree of ruggedness.

5. Health and safety criteria, including consideration of adverse explosive, mechanical, and biological effects.

6. Security criteria.

7. Media format for delivered software.

3.2.3 Reliability

Reliability is stated in quantitative terms. The conditions under which the reliability requirements must be met are explicitly defined.

3.2.4 Maintainability

This paragraph specifies maintainability requirements. The requirements apply to maintenance in a planned maintenance and support environment and are stated in quantitative terms. Examples are:

1. Time (e.g., mean and maximum downtime, mean time between failures, reaction time, turnaround time, mean and maximum time to repair, mean time between maintenance).

2. Rate (e.g., maintenance man-hours per specific maintenance action, maintenance hours per operating hour, frequency of preventive maintenance).

3. Maintenance complexity (e.g., number of people and skill levels, variety of support equipment).

Figure 9.7 *Continued.*

3.2.5 Availability

The degree to which the system is available, e.g., 24 hours per day, 7 days per week.

3.2.6 Environmental Conditions

This paragraph includes environmental conditions to be encountered during operation and/or storage of the system equipment. The following subjects should be considered: natural environment (wind, rain, temperature, etc.); induced environment (motion, shock, noise, etc.); electromagnetic signal environment.

3.2.7 Transportability

This paragraph includes requirements for transportability which are common to all system equipment to permit installation and logistic support. All system equipment that will be unsuitable (due to operational or functional characteristics) for normal transportation methods is identified.

3.3 Allocation of System Functions

This paragraph allocates the system functions described above to hardware, software or other system elements. Minimum system design and construction standards are also specified.

3.3.1 Division of Hardware and Software

This paragraph contains a list of all system functions and the allocation of either hardware or software.

3.3.2 Materials, Processes and Parts

This paragraph specifies requirements that govern the use of materials, parts, and processes. It contains specifications for particular materials and processes to be utilized in the design of system equipment. In addition, requirements for the use of standard components and parts for which a qualified products list has been established are specified.

3.3.3 Workmanship

This paragraph contains workmanship requirements for equipment to be produced during system development. Requirements for manufacture by specified production techniques are also discussed.

3.3.4 Interchangeability

This paragraph specifies the requirements for system elements that are interchangeable and replaceable. Entries in this paragraph are for the purpose of establishing a condition of design, and are not to define the conditions of interchangeability.

3.3.5 Safety

This paragraph specifies safety requirements that are basic to the design of the system. Equipment characteristics, methods of operation, and environmental influences are considered.

3.3.6 Human Engineering

Human engineering requirements for the system are specified and applicable documents (e.g., regulations) included by reference. Particular attention is paid to areas in which the effects of human error would be particularly serious.

3.4 Logistics

This paragraph considers the logistical characteristics that are required for the implementation, support and maintenance of the system.

3.4.1 Maintenance

This paragraph includes consideration of factors such as: 1) use of test equipment; 2) repair versus replacement criteria; 3) organizational levels of maintenance; 4) maintenance and repair cycles; 5) accessibility, and 6) distribution and location of spare parts.

3.4.2 Support

This paragraph specifies the support required for the system. Considerations include vendor hardware and software support and maintenance response requirements.

3.4.3 Facilities and Equipment

This paragraph specifies the impact of the system on existing facilities and facility equipment. It also describes requirements for new facilities, auxiliary equipment or software to support the system. Software development facilities and test facilities are also included.

3.5 Personnel, Training and Documentation

This paragraph specifies the training requirements for the system and includes: 1) how training is to be accomplished (e.g., school, on the job); 2) identification of equipment or software that will be required for training purposes; and 3) course material and training aids. Additionally, a training schedule and location are specified.

3.5.3 Documentation

This paragraph specifies the documentation required for each system element. References to standards are acceptable.

Figure 9.7 *Continued.*

4.0 QUALITY ASSURANCE

Provisions and criteria established in this section are used to verify functional and performance requirements specified in Section 3. Requirements for test/validation are defined. The following topics should be considered:

1. Collection and recording of data during all testing for use as part of the reliability analysis;

2. engineering evaluation and test requirements in direct support of design and development activity;

3. integration testing such as continuity checking, software/hardware interface mating, and function operation in the installed environment; support equipment compatibility and documentation verification;

4. formal test and validation of functional and performance characteristics.

5.0 NOTES

This section contains background information and other material pertinent to a complete description of the system.

6.0 APPENDIX

This section contains supplementary material that is part of the specification but has been appended for convenience in specification maintenance.

Figure 9.7   *Continued.*

Item	Req'd	Not Req'd	Incor- porated
General and Miscellaneous			
Provides for precluding dependence on administrative procedures			
Provides for using information control concept for deriving the authorization code for the activation of the authorization device			
Provides that the software contains only features or capabilities required by the system, and that it does not contain additional capabilities, e.g., testing, troubleshooting, etc.			
Provides for positive control of system safety-critical functions at all times			
Provides for safety-critical subroutines and subprograms to include "Come From" checks to verify that they are being called from a valid calling program			
Separation of Commands, Functions, Files, and Ports			
Provides for using separate launch authorization and separate launch control functions to initiate a missile launch			
Precludes the ground ordnance enabling arming code from being same as the enable launch authorization code			
Provides for requiring separate "arm" and "fire" commands for ordnance initiation			

Figure 9.8   Example of a software system safety checklist.

Item	Req'd	Not Req'd	Incor-porated
Separation of Commands, Functions, Files, and Ports			
Precludes using input/output ports for both critical and noncritical functions			
Provides for sufficient difference in addresses for critical input/output ports versus noncritical ports that a single address bit failure does not allow access to critical functions or ports			
Provides for having files that are unique and have a single purpose			
Interrupts			
Provides for defining specific interrupt priorities and responses			
Provides for software system management of interrupt control so as to not compromise safety-critical operations			
Shutdown, Recovery, and Safing			
Provides for a fail-safe recovery from inadvertent instruction jumps			
Shutdown provisions are included in software upon detection of unsafe conditions			
Provides for the system reverting to a known predictable safe state upon detection of an anomaly			
Provides for software safing of safety-critical hardware items			
Provides for an orderly system shutdown as the result of a command shutdown, power interruptions, or other failures			
Requires that the software be capable of discriminating between valid and invalid external interrupts and shall recover to a safe state in the event of an erroneous external interrupt			
Provides for entry into a safe state in the event of erroneous entry into a critical routine			
Protects against out-of-sequence transmission of safety-critical function messages by detecting any deviation from the normal sequence of transmission. When this condition is detected, the software terminates all transmissions, recycles to a known safe state, and displays the existing status so the operator can take compensatory action			

Figure 9.8 *Continued.*

Item	Req'd	Not Req'd	Incor-porated
Shutdown, Recovery, and Safing			
Provides for initializing all unused memory locations to a pattern, that if executed as an instruction, will cause the system to revert to a known safe state			
Provides for identifying safing scenarios for safety-critical hardware and including them into the decision logic			
Provides for the capability of reversing or terminating launch authorization and ordnance arming functions			
Preventing, Precluding, and Disallowing Actions			
Provides for preventing inadvertent generation of critical commands			
Provides for disallowing coexistence of potentially hazardous routines			
Provides for preventing bypass of safety devices during test			
Following computer memory loading, automatic control is prevented until all data is loaded and verified			
Precludes inadvertent operation of data entry control to critical routines			
Provides for precluding a change in state if data synchronization is lost			
Provides for prevention of a hardware failure or power interruption from causing a memory change			
Provides for prevention of memory alteration or degradation over time during use			
Provides for program protection against unauthorized changes			
Provides for not allowing the safety-critical time limits in decision logic to be changed by the console operator			
Provides for preventing inadvertent entry into a critical routine			
Provides for not allowing a hazardous sequence to be initiated by a single keyboard entry			
Prohibits transmission of any critical command found to be in error and notifies the operator of the error			

Figure 9.8 *Continued.*

Item	Req'd	Not Req'd	Incor- porated
Shutdown, Recovery, and Safing			
Provides that controlling or monitoring of nuclear weapons be incapable of bypassing operator control of safety-critical functions			
Provides for disallowing use of workaround procedures when reverting to a safe configuration after the detection of an anomaly			
Provides for not using a "stop" or "halt" instruction or causing a CPU "wait" state. The CPU is always executing, whether idling with nothing to do or actively processing.			
Provides for detection and termination of commands requesting actions beyond the performance capability of the system			
Provides for disallowing performance of a potentially hazardous routine concurrently with a maintenance action			
Memory, Storage, and Data Transfer			
Precludes storage, in usable form, of information required to cause initiation of a safety-critical function			
Provides for self-test capability to assure memory integrity			
Provides for prevention of a hardware failure or power interruption from causing a memory change			
Provides for prevention of memory alteration or degradation over time during use			
Provides for erasure or obliteration of clear text secure codes from memory			
Provides for limiting control access to storage devices memory			
Provides for protecting the accessibility of memory regions dedicated to critical functions			
Provides for having safety-critical operational software instructions resident only in nonvolatile read-only memory			
Provides for not using scratch files for storing or transferring safety-critical information between computers			
Provides that remote transfer of data cannot be accomplished until verification of data to be transferred is accomplished and authorization to transfer the data has been provided by the operator(s)			

Figure 9.8 *Continued.*

Item	Req'd	Not Req'd	Incor- porated
Memory, Storage, and Data Transfer			
Provides for self-test capability to assure memory integrity			
Verification and Validation Checks			
When a test specifies for the removal of safety interlocks, the software provides for verification of reinstatement of these safety interlocks at the completion of the testing			
Provides for verification and validation of status flags			
Requires that critical data communicated from one CPU to another be verified prior to operational use			
Provides for software validation of critical commands			
Provides for verification of the existence of prerequisite conditions prior to command issuance in accordance with predefined operational requirements			
Provides for verification of the results of safety-critical algorithms prior to use			
Provides for verification of safety-critical parameters or variables before an output is allowed			
Decisioning verifies the sequence and logic of all safety-critical command messages and rejects commands when sequence or logic is incorrect			
Provides that remote transfer of data cannot be accomplished until verification of data to be transferred is accomplished and authorization to transfer the data has been provided by the operator(s)			
Provides that all operator actions that set up safety-critical signals are verified by software based on control device positions			
Provides for control of analog functions having feedback mechanisms that provide positive indications of the function having occurred			
Provides for verification and validation of the prompt for the initialization of a hazardous operation or sequence of hazardous operations			
Provides for verification of accomplishment of each step of a hazardous operation, or sequence of hazardous operations, by setting of a dedicated status flag prior to proceeding to and initiating the next step in the operation or series of operations			
Provides for verification/validation of all critical commands prior to transmission			

Figure 9.8   *Continued.*

Item	Req'd	Not Req'd	Incor-porated
Logic, Structure, Unique Codes, and Interlocks			
Provides for identification of flags to be unique and single purpose			
Provides for using unique arming codes to control critical safety devices			
Provides for inclusion of system interlocks			
Provides for using a minimum of two separate independent commands to initiate a safety-critical function			
Provides for the majority of safety-critical decisions and algorithms to be contained within a single (or few) software development module(s)			
Provides for single CPU control of a process which can result in major system loss, system damage, or loss of human life to be incapable of satisfying all of the requirements for initiation of the process			
Requires that decision logic using registers which obtain values from end-item hardware and software not be based on values of all "ones" or all "zeroes"			
Requires that decision logic using registers which obtain values from end-item hardware and software use specific binary patterns to reduce the likelihood of malfunctioning end-item hardware/software satisfying the decision logic			
Provides for cooperative processing between launch control point and missile computer(s) to process safety-critical functions			
Provides for having safety-critical modules with only one entry and one exit point			
Provides for having files that are unique and have a single purpose			
Provides for not having operational program loads contain unused executable code			
Monitoring and Detection			
Provides for inclusion of monitoring of safety devices			
Provides for detection of inadvertent computer character outputs			
Provides for detection of errors during computer memory loading to terminal loading process			
Provides for detection of unauthorized operation of data entry control			

Figure 9.8 *Continued.*

Item	Req'd	Not Req'd	Incorporated
Monitoring and Detection			
Provides for identification of safety-critical functions requiring continuous monitoring			
Provides for detection of improper processing that could degrade safety			
Provides for detection of a fault having the potential of degrading safety			
Provides for detecting a predefined safety-critical anomaly and informing the operator what action was taken			
Requires that the software be capable of discriminating between valid and invalid external interrupts and shall recover to a safe state in the event of an erroneous external interrupt			
Provides for detection of improper sequence requests by the operator			
Provides for detection of inadvertent transfer of safety-critical routines			
Provides for detection and termination of commands requesting actions beyond the performance capability of the system			
Reasonableness Checks			
Provides for software system reasonableness checks of all safety-critical inputs			
Provides for performing parity on other checks, requiring two decisions, before providing an output			
Initialization, Timing, Sequencing, and Status Checking			
Provides for a status check of critical system elements prior to executing a potentially hazardous sequence			
Provides the proper configuration of inhibits, interlocks, safing logic, and exception limits at initialization			
Provides for issuance of good guidance signal subsequent to satisfaction of performance of flight safety checks			
Provides for timing sufficiency of commands relative to response to detected unsafe conditions			
Provides for software initialization to a known safe state			
Provides for performing a status check of safety-critical elements prior to executing a potentially hazardous sequence			

Figure 9.8   *Continued.*

Item	Req'd	Not Req'd	Incor-porated
Initialization, Timing, Sequencing, and Status Checking			
Provides that all critical timing relative to hazardous operations processing is automated			
Provides for employing time limits for operations impacting system safety and having these time limits included in decision logic			
Protects against out-of-sequence transmission of safety-critical function messages by detecting any deviation from the normal sequence of transmission. When this condition is detected, the software terminates all transmissions, recycles to a known safe state, and displays the existing status so the operator can take compensatory action			
Provides for initializing all unused memory locations to a pattern, that if executed as an instruction, will cause the system to revert to a known safe state			
Provides for matching of observed flight terrain being matched to computer-stored flight terrain map prior to issuance of critical commands (e.g., arming, fire, climb, descend, etc.)			
Applies use of software timing coincident with hardware timing to prevent initiation of safety-critical functions			
Provides for verification and validation of the prompt for the initialization of a hazardous operation or sequence of hazardous operations			
Provides for verification of accomplishment of each step of a hazardous operation, or sequence of hazardous operations, by setting of a dedicated status flag prior to proceeding to and initiating the next step in the operation or series of operations			
Operator Responses and Limitations			
Requires an operator response for initiation of any potentially hazardous sequence			
Provides for not allowing the safety-critical time limits in decision logic to be changed by the console operator			
Provides for concise definition of operator interactions with the software			
Provides for operator cancellation of current processing in a safe manner			
Requires that an operator cancellation of current processing be verified by an additional operator response			

Figure 9.8 *Continued.*

Item	Req'd	Not Req'd	Incorporated
Operator Responses and Limitations			
Provides that controlling or monitoring of nuclear weapons be incapable of bypassing operator control of safety-critical functions			
Requires that the system responds to predefined safety-critical anomalous conditions by notifying the operator of the condition and identifying the action taken			
Provides that upon safing the system, the resulting system configuration or status be provided to the operator and await definition of subsequent software activity			
Provides that remote transfer of data cannot be accomplished until verification of data to be transferred is accomplished and authorization to transfer the data has been provided by the operator(s)			
Provides that operator control of safety-critical functions is maintained under all circumstances throughout weapon system operation			
Provides that all manual actions that set up safety-critical signals are verified by software based on control device positions			
Operator Notification			
Requires that an override of a safety interlock be identified to the test conductor by a display on the test conductor's panel			
Provides for generation of critical status to operator			
Provides to operator identification of overrides to safety interlocks			
Provides for software indication if unauthorized action has taken place			
Provides for the system informing the operator of the anomaly detected			
Provides system configuration status to operator upon safing of safety-critical hardware items			
Provides for positive reporting of changes of safety-critical states, e.g., absence of an armed indication does not constitute a safe condition			
Provides for detecting a predefined safety-critical anomaly and informing the operator what action was taken			
Provides for the software system to display safety-critical timing data to the operator			

Figure 9.8 *Continued.*

Item	Req'd	Not Req'd	Incor-porated
Operator Notification			
Provides for the software systems to indicate to the operator the currently active operation(s) and function(s)			
Provides for identification to the operator that a safing function execution has occurred; provides the reason for the execution with a description of the safing action taken			
Provides for notification of improper keyboard entries by the operator			
Prohibits transmission of any critical command found to be in error and notifies the operator of the error			
Provides that upon safing the system, the resulting system configuration or status be provided to the operator and await definition of subsequent software activity			

Figure 9.8 *Continued.*

9.3 SOFTWARE REQUIREMENTS DURING CODING AND INTEGRATION

During the coding and integration stage, plans are made to prevent coding errors. One of the best ways to reduce errors is to standardize the programming language, but there has been only limited acceptance of a standard language. The Pentagon standardized on the Ada language for all weapon systems so that reliable software components can be put together, just as in hardware systems, but it has limited the ability of military systems to work with commercial software.

The challenge in design assurance at code level is to integrate the following scenarios.

Input is true; software recognizes it as true. (No problem)

Input is true; software recognizes it as false. (Can be a big problem, need fallback mode)

Input is false; software recognizes it as false. (No problem)

Input is false; software recognizes it as true. (Can be a big problem, need fallback mode)

9.3.1 Coding Errors

Many types of errors are rampant in codes. A programmer can remove many errors by searching for them by type. Major error categories are the following:

1. *Computational Errors*: Result from coded equations, algorithms, and models.
2. *Configuration Errors*: Code is incompatible with the operating system or the application software. This generally occurs after a modification of the software.

3. *Data Handling Errors*: Made in reading, writing, moving, storing, and modifying data.

4. *Documentation Errors*: Mistakes in software descriptions, user instructions, and other documents.

5. *Input/Output Errors*: Result from input and output code.

6. *Interface Errors*: Occur in (1) routine-to-routine interface, (2) routine-to-system software interface, (3) file processing, (4) user interface, and (5) database interface.

7. *Logic Errors*: Design errors.

8. *Operating System Errors*: Occur in operating system software, the compiler, the assembler, and the special utility software.

9. *Operator Errors*: Human-induced errors.

10. *Database Errors*: Residual errors in the database, usually from the user.

11. *Recurrent Errors*: Duplicates of errors, usually arising when problems are reopened.

12. *Requirements Compliance Errors*: Software fails to provide capability specified in the requirements specification.

13. *Definition Errors*: Improper definition of variables or constants.

14. *Intermittent Errors*: Caused by temporary faults in the hardware or software.

9.3.2 Quantifying Software Errors

Demand for higher software reliability has created the need for measurement models. For hardware, data banks can offer some insight into reliability prediction, but software reliability is heavily process- and human-dependent. Little has been published on predicting software reliability from historical data. However, at least 67 mathematical models for software reliability have been published. Almost all are variations of two curves (Fig. 9.9) [14].

The experience of Sagols and Albin [15] suggests that the first model applies to simpler software. The second model is for complex software where the learning process is longer. Some software data do not fit either of these models. In such cases, the most common approach is to use the data on number of errors at time *t* to predict the number of errors remaining. Some compute mean time between failures based on the exponential model. This is a useless measure because most users do not understand that if MTBF is 200 hours, there is a 67% chance of failure. Nobody in his/her right mind wants to have such a high probability of failure. Reference 2 classifies some models as given below. Readers interested in more details on models should consult Reference 15.

Reliability Measurement Models

- Hecht measurement model
- Nelson model

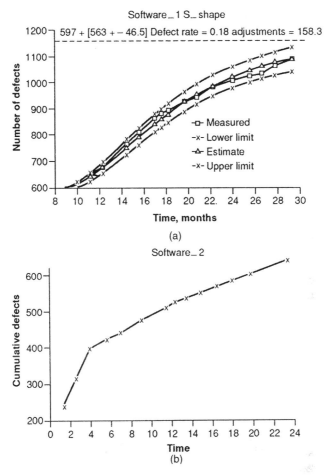

Figure 9.9 Two basic models of software reliability: (a) S-shaped curve and (b) standard exponential growth with negative coefficient.

Reliability Estimation Models

- Reliability growth model (Application of Duane model, p. 68)
- Mills model
- Rudner model
- Musa model
- Jelinski–Moranda deeutrophication model
- Jelinski–Moranda geometric deeutrophication model
- Jelinski–Moranda geometric Poisson model
- Schick–Wolverton model
- Shooman exponential model

- Weibull (Wagoner) model
- Goel–Okumoto bayesian model
- Littlewood–Verrali bayesian model
- Shooman–Natrajan model
- Shooman–Trivedy Markov model
- Shooman micromodel
- Littlewood Markov model
- Littlewood semi-Markov model
- Moranda a priori model
- Hecht estimation model

Reliability Prediction Models

- Motley and Brooks model
- McCall, Richards, and Walters model
- Halstead model

These models mostly predict the fault density. There is a way to assess reliability. Start with a simple model:

$$\text{System reliability} = (\text{hardware reliability}) \times (\text{software reliability})$$

One can use a Weibull plot for software and hardware failures together to assess system reliability. Then plot a Weibull curve only for the hardware failures to assess hardware reliability. Since the only unknown in the equation is the software reliability, its value can be computed from the equation.

If we want to expand this model, we can include interface reliability into the equation. If the software is interactive, we can include human reliability.

9.3.3 Coding Error Prevention

There are many approaches to reducing coding errors: high-level languages; redundant algorithms; microcoding for complex algorithms; multiuser support for editing, compiling, and unit testing; compiler conformance to American National Standards Institute (ANSI) document X3.9-1978; user support in establishing detailed requirements, testing, and acceptance; error detection codes; independent verification and validation; system integration testing; independent testing; fault injection testing; use of a software support library; configuration control; thorough software development plans; rigidly controlled requirement specifications; rigidly controlled software functional specifications; requirements traceability matrixes; rigidly controlled detailed design specifications; structured analysis tools; program specification language; program design language; automated test tools; computer-aided software engineering; peer walk-throughs; progress reviews; quality audits;

software problem reports; specification change notices; software requirements reviews; test readiness reviews; extensive interface reviews; and human interface requirements.

Fortunately, there are several software development tools available that can detect features of a software language. (An example is a feature Utility in the UNIX operating system, which detects bugs in C language.) The tools perform much more thorough checks than a compiler can. Some detect bugs in the program files. They identify unreachable statements, loops not entered at the top, variables declared and not used, and inconsistencies in function calls. With the current trend to portability of software to new platforms, such tools have started playing a very important role in high-quality software development.

A good rule of thumb for avoiding many coding errors is: "Those who design software should not write the code; those who write the code should not do the testing." Another good rule often practiced at IBM is to develop test with the customer for worst-case scenarios. Usually this points to the missing requirements in the specification.

Example 9.1   Some coding errors can be prevented by using a simple integration certification form (Fig. 9.10). This makes sure the integration *is* done properly. The important coding errors should be caught in the process.

Example 9.2   To assure that codes will not cause any major mishaps, the review team should identify the undesirable paths that should not occur and the important paths that must occur. These paths are entered in a matrix along with test cases to test them. Such a matrix is shown in Fig. 9.11.

Processor(s): _____ DP _____

Unit name: _____ GP-PART-UPDATE _____   Release/version: __1.0___

Test certified by: _____   Date: _____

Testing: Yes   No

1.   Are all necessary elements present?   ___   ___

    a.   Software design notes
    b.   Test cases/scripts
    c.   Test drivers?

2.   Do the test cases adequately test the program   ___   ___
    logic or intergration dataflow?

3.   Is the data used sufficiently close to real-   ___   ___
    world data?

Comments:

Figure 9.10   Example of a certification form for test integration.

Unit Name **GP-PART-UPDATE** Task **DP** Date _____

1.0 — Test Certifications →

Path	Path 10	Path Condition	1	2	3	4	5	6	7	8	9	10	11	12	13	14	15	16
									Test Case Coverage								
1	DF-T	Do for "NUM-XFER" entries	X	X														
2	IF-T	Found HDR entry in XFER table	X	X														
3	IF-T	Found FILE-ID, FILE-TYPE match	X	X														
4	IF-F	Match not found																X
5	IF-T	Found TRLR entry in XFER table	X	X														
6	IF-T	Found FILE-ID, FILE-TYPE match	X	X														
7	IF-F	Match not found																X
8	IF-F	No HDR or TRLR entry																X
9	DF-F	All entries complete	X	X														
10	IF-T	Error from GP-DIR-UPD																X
11	IF-F	No error from GP-DIR-UPD	X	X														
12	IF-T	FILE-TYPE "UPD"	X	X														
13	IF-F	Invalid FILE-TYPE																X
14	IF-T	Error in FMBPEN																X
15	IF-F	No error in FMPPEN	X	X														
16	IF-T	Error in FMP position																X
17	IF-F	No error in FMP position	X	X														
18	DF-T	Do for "NOM-XFER" entries	X	X														
19	IF-T	Found data entry in XFER-TAB	X	X														
20	IF-T	Found FILE-ID, FILE-TYPE match	X	X														
21	DF-T	Do for all words in record	X	X														
22	DF-F	All words complete	X	X														
23	IF-T	Error from FMPWRITE																X
24	IF-F	No error from FMPWRITE	X	X														
25	IF-T	Error from FMPCLOSE																X
26	IF-F	No error from FMPCLOSE	X	X														
27	IF-F	No match of FILE-ID, FILE-TYPE																X
28	IF-F	No data entry in XFER-TABLE																X
29	DF-F	All num-XFER entries complete	X	X														
30	IF-T	Error from GP-DIR-UPD																X
31	IF-F	No error from GP-DIR-UPD	X	X														
32	IF-T	TABLE-ID exists	X															
33	IF-F	No TABLE-ID		X														
34	DF-T	Do for delete "NUM-XFER" entries	X	X														
41	DF-F	All entries deleted	X	X														

Figure 9.11 Example of a test matrix for desired and undesired paths.

9.4 SOFTWARE TESTING

Tests are conducted for quality, reliability, safety, and maintainability. Tests may also be done for overall qualification. In complex software, there should be a separate test for internal and external interfaces.

9.4.1 Testing for Quality

Tests for quality are aimed at measuring the accuracy and precision with which the software performs. Acceptance is based on the software's ability to perform according to specifications. Some common quality control tests are defined below.

1. *Black Box Testing*: This testing strategy derives test data solely from the requirements specification.

2. *White Box Testing*: This is a complement of black box testing—a strategy that derives test data from knowledge of the program's internal structure.

3. *Gray Box Testing*: This strategy derives test data by combining elements of black box and white box testing.

4. *Unit Testing*: Individual testing of each module by its implementing programmer to verify its correctness.

5. *Validation Testing*: A formal test and evaluation of the software to demonstrate its conformance to the requirements specification.

6. *Boundary Value Testing*: This testing strategy derives test data from boundary conditions, that is, situations directly on, just above, and just below the boundaries of input and output equivalence classes.

9.4.2 Testing for Reliability

Since reliability tests are of very long duration, mathematical models are often used as part of either software field assessment or acceptance testing. In many military contracts, Mil-Std-781 [16], which is based on an exponential probability distribution, is also used for acceptance testing for hardware as well as the software. In some contracts, if the cause of error is not isolated in hardware or software, it counts as two errors.

Testing can be accomplished in many ways. Individual modules may be tested with stubs as the software develops and integrated later. At the system level, testing may be done bottom–up or top–down or may start in the middle (sandwich testing). Myers [17] suggests several such tests for quality and reliability.

9.4.3 Testing for Maintainability

Most maintainability parameters are measurable, such as modularity, testability, flexibility, and transportability. The tests have to be developed from the customer requirements, since very little is published in this area.

9.4.4 Testing for Software Safety

Special safety tests are done to verify the performance of safety-related test descriptions, procedures, and cases. They include verification of fallback modes for inhibits, interlocks, and traps. Tests are conducted under specified and abnormal conditions. The tests are documented with special care because they can be summoned in the courts in case of injuries or deaths.

Special consideration should be given to testing of interfaces: the hardware–software interface, software–software interface, and software–human interface. The inputs to safety tests come from various hazard analysis techniques covered in this chapter. Make sure to test rare events thoroughly such as for the air bag in automobiles. The code may be exercised after 17 years of use. In that case, software should monitor hardware performance throughout the life.

Safety tests should go beyond normal testing to allow for safely margins. The following should be developed [18]:

- Tests to verify that the software responds correctly and safely to credible single and multiple failures (and faults)
- Tests for operator input errors and component failures
- Tests to assure that errors in data received from external sensors or other software processes will be handled safely
- Tests for failures associated with entry into, and execution of, safety-critical software components
- Tests for negative and no-go testing
- Tests to assure that the software performs only those functions for which it was intended, and no extraneous functions
- Test to verify fallback modes in case of overloads

9.4.5   Testing for Overall Qualification

The most important test is the fault injection (also called fault seeding) test. In this test one also physically disables components and interfaces to assure a desirable fallback mode. Qualification testing is generally done in stages. It starts with unit testing and goes all the way to operational testing. Howley [19] suggests the following tests. Not all tests are required for every project; the selection will depend on the test objectives (e.g., validating logic) and the design or operational performance capabilities.

1. *Checkout*: Involves compiling or assembling and executing the program, ascertaining performance characteristics, and performing error detection.
2. *Unit/Module Testing*: Includes verifying design requirements and performing error detection.
3. *Integration Testing*: Verifies the compatibility of composite program elements.
4. *Computer Program Testing*: Consists of verifying the functional characteristics of the program, demonstrating the validity of processing results, and establishing the validity of specifications. This level of testing includes verification and formal validation. Acceptance and quality assurance testing may be at this level or the system test level.
5. *System Testing*: Involves verifying the compatibility and joint performance of the software and hardware components of the total system. This level includes acceptance and quality assurance testing and system performance demonstrations.
6. *Operational Testing*: Involves providing realistic data to the computing system in the operational environment, under full operational

demands. These tests are referred to as *operational soaks* or *operational demonstrations.*

9.5 SOFTWARE PERFORMANCE IN THE USE STAGE

Since software performance characteristics are still vague, a good percentage of performance is improved in the use stage. In fact, many bugs are introduced as a result of poor documentation, poorly structured software, and vague specification. At least five areas are critical in the use stage:

- *Failure Logs*: To use lessons learned on future projects.
- *Failure Management*: To assure there is a closed-loop system for removing the bugs permanently.
- *Configuration Control*: Formal procedures to control changes made in software. The procedures assure that the basic structure of the software design is not changed and that all changes are accounted for.
- *Customer Support*: To assure quality of services.
- *Software Maintenance*

The last item is a new and complex area. The broadest definition is offered by Morton [20]: "Maintenance is any change to an existing program." This view is helpful because the source of the problem is change, not the type of change. Morton's definition of maintenance includes:

- Major enhancements
- Bug fixes
- Local adaptation of vendor-supplied systems
- Installation of vendor-supplied systems
- Adaptation to environmental changes such as new hardware

For the purpose of configuration control, Morton also suggests the following guidelines:

- Any change in the functional capabilities or performance specifications is a requirements change.
- Any other user-visible change (other than fixing a bug) is a change in the functional specifications.
- Any change in the allocation of functions to software components or any change in interfaces between components is an architectural change.
- Any change in the design of a module, which is limited to only one module, is a module design change.
- Any change to the code, which does not also change the module design, is a coding change.

- Any bug after testing requires two fixes, one in the system (at whatever levels may be appropriate) and one in the test.

Reference 21 contains additional information on software maintenance.

9.6 TOPICS FOR STUDENT PROJECTS AND THESES

1. Differentiate between software quality and software reliability. Summarize various approaches. Suggest your own version if the current definitions are unsatisfactory.
2. Summarize the techniques for achieving software maintainability goals. If the present techniques are inadequate, suggest an alternative technique.
3. Write a short report on design approaches for fault tolerance for human errors.
4. Summarize reliability measurement models. Recommend a model (or models) and justify your selection.
5. Develop a model for predicting software reliability which can be applied before the software is coded.
6. Are the current software safety approaches adequate? If they are adequate, justify. Propose your own approach.
7. Explain your views on approaches to software logistics analysis.
8. Can FMEA be applied to software alone? If so, perform this analysis on a portion of software.
9. Can fault-tree analysis be applied to software alone? Summarize your views with an example.
10. What kind of structured programming is best suited for software maintenance? Present your views.
11. Select a complex automated system and conduct a system hazard analysis and provide supportive software hazard analysis.
12. Chose a high-level software programming language (e.g., Ada, C, C++), and discuss pros and cons from a system safety view.

REFERENCES

1. U.S. Army Solicitation No. DAH CO6-85-R-0018.
2. E. C. Soistman and K. B. Ragsdale, Impact of Hardware/Software Faults on System Reliability, Report No. OR 18,173, Griffiths Air Force Base, Rome Air Development Center, Rome, NY, 1985.
3. *Defense System Software Development*, DoD-Std-2167A, 1986.

4. *System Safety Program Requirements*, Mil-Std-882B, Naval Publications and Forms Center, Philadelphia. Mil-Std-882E is in the process.

5. General Electric Co., *Software Engineering Handbook*, McGraw-Hill, New York, 1986.

6. F. Brooks, No Silver Bullet: Essence and Accidents of Software Engineering, *Computer*, April 1987.

7. *A Standard Classification for Software Errors, Faults and Failures*, IEEE Draft Standard No. P1044/D3, IEEE, New York, December 1987.

8. D. Raheja, Software FMEA: A Missing Link in Design for Robustness, SAE World Congress Conference, 2005.

9. R. C. Clardy, Sneak Analysis: An Integrated Approach. In: *International System Safety Conference Proceedings*, 1977.

10. W.L. Chenoweth, Oral Presentation, In: *IEEE Conference on Computer Assurance*, IEEE Computer Software Assurance Conference, COMPSAC 86, Washington, DC, 1986.

11. Software System Safety Handbook, AFISC SSH 1-1, U.S. Air Force Inspection and Safety Center, Norton Air Force Base, Calif., 1985.

12. N. Leveson, Safety Analysis Using Petri Nets, *IEEE Transactions on Software Engineering*, 1986.

13. S. Mattern, Confessions of a Modern-Day Software Safety Analysis. In: *Proceedings of the Ninth International System Safety Conference*, Long Beach, California, 1989. Group Meeting, June 1989.

14. A. L. Goel and K. Okumoto, A Time Dependent Error Rate Model for Software Reliability and Other Performance Measures, *IEEE Transactions on Reliability*, no. R-28, pp. 206–211, 1979.

15. G. Sagols and J. L. Albin, Reliability Models and Software Development, A Practical Approach. In: *Software Engineering: Practice and Experience*, E. Girard (Ed.), Oxford Publishing, Oxford, UK, 1984.

16. *Reliability Qualification and Production Approval Tests*, Mil-Std-781, Naval Publications and Forms Center, Philadelphia.

17. G. Myers, *Software Reliability: Principles and Practice*, John Wiley & Sons, Hoboken, NJ, 1976.

18. Mil-Std-882C.

19. P. P. Howley, Jr., A Comprehensive Software Testing Methodology. In: *IEES Workshop on Software Engineering Standards*, San Francisco, 1983.

20. R. P. Morton, The Application of Software Development Standards to Software Maintenance Tasks. In: *IEEE Workshop on Software Engineering Standards*, 1983.

21. D. H. Longstreet, *Software Maintenance and Computers*, IEEE Computer Society Press, New York, 1990.

CHAPTER 10

SYSTEM EFFECTIVENESS

10.1 INTRODUCTION

There is a saying: "Behind every technical disaster *is* a management shortcoming." Accident investigators who use MORT (management oversight and risk tree) often conclude this observation. MORT training originated at the Department of Energy (DOE). For a textbook see the Further Reading section at the end of this chapter.

Quality experts say management is responsible for 85% of quality problems. Reliability experts say the cause of unreliability is that there is never enough budget and time for thoroughly analyzing designs. The fact that maintenance costs usually are high suggests that management should get strongly involved in integrating such issues. Product costs can be cut drastically if waste in manufacturing and excess field failure costs are avoided. Avoidance of these costs also results in lower downtime costs to the customer. With integration of assurance technologies, both the supplier and the customer benefit. That is the secret of market leadership.

Typical situations mentioned below highlight the dollars lost by the companies. If these dollars were saved, companies might not need loans from banks; more capital would be generated internally than a bank could offer. Companies spend tremendous amounts of money for burn-in and screening tests instead of solving the problems. This is just one example.

Another example is paying for warranty fixes. One midsize company spends about $60 million per year to pay for poorly designed products. Yet another example is the safety recalls. The Ford sports utility vehicle recall cost $9 *billion*.

Assurance Technologies Principles and Practices: A Product, Process, and System Safety Perspective, Second Edition, by Dev G. Raheja and Michael Allocco
Copyright © 2006 John Wiley & Sons, Inc.

Everyone recognizes that it is cheaper to solve problems than to test and inspect products forever.

A lack of teamwork and poor integration prevents this from happening. Avoiding unnecessary tests can save millions of dollars. Moreover, good tests are often overlooked because management is already burdened with test costs. Waste because of inadequate manufacturing processes falls into the same category. Such unnecessary costs in some companies amount to 10–30% of sales, according to George Fisher, ex-chairman of Motorola.

Many companies waste enormous amounts of money because of ignorance in integration among the technical departments. In one company, design engineers conducted all kinds of tests to demonstrate high reliability and durability in the products, but the manufacturing department rarely evaluated with such thoroughness. In another situation, both the supplier and the customer knew that expensive and useless tests were being performed on a system but no action was taken for years. The result was that many engineering dollars were put into good designs, which the manufacturing department could never deliver. The problems are often known but no one seems to be accountable or has the time to integrate.

Assurance technologies are not independent of each other. Information on reliability is used for maintainability, quality, and logistics analyses. Test programs are developed from various analyses. Almost every product requires trade-offs between the following:

- Reliability and maintainability
- Functional performance and reliability
- Number of functions and reliability (e.g., in a microprocessor)
- Maintainability and maintenance
- Hardware and software
- Component tolerances and reliability
- Human factors and design cost
- Operability and reliability
- Operational readiness and inventory cost

It is management's responsibility to provide the environment that encourages free discussion of these trade-offs. This will require team effort in various analyses and the budget. But there is a potential payoff of $40–100 for every extra dollar in the budget.

10.2 SYSTEM EFFECTIVENESS PRINCIPLES

For a system to perform at expected levels, the reliability, *maintainability*, safety, and quality functions need to work in harmony. If the equipment breaks down, it must be repairable quickly so that the system is available to the user. From a customer's perspective, a system's availability is more important. For instance, a commuter used to taking a train at 6 a.m. and again at 6 p.m. expects the train to arrive at those hours. If the train breaks down every night but is repaired before 6 a.m., these failures are of little concern to the commuter. Availability can be improved either by not allowing the train to fail or by designing it in such a way that it can be repaired very fast. At the user level, availability replaces reliability. It is defined as the percent time the system is in working condition when the user needs it. Power companies usually measure their performance in terms of availability. One utility's rate increases are tied to availability; customers pay less if they experience higher than usual loss of power.

Availability can also be defined as a measure of the degree to which an item is in an operable and committable state at the start of the mission, when the mission is called for at an unknown (random) point in time [1]. The technical definition of availability in terms of mean time between failures (MTBF) and mean time to repair (MTTR) is covered in Chapter 4.

Availability is actually only part of the customer's needs. As the system grows in complexity, assurance technologies such as reliability, maintainability, and quality assurance begin to look like the pieces of a puzzle called *system effectiveness*. Engineers seem to apply these technologies to hardware system effectiveness alone or to software system effectiveness alone, but the truth is that performance also depends on the integration of software and hardware effectiveness along with that of the humans who operate the system. This broad picture is often overlooked. Someone who has the responsibility for the total picture (that means top management) should also be accountable for overall effectiveness. Table 10.1 [2] gives a typical breakdown of hardware, software, and other causes of outages in software-controlled systems. The *Other* category includes several causes including humans.

TABLE 10.1 System Outages in Minutes by Causes

Month	Hardware	Software	Other	Total
1	43	102	195	340
2	28	121	0	149
3	133	39	21	193
4	31	81	119	231
5	261	128	411	800
6	76	109	916	1101
7	284	44	7	335
8	0	209	0	209
9	17	181	46	244
10	13	158	34	205
11	3	107	36	146
12	154	100	48	302
Mean	86.9	114.9	152.8	354.6
Percentage	24.5	32.4	43.1	100.00

The numbers indicate that software causes more outages than hardware causes. The largest percentage of failures (43.1%) is unaccounted for. In other words, no one is responsible for almost half the system problems. This makes accountability for system availability more important than stand-alone technology.

The first element to be considered in system effectiveness is availability. System availability can be modeled in terms of component availabilities A:

$$A_{\text{system}} = A_{\text{hardware}} \times A_{\text{software}} \times A_{\text{human}} \times A_{\text{interface}}$$

Here, A_{human} does not merely mean that a human is available to do the work; it means that a trained and alert human who knows what to do during emergencies is available. $A_{\text{interface}}$ refers to the proportion of time the interfaces between the hardware, software, and human are in usable condition.

The next item to be integrated is dependability. The system is not acceptable if it is not dependable, and management therefore cannot trust it. It also has to be capable of performing missions correctly throughout the intended period. When these principles are integrated, the result is system effectiveness. Reference 3 contains models and their sources. The information below is derived from this reference.

System effectiveness may be defined [4] as "the probability that the system can successfully meet an operational demand within a given time when operated under specified conditions." For a single-shot device, the system effectiveness can be "the probability that the system (such as a parachute) will operate successfully when called upon to do so under specified conditions." The Mil-Std-721[1] definition is "a measure of the degree to which an item can be expected to achieve a set of specific mission requirements and which may be expressed as a function of availability, maintainability, dependability, and capability." The WSEIAC model (Weapon System Effectiveness Industry Advisory Committee) developed by the U.S. Air Force Systems Command seems to be accepted by most practitioners. The U.S. Navy has also developed a model based on the WSEIAC model, as has the U.S. Army. These models are presented in Fig. 10.1. For more details, see References 5, 6, and 7.

Some features shown in these models were developed for special applications. Others are unique to the organization. However, two commonly used terms can be defined:

Capability: The probability that the system's designed performance level will allow it to meet mission demands successfully provided that the system is available and dependable (Mil-Std-721). Capability includes elements such as consistency of power supply, fidelity, and precision.

Dependability: A measure of the item's operating condition at one or more operating points during the mission, including the effects of reliability, maintainability, and survivability, given the item's condition at the start of the mission (Mil-Std-721). Dependability includes safety and ease of maintenance.

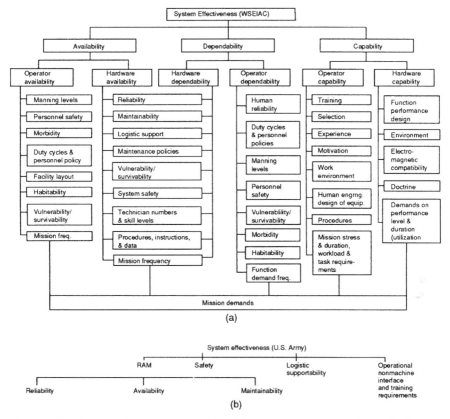

Figure 10.1 System effectiveness models: (a) U.S. Air Force system effectiveness (WSEIAC) model and (b) U.S. Navy reliability, availability, and maintenance (RAM) and system effectiveness model.

10.3 IMPLEMENTING THE PROGRAMS

It is obvious that, if there is no plan to carry out certain activities, there will be no budget for them. Without a budget the tasks will not be done or will not be done on time. This affects engineering morale and productivity. Many so-called total quality management programs have failed because there were no real resources behind them.

Table 10.2 presents programs for reliability (Mil-Std-785) [8] and Table 10.3 presents programs for maintainability (Mil-Std-470) [9]. Chapter 5 includes programs for system safety. Similar programs can be constructed for quality assurance from DoD directive 5000.51-G [2] or from Mil-Q-9868 [10], and for logistic engineering from Mil-Std-1388-2A [11]. These programs are not exhaustive; if a certain task is of proper value, it should be added to the program.

The reliability and maintainability programs have some similar tasks. A good program manager will group the similar and related tasks together and have

TABLE 10.2   Reliability Program (Mit-Std-785)

		Program Phase		
Task		Concept	Validation	FSD	Production
101	Reliability Program Plan	X	X		
102	Monitoring Subcontractors		X	X	X
103	Program Reviews		X	X	X
104	Failure Reporting Analysis, and Corrective Action System (FRACAS)		X	X	X
105	Failure Review Board (FRB)		X	X	X
201	Reliability Modeling	X	X		
202	Reliability Allocation	X	X		
203	Reliability Predictions	X	X		
204	Failure Modes, Effects, and Criticality Analysis (FMECA)	X	X		
205	Sneak Circuit Analysis (SCA)		X		
206	Parts Circuits Tolerance Analysis		X		
207	Parts Program		X	X	
208	Reliability Critical Items		X	X	
209	Effects on Storage, Handling, Packaging, Transportation, and Maintenance		X		
301	Environmental Stress Screening		X	X	
302	Reliability Growth Testing		X		
303	Reliability Qualification Test Program		X	X	
304	Production Reliability Acceptance Tests				X

them analyzed together. For example, FMECA is used for reliability as well as maintainability, but for different objectives. About 80% of the analysis will be common.

The results of the FMECA are also necessary for logistics analysis. It will be highly productive if the design engineer, reliability engineer, logistics engineer, and maintainability engineer conduct the analysis as a team. More and more companies are finding that the safety engineer should also be a team member.

A company cannot implement the complete intent of these programs on every product. The programs must be tailored for the project. Tailoring is explained by the following illustration. Consider a large dining table, with 14 legs. The table may be safe and reliable, but it may not be cost-effective. With four legs, however, it may provide the desired level of safely and reliability at a reasonable cost. *Tailoring* means picking out the right tasks with due consideration to cost and schedule constraints. It also means that, if a new task will make the product much safer, it should be added to the program. If the dining table had only three legs, it might be wise to add a fourth leg.

A major aspect of implementing the tasks is to perform them efficiently by using the principles of concurrent engineering. Figure 10.2 gives an example of tasks and their concurrent schedules.

TABLE 10.3 Maintainability Program (Mit-Std-470)

	Program Phase				
Task	Concept	Validation	FSD	Production	Operating System Development
101 Maintainability Program Plan		X	X	X	X
102 Monitor/Control of Subcontractors and Vendors			X	X	
103 Program Reviews	X	X	X	X	
104 Data Collection, Analysis, and Corrective Action System			X	X	
201 Maintainability Modeling	X		X		
202 Maintainability Allocations	X		X		
203 Maintainability Predictions			X		
204 Failure Modes and Effects Analysis (FMEA) Maintainability Information			X		
205 Maintainability Analysis	X	X			
206 Maintainability Design Criteria			X		
207 Preparation of Inputs to Detailed Maintenance Plan and Logistics Support Analysis (LSA)			X		
208 Maintainability Demonstration (MD)			X		

10.4 MANAGING BY LIFE-CYCLE COSTS

A consumer rarely buys an automobile because it has the lowest price. Instead, a prospective buyer subconsciously looks at the costs of repair, maintenance, and fuel consumption. Decisions are usually based on life-cycle costs. Managers face similar challenges in making investments, which may not provide immediate benefits.

Unfortunately, many times they are not convinced there are enough benefits. They overlook many hidden ones. For example, capital equipment may be bought without evaluation of downtime costs. Higher-priced equipment with enormous savings in downtime may be rejected.

The concept of life-cycle costs in making engineering decisions has found wide token acceptance, but very few are willing to take it seriously. Two big reasons are

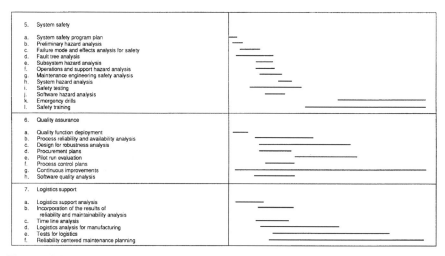

Tasks	Design phase	Manufacturing phase	Test phase	Use phase

1. Management
   a. Program plan
   b. Budgets

2. General
   a. Preliminary design review
   b. Intermediate design reviews
   c. Critical design reviews

3. Reliability
   a. Reliability prediction allocation and modeling
   b. Failure mode and effects analysis
   c. Worst case analysis
   d. Parts tolerance analysis
   e. Sneak circuit analysis
   f. Parts program
   g. Analysis of storage, handling, packaging, and transportation
   h. Reliability growth tests
   i. Reliability qualification tests
   j. Production and screening tests
   k. Process reliability
   l. Software reliability analysis
   m. Human factors analysis for reliability

4. Maintainability
   a. Maintainability analysis
   b. Maintainability modeling and procedures and allocations
   c. Failure mode and effects analysis for maintainability
   d. Testability analysis
   e. Human factors analysis for maintainability
   f. Maintainability demonstration

5. System safety
   a. System safety program plan
   b. Preliminary hazard analysis
   c. Failure mode and effects analysis for safety
   d. Fault tree analysis
   e. Subsystem hazard analysis
   f. Operations and support hazard analysis
   g. Maintenance engineering safety analysis
   h. System hazard analysis
   i. Safety testing
   j. Software hazard analysis
   k. Emergency drills
   l. Safety training

6. Quality assurance
   a. Quality function deployment
   b. Process reliability and availability analysis
   c. Design for robustness analysis
   d. Procurement plans
   e. Pilot run evaluation
   f. Process control plans
   g. Continuous improvements
   h. Software quality analysis

7. Logistics support
   a. Logistics support analysis
   b. Incorporation of the results of reliability and maintainability analysis
   c. Time line analysis
   d. Logistics analysis for manufacturing
   e. Tests for logistics
   f. Reliability centered maintenance planning

Figure 10.2 Master schedule of concurrent engineering tasks. Note that the manufacturing and test activities start during the design phase.

(1) the short-term profits are reduced because the investment does not yield immediate gains—Wall Street does not like that; and (2) the investment is made by the design department but the gains are enjoyed by other departments such as manufacturing, marketing, and maintenance, and by the customer. Smart companies that take advantage of the benefits of the concept capture higher market shares and profits in the long run. Shortsighted companies sooner or later go out of business, unless they have aggressive new product programs.

Chapter 4 gives examples of analyzing life-cycle costs. Another example is presented below to emphasize that such an approach is not only meaningful for selecting engineering options but also for making higher-level decisions. The life-cycle costs model generally contains current costs and future costs, to the supplier as well as to the buyer. The best situation is when the costs are reduced for the supplier as well as for the customer.

Example 10.1   A product contained a plastic gear on a plastic shaft. The product failed because the gear moved on the shaft. This was the result of expansion and contraction caused by changing environments. The analysis showed that the problem could be permanently designed out if the gear and the shaft were molded as one part. Even though the change made logical sense, the solution could not be implemented. The argument was that, since the investment in the existing process was already recovered, why *pay* $2 million for a new process. The life-cycle cost analysis, however, surprised the company. It discovered that the investment of $2 million would produce savings of more than $20 million over the next 10 years in terms of reduced warranty and related costs. This is another example of how life-cycle savings can be shared by the supplier as well as the customer. The customer benefits, of course, would be in terms of reduced failures.

10.5 SYSTEM EFFECTIVENESS MODEL

As pointed out earlier, system effectiveness is a function of availability, dependability, and capability. Therefore, a model for system effectiveness may be written as a combination of probabilities:

$$P(\text{SE}) = P(\text{A}) \times P(\text{D}) \times P(\text{C})$$

where SE, A, D, and C indicate that the probabilities are those of system effectiveness, availability, dependability, and capability.

One more term can be added to the system effectiveness model: operational readiness. This accounts for readiness of operating personnel and materiel, as in the U.S. Navy model in Fig. 10.1. The literature does not define operational readiness clearly. But the authors recommend including the readiness of the operating crew and of support services such as ambulance, fire control, and rescue squads. The picture of assurance technologies now seems complete!

10.6 AUTHORS' RECOMMENDATIONS

A major division of a large corporation, which seemed on the verge of going out of business, was turned around and instead achieved the number one spot as market leader in quality in less than 2 years. This took place in 1974. For years, many members of the corporate elite refused to believe the principles applied to the turnaround—until Japan used them on a massive scale. Some principles may still seem controversial, but the authors are willing to share with the reader the truth, and nothing but the truth. The principles are embodied in the ten commandments of good product assurance management:

1. You shall always assess competitors' weaknesses and strengths to plan where you want to be. A great company always stays ahead of the competition. By the time the competition catches up, you are 5 years ahead.

2. You shall always make sure all the specifications are thoroughly reviewed for reliability, maintainability, safety, quality, and logistics support before detailed design work starts. This must result in many changes in the system specification.

3. You shall minimize life-cycle costs by making many design changes before going into production. This will require more fire prevention but much less firefighting. Prevention is a one-time cost. Firefighting may go on for years.

4. You shall use concurrent engineering principles on the right things such as challenging specifications and analyses in a cross-functional team, to make products faster, better, and cheaper.

5. You shall always review the manufacturing process as thoroughly as the product design before the process is installed.

6. You shall not use control charts in manufacturing until you have demonstrated that the problem cannot be eliminated completely. All inspection, including control charts, should be eliminated as soon as possible.

7. You shall not perform burn-in and screening on 100% of devices unless you have demonstrated that the causes of failures cannot be eliminated.

8. You shall try to build customer trust and increase market share by making a goal of zero mission stopping failures and zero fatalities.

9. Top management should make sure the bad news reaches them as fast as the good news.

10. For innovative companies, the lower the failure rate, the higher the profit. The authors of this book recommend 500% return on every failure and accident prevented. Safety is a business case, not a cost.

10.7   SYSTEM RISK AND EFFECTS ON SYSTEM EFFECTIVENESS

If a system has not been designed appropriately from a system safety perspective, there can be system accidents. Bottom line is: if safety has not been adequately considered within the design other system engineering efforts may be futile. Many entities that make or operate products, processes, or systems cannot withstand many catastrophic events. Both new and existing products, processes, and systems can be enhanced for system safety. In applying concepts of risk management and system safety, the first step involves the identification of system and synergistic risks.

10.7.1 System Accidents

Perrow discussed the concept of system accidents[1] in his book, *Normal Accidents*. Since the early 1970s, safety professionals were aware of a similar concept of a system accident; however, it was referred to as multicausal events. It became a common understanding that accidents involving systems were the result of many hazards. Benner and Hendrick in their book, *Investigating Accidents with STEP* [12, p. 27], defined accidents as a process. Such a process should be equated to failure states rather then success states. Hammer [13, pp. 62–64], also in the 1970s, provided support to the concept of multicausal events. He had equated the potential accident and hazard analysis, by providing the explanation of the concept of initiating, contributory, and primary hazards. It was noted that there are initiators, sequenced contributory hazards, and primary hazards, which were defined as the outcomes.

10.7.2 Complex System Risks

System accidents are never the result of a simple single failure, or a behavior deviation, or a single error. Although simple adverse events still do occur, system accidents are the result of many initiators and contributors, combinations of errors, failures, and malfunctions. It is not easy to foresee such an adverse sequence or to connect the dots while evaluating multicontributors within adverse events, identifying initial events, and identifying subsequent events to the final outcome. System risks can be unique, undetectable, unperceived, unapparent, and very unusual. A novice investigator, analyst, or outside party can question the credibility of such diverse events.

10.7.3 Synergistic Risks

In addressing risk management, there are many other risks that are not apparently directly related to safety, depending on points of view. There are interactions between other risks as a result of adverse effects—synergistic risks that can result in harm. Consider the following examples:

- A security-related risk of a hacker gaining access into a control system of a particular hazardous process and altering safety parameters within the control system
- The loss of key personnel at a critical time during contingency or system recovery

---

[1]Dr. Perrow in 1984 in his book *Normal Accidents* indicated and enhanced the multi-leaner logic discussion with the definition of a *system accident*—system accidents involve the unanticipated interaction of multiple failures.

- Physical damage of a landline that interrupts the transfer of safety-critical information
- The loss of valuable papers and records that contain important safety-critical data
- Inadequate resources to maintain a safety-critical product, process, or system
- The compression of scheduling to save cost adversely affecting a safety-critical product, process, or system
- Interruptive events such as loss of power, transients, power spikes, fires, physical damage, floods, storms, earthquakes
- Inappropriate decisions, omissions, and oversights
- Adverse public relations
- Strike, civil disorder, vandalism, or malicious acts
- Negative political influence
- Inadequate system requirements involving reliability, maintainability, quality, human factors, and logistics
- Poor company morale
- Layoffs
- Poor management and administration, inappropriate or conflicting objectives
- Civil legal action
- Poor communications
- Limited knowledge of system operations and complexities
- Poor system integration
- Poor system documentation and configuration control
- Inaccurate depiction of system, models, visualizations
- Poor planning and procedures

10.7.4   Controlling Risks with Effective System Safety Requirements and Standards

System and synergistic risks may be inherent in a design if they were not initially considered. Designers are generally concerned with meeting a customer's needs; however, in many situations neither the customer nor the designer may be aware of the synergistic or system risks associated with the concept. Experience shows that over 50% requirements are either not defined or not articulated clearly by the customer, or they are vague.

It takes some skill, knowledge, and experience to be able to identify complex system safety needs—such as the mitigation of system or synergistic risks. One must read this entire book to be reminded of most of the requirements. As a result of hazard analyses and risk assessment, hazard controls, recommendations, and precautions are defined. Formally, these mitigations are to be in the form of system requirements forming standards. To accomplish this, the safety analyst slightly lags behind the design process in order to conduct analysis and define mitigations. The analyst works within a concurrent engineering team, which is comprised of product assurance and other system specialists. This is to assure that system-level specifications are integrated and that they are not conflicting.

The design process is iterative and involves general activities [14, pp. 2–5]:

- Designers must have knowledge and understand their customer's needs.
- From a system safety view, a preliminary hazard list, preliminary hazard analysis (PHA), and initial risk assessment are to be conducted to identify risks, preferably by brainstorming with the customer or stakeholder.
- Designers must define the problem to be solved to satisfy the needs.
- Based on known information, a system-level PHA is needed to further identify the synergistic and system hazards. Hazards are to be mapped or integrated into risks, defining potential event sequences, via initial risk assessment.
- Designers should conceptualize the solution through synthesis.
- As an output of hazard analysis, hazard controls are defined, along with recommendations and precautions—the mitigations that will eliminate or control the identified risks. Have several ways to mitigate in order to choose the most intelligent solution.
- Initial analyses, studies, and tests are to be conducted to improve hazard analysis and risk assessment.
- Designers should perform analysis to optimize the proposed solution. Other analyses, studies, and tests are to be conducted to validate and refine the hazard controls, recommendations, and precautions. A determination is made to assure that the identified risks will be eliminated or controlled to an acceptable level.
- Designers should check the resulting design solution to determine if the original needs have been met. Continue hazard analysis and risk assessment iteratively until the design is fully mature and all life-cycle risks have been identified, eliminated, or controlled.
- Designers should conduct hazard tracking and risk resolution efforts, which will verify that all hazard controls, recommendations, and precautions have been incorporated into the design, as well as the life cycle.
- Designers should validate the solution to make sure there are no side effects.

10.7.5 Effective System Safety Requirements and Standards

Once system safety needs have been identified, the objective is to design a system with acceptable risks via the application of hazard controls; these controls are to be transformed into requirements and standards. Within axiomatic design,[2] a functional requirement is the characterization of a design goal [14]. It adequately

---

[2]Axiomatic design is a means to establish a scientific basis for design and to improve design activities by providing the designer with a theoretical foundation based on logical and rational thought processes and tools. Axiomatic design provides an additional context—a different way of looking at design rather than singularly from a functional view.

describes what is to be achieved to satisfy needs. Design parameters or physical design requirements define how the goal is to be achieved. Design goals are also subject to constraints, which provide bounds on the acceptable design solutions. There are also process variables and process requirements to enable the design, such as processing methods. Needs and goals are defined within functional requirements. How the needs are met is defined within design requirements. Process requirements provide the means to accomplish the design.

10.7.5.1 Standards and Federal Codes

A standard is a set of rules, conditions, or requirements that define and specify design, performance, procedures, or operations. For discussion, a safety standard is comprised of a set of system safety requirements, to assure that risks have been identified, eliminated, or controlled to an acceptable level. Thus far, functional, design, and process requirements have been discussed. Administrative controls such as safety operating procedures can be considered process requirements.

Usually the federal standards address passive safety—protection in the event of an accident. Active safety is just as important. For example, some automobiles have gone backward when the driver wanted to go forward. Federal standards cannot regulate such unusual events.

A code can be a specific set of related regulations, such as the Life Safety Code of the National Fire Protection Association. A set of laws that have been defined as regulations are also considered codes, such as The U.S. Code of Federal Regulation. The Code of Federal Regulation is a codification of the general and permanent rules published in the *Federal Register* by the executive departments and agencies of the U.S. Federal Government.

Conformance to safety-related codes and standards is required to meet minimal levels of protection or risk mitigation. However, meeting codes and standards does not totally assure acceptable risk. As discussed, appropriate hazard analysis and risk assessment enables the development of specific safety requirements. Generalized codes and standards may be the result of consensus, bias, and past catastrophic accidents.

Consensus and bias can adversely affect a standard. General requirements may not be adequate for a particular unique system or synergistic risk. Still it may be required that within a particular safety program an inventory of applicable safety-related codes should be maintained along with the appropriate documentation as to how the code has been met.

10.7.5.2 General System Safety Requirements

Prior to initial design, general requirements[3] may be appropriate in order to apply basic principles or axioms. The essence of a safety program should involve applying system safety principles to new systems, existing systems, processes, procedures, tasks, or

---

[3]As part of Mil-Std 882 series, general system safety requirements have been derived. These requirements have become axioms in the application of system safety.

operations. The following general system safety requirements should be considered and defined within the program:

- Eliminate risks early via design, material selection, and/or substitution.
- When hazardous components must be used, select those with the least risk throughout the life cycle of the program.
- Design software-controlled or software-monitored functions to minimize initiation of hazardous events, through software controls such as error source detection, modular design, firewalls, command control response, and failure detection.
- Design to minimize risk created by human error in the operation and support of the program.
- Consider alternate approaches to minimize risk, such as interlocks, fail-safe design, and redundancy.
- Provide design protection from uncontrolled energy sources (e.g., physical protection, shielding, grounding, and bonding).
- Locate equipment, devices, and subsystems so that access during operations, servicing, maintenance, installation, and repair minimizes exposure to associated risks.
- Categorize risks associated with the system and its performance.
- Identify the allocation of safety requirements to specification traceability and assess the effectiveness of this allocation on risk control.
- Apply the system safety precedence: select lower risk designs, items, or approaches. If this is not possible, remove risks via design or provide safety devices or warning devices. Lastly, apply administrative procedures to control the risk to an acceptable level.

10.7.5.3 Derived Requirements Derived safety requirements are developed as an output of safety-related experiences, accident analyses, system safety analyses, studies, assessments, safety reviews, surveys, observations, tests, and inspections. Safety-related risks are either eliminated or controlled via the application of derived requirements. A process for developing and implementing derived requirements should be described within the safety program.

10.7.5.4 Requirements Testing Depending on the system complexity, it may or may not be possible to test all design requirements; for example, consider a large major automated system with extensive software and millions of lines of code. From a system safety point of view, it remains important to verify that safety requirements have been implemented in a design. Requirements for safety devices, structural integrity, built-in testing, detection capabilities, and system reliability, for example, should be tested. Special safety tests may be conducted to verify system operations. Requirements testing is an important aspect associated with hazard tracking and risk resolution.

In some cases, a test safety analysis may be needed when there are safety-related risks linked with a particular test. The output of such an analysis will provide the hazard controls or safe operating procedures for the test itself.

10.7.5.5 Requirements Development There are existing hazards associated with operations and there are also newly identified hazards due to changes within a system. Changes in procedures or systems can introduce additional hazards. As new systems are developed, hazard analysis is to be conducted to identify new risks. As a result of these efforts, both administrative and engineering controls are developed to mitigate risk, thereby providing new safety requirements.

As discussed, additional derived safety requirements are developed as an output of safety-related experiences, accident analyses, system safety analyses, studies, assessments, safety reviews, surveys, observations, tests, and inspections.

10.7.5.6 Requirements Compliance Specific verification techniques as to how compliance to safety-related requirements are met should be defined. Considering compliance with safety requirements, safety verification is needed to assure that system safety is adequately demonstrated and all identified risks that have not been eliminated are controlled. Risk controls (mitigation) must be formally verified as being implemented. Safety verification, for example, is accomplished by the following methods:

Inspection
Analysis
Demonstration
Test
Implicit verification

It should be noted that no single method of verification indicated above provides total assurance. Safety verification is conducted in support of the closed-loop hazard tracking and risk resolution process.

Hazard control analysis or safety validation considers the possibility of insufficient control of the system. Controls are to be evaluated for effectiveness. They are to enhance the design. Keep in mind that system safety efforts are not to cause harm to the system, and that any change to a system must be evaluated from a system and synergistic risk viewpoint.

10.7.5.7 Requirements Revision As systems change, the risks associated with the changes are identified and the safety-related requirements may have to be revised. Further consider the development of a new system; as the system design matures, so will the requirements. For complicated systems, the requirements revision process can be laborious and extensive. Formal reviews and meetings may be required. Any changes to a system, subsystem, process, procedure, task, or activity related to system safety should be evaluated and requirements revised as needed. This effort should also be documented within the safety program.

10.7.5.8 Requirements Traceability   For large complex systems, requirements traceability is important. Safety requirements at first can be written at a high level of abstraction, since initially a system may be at a concept stage of development. The same situation may be true for the development of almost any complex process, procedure, function, activity, or task. Consequently, requirements are written to a specific level of abstraction. As the system matures, so will its associated requirements. Furthermore, to meet the objective of the high-level requirement, due to the nature of the design, it may take many lower-level requirements to actually implement the requirement. The concept of traceability involves providing (or proving) the connectivity between the higher-level and lower-level requirements. This system engineering or management process should also be documented within the program. The functional requirements, design, and process attributes are to be iteratively developed as the design matures. The high-level general requirements must be traced to lower-level specific requirements. There must be evidence of connectivity between high-level and low-level requirements. This means that every engineering change should point to what other portions of software, hardware, and interfaces are affected. This makes verification and validation more effective.

10.7.5.9 Requirements Documentation   Activity associated with system safety requirements should be formally documented. An overall process or procedure should be defined and criteria established for specific types of reports, forms, checklists, requirements models, and testing results, for example. For large entities, formal requirements programs are implemented. For large complex programs, automated programs and models are used to enable traceability and the documentation of requirements. Commercial software tools such as Rational Rose and Perfect Developer are available for this purpose.

10.7.5.10 Requirements Language   It is important that requirements are stated in a way that can be verified, usually by test if appropriate. However, as indicated, there are other ways in which a requirement can be verified by inspection, analysis, demonstration, test, and implicit verification. There are many situations during initial design stages where high-level safety requirements include the need to conduct particular studies, analyses, tests, modeling, and simulation. Eventually, the goal should be for the lower-level requirements to be very specific and testable.

Contractually and from a government context, requirements must be written with specific language conventions. Requirement statements containing the word "shall" are mandatory and contractually binding.[4] Statements that contain the word "should" or "is preferred" or "is permitted" are guidelines that are desirable and recommended, and not mandatory. Statements that contain the word "may" are optional.

10.7.5.11 Redundant Requirements   Requirements must be independent from each other such that each requirement at the lower level is functionally

---

[4]There is an apparent movement away from the conventions especially using the "shall" statement, in the application of plan language and active rather than passive voice.

independent or there will be conflict and a poor design will result. There are many ways to state a particular need or indicate what will be accomplished to meet that need. For complex systems, there may be many lower-level requirements to meet a particular higher-level requirement.

10.7.6 Additional System Effectiveness Models

As discussed within this chapter, there are many ways to assess the effectiveness of systems, and consequently the models depend on different abstractions that are used to decompose systems. Ebeling [15, pp. 148–149] in his 1997 publication discusses a system effectiveness model with three levels. At the top, system effectiveness is indicated. The second level shows operational readiness, mission availability, and design adequacy. Under mission availability, reliability and maintainability are subsets. The three midlevel components are defined to be probabilities that are assumed to be independent. Consequently,

$$\text{System effectiveness} = (\text{operational readiness}) \times (\text{mission availability})$$
$$\times (\text{design adequacy})$$

10.7.7 Other Indicators of System Effectiveness or Success

There are many methods that measure system success both quantitatively and qualitatively. Apparently there is a need to develop a systematic, integrated, and sound way of evaluating systems. Although econometrics and system costs analysis may be appropriate ways of determining system success, consider the integration of other factors, such as:

Liability risk
Professional liability
Criminal liability
System robustness
System extensiveness
System trends
System risk tolerance
Return on investment
System complexity
System coherence
System stability
System trust
Software complexity
System adaptability
System growth

10.8   TOPICS FOR STUDENT PROJECTS AND THESES

1. Develop a model for system effectiveness other than the one described in this chapter. Include operational readiness.
2. Suggest your approach for an operational readiness model.
3. Suggest your approach to integration of assurance technologies. Consider cost, schedule, and performance in your approach.
4. Search for the best life-cycle model. Suggest your approach.
5. Discuss the role of management in system effectiveness.
6. Select a complex system and conduct a system hazard analysis. Identify synergistic and system risks. Develop safety requirements to eliminate or control risk.
7. Develop an integrated method to evaluate system success.
8. Develop an integrated model to evaluate system effectiveness.
9. Describe how system safety can enhance system effectiveness.
10. Select a complex system and develop a system engineering program and detail how concurrent engineering can be accomplished with the various disciplines integrating.

REFERENCES

1. *Definitions of Effectiveness Terms for Reliability, Maintainability, Human Factors, and Safety*, Mil-Std-721.
2. DoD Directive 5000.51-G.
3. F. A. Tillman, C. L. Hwang, and Way Kuo, System Effectiveness Models: An Annotated Bibliography, *IEEE Transactions on Reliability*, no. R-29, October 1980.
4. ARINC Research Corp., *Reliability Engineering*, Prentice-Hall, Englewood Cliffs, NJ, 1964.
5. Weapon Systems Effectiveness Industry Advisory Committee, AFSC-TR-65, vols. I, II, and III, U.S. Air Force Systems Command, Andrews Air Force Base, MD, 1966.
6. *Naval System Effectiveness Manual*, Technical Document 251 (NBLC/TD261), Naval Electronics Laboratory Center, San Diego, 1973.
7. *Product Assurance—Army Materiel Reliability, Availability, and Maintainability*, AR 702-3, Department of the Army, Washington DC, 1973, revised 1976.
8. *Reliability Program for Systems and Equipment Development and Production*, Mil-Std-785.
9. *Maintainability Program Requirements*, Mil-Std-470.
10. *Quality Program Requirements*, Mil-Q-9858.
11. *Logistic Support Analysis*, Mil-Std-1388; and *Logistic Support Analytic Records*, Mil-Std-1388-2A.
12. L. Benner and K. Hendrich, *Investigating Accidents with STEP*, Marcel Dekker, New York, 1987.

13. W. Hammer, *Handbook of System and Product Safety*, Prentice-Hall, Englewood Cliffs, NJ, 1972.

14. Suh Pyo Nam, *Axiomatic Design: Advances and Applications*, Oxford University Press, New York, 2001.

15. E. C. Ebeling, *An Introduction to Reliability and Maintainability Engineering*, McGraw-Hill, New York, 1997.

FURTHER READING

Anenault, J. L., and J. A. Roberts (Eds.), *Reliability and Maintainability of Electronic Systems*, Computer Science Press, Potomac, MD, 1980.

Booth, G. M., *The Design of Complex Information Systems*, McGraw-Hill, New York, 1983.

Chapanis, A., *Human Factors in System Engineering*, John Wiley & Sons, Hoboken, NJ, 1996.

Electronic Reliability Design Handbook, Mil-Hdbk-338.

Farry, K. A., and I. F. Hooks, Customer-Centered Products: Creating Successful Products Through Smart Requirements Management, AMACOM, American Management Association, 2001.

Hammer, W., *Handbook of System and Product Safety*, Prentice-Hall, Englewood Cliffs, NJ, 1972.

Johnson, W., *MORT Safety Assurance Systems*, Marcel Dekker, New York, 1980.

Proceedings of the Reliability and Maintainability Conference, annual publication, IEEE, New York.

Sheridan, T. B., *Humans and Automation: System Design and Research Issues*, John Wiley & Sons, Hoboken, NJ, 2002.

Suh Pyo Nam, *Axiomatic Design: Advances and Applications*, Oxford University Press, New York, 2001.

Yourdon, E. *Modern Structured Analysis*, P.T.R. Prentice-Hall, Englewood Cliffs, NJ, 1998.

CHAPTER 11

MANAGING SAFETY-RELATED RISKS

11.1 ESTABLISH THE APPROPRIATE SAFETY PROGRAM TO MANAGE RISK

In establishing a safety program, there are many facets to contemplate involving standards, the human element, and the complexity of risk. Standards, codes, requirements, practices, rules, protocols, and procedures for particular industries have become extensive. Compliance and assurance engineering are important elements in modern safety programs. The human side of the equation remains a most complex consideration, as initiators or contributors in previous accidents or possible initiators or contributors in potential future accidents. Reliability, maintainability, quality, logistics, human factors, software performance, and system effectiveness are also important elements. Understanding the potential safety-related risks involves the application of the various assurance aspects of this book. It appears that the idealistic goal of system safety may be a very long-term objective—to design systems free of risks. These observations lead to a single conclusion: safety programs are needed in order to integrate, implement, and manage the complications.

11.1.1 Specific Safety Programs

Specific safety programs have been developed for any safety-related risk. They are comprised of specific standards, codes, requirements, practices, rules, protocols,

Assurance Technologies Principles and Practices: A Product, Process, and System Safety Perspective, Second Edition, by Dev G. Raheja and Michael Allocco
Copyright © 2006 John Wiley & Sons, Inc.

practices, and procedures. Guidelines and recommended practices have been established that detail program content, elements, and protocols. Examples of specific safety programs (or subprograms) are process safety, product safety, life safety, weapon safety, fleet safety, ergonomics, industrial hygiene, aircraft/flight safety, environmental protection, highly protective risks, fire safety, radiation safety, and construction/facility safety.

11.2   PROGRAMS TO ADDRESS PRODUCT, PROCESS, AND SYSTEM SAFETY

Generally, system safety principles can be universally applied when designing and implementing product and process safety programs. There are similarities between these programs or common safety program elements: hazard analysis, systemized safety training, communication, accident/incident investigation, safety review, inspection, evaluation, observation, peer review, data collection, resource allocation, monitoring, and maintenance of program. Specific standards, codes, requirements, practices, rules, protocols, and procedures for particular product, process, and system safety program applications are also relevant.

11.2.1   Product Safety Management

In the 1970s, Product Safety became important due to large product liability judgments against manufacturers of consumer products in the United States. Insurance companies developed criteria for Product Liability Claims Defense (PLCD) Programs[1] to control such potentially large losses associated with product liability. Loss Control engineers were trained in product safety in order to consult with the insured and assist in program development. Extensive audit checklists were developed from the criteria required for product safety. Judgment was made as to meeting, exceeding, or not meeting the product safety criteria. The PLCD audit, for example, addressed the following:

Loss Control Program: The establishment of a Loss Control Program,[2] which addressed product safety. The program is to be inclusive of all activities conducted within an organization, which addressed product safety.

Policy Statement: The publishing of a Corporate Policy Statement signed by the CEO or COO. The Policy Statement is a widely publicized document of formal statement, as a matter of record, regarding top management's commitment to state-of-the-art product safety and the importance of product safety during product life cycle.

---

[1]Aetna Causalty developed the Product Liability Claims Defense Program; Michael Allocco was a Loss Control engineer for Aetna and received extensive training in product safety in the 1970s.
[2]Total Loss Control is the concept of holistically applying hazard controls considering a specific exposure throughout the organization.

Product Safety Officer: The appointment of a Product Safety Officer with defined duties and responsibilities in product safety. The Product Safety Officer should be an executive of corporate management who has direct access to the CEO or COO.

Policies, Procedures, and Protocols: The completion of formal policies, procedures, and protocols throughout the organization associated with product safety. Specific assignment of responsibility and authority should be explicitly defined and established at the executive level and integrated throughout the organization and emphasized in position descriptions, including performance objectives.

Product Safety Committee: The establishment of a product safety committee, which is comprised of representatives from legal, engineering, design, manufacturing, quality, facilities, marketing, public relations, and operations. Each discipline should have specific responsibilities and duties applicable to product safety.

Hazard Analysis and Risk Assessment: The implementations of safety requirements to conduct hazard analysis and risk assessment of all products to identify safety-related risks. Standards on how to conduct analyses and risk assessment are to be defined. Concepts associated with system safety are included, such as designing or engineering-out hazards.

Design Review and Safety Review: The requirements to conduct formal design review and safety reviews where needed. These reviews involve the assessment of the product from a systems view. This included evaluation of parts, materials, components, configurations, subsystems, prototypes, mockups, breadboards, packaging design, shipping, storage, handling, transport, disposal, and labeling (instructions and cautions or warnings) to identify, evaluate, and control product safety-related risks. The risks associated with the life cycle of the product were identified, eliminated, or controlled to an acceptable level.

Testing: The standards and requirements to conduct safety testing are to be implemented. Field-testing or survey activities are to be defined, evaluated, and examined to identify product safety risks. Consideration is also given to the potential misuse of the product, product suitability, form-fit or function of the product.

State-of-the-Art in Design: The requirements associated with the state-of-the-art in design are identified and applied. Best practices are also to be established concerning protection against risks. Legal concepts are also defined; concepts of reasonable care or great care are to be addressed.

Training: A formal training program in product safety is also required involving policies, procedures, and protocols throughout the organization associated with product safety. Training associated with codes, standards, or requirements is also established. Training to conduct analysis, risk assessment, and safety review are required as well. Criteria should also include customer, subcontractor, shipper, retailer, and distributor training.

Quality Assurance and Process Control: Quality assurance is another important element within product safety. Quality is the conformance to a set of requirements that, if met, results in a product that is fit for its intended use (see discussions in Chapter 6). Quality standards include design, quality of processes, and quality of services. Specific standards address the product developmental phases, from material selection, incoming inspection, parts selection, assembly, manufacturing, product review, statistical control, maintenance, process control, storage, handling, shipping, distribution, logistics, product performance monitoring, customer feedback, product recall, and product disposal.

Documentation and Records: Documentation requirements are also considered necessary. Records should be kept for time periods that exceed the product life cycle. Records include product developmental information, initial product testing, product safety program documentation, analyses, inspection results, marketing tests, customer feedback, recall information, process documentation, and safety review minutes. Note that as a result of corporate mergers, the existing product liability is assumed. Consideration should be given to the review of past products, expected product life of the previously manufactured products, product safety-related records, documentation, and past claims and losses.

Communication: Requirements for product safety-related communication are also needed. Criteria associated with the appropriate (safe) use of the product are to be defined and evaluated. Consideration should include appropriate instruction on assembly, the intended use of product, maintenance and upkeep, inspection requirements, packaging and handling, distributor requirements, advertisement, product recall information, misuse potential, conformance to applicable installation codes and standards, inadvertent damage, disassembly, and use of tools and testing equipment. Criteria are also needed concerning the communication related to hazard notification, development of cautions and warnings, emergency first aid, nature of hazard, and misuse potential; consideration involving miscommunication, procedure deviation, language use, stereotyping, social convention, conventional use, and human factors.

Product-Related Hazards and Risks: Sources of information within the United States concerning product-related hazard and risks can be acquired from the Consumer Product Safety Commission, the National Highway Traffic Safety Administration, and the Food and Drug Administration.

11.2.2 Process Safety Management

The U.S. Occupational Safety and Health Standard 1910.119 addresses process safety management of highly hazardous chemicals. Consult the standard for explicit information on process safety. The following discussion provides an overview for general background information.

Scope of Standard   The standard contains requirements for preventing or minimizing the consequences of catastrophic releases of toxic, reactive, flammable, or explosive chemicals. These releases may result in toxic, fire, or explosion hazards.

Hazard Analysis   According to the standard, the employer shall perform an initial process hazard analysis (hazard evaluation) on processes covered by the standard. The process hazard analysis shall be appropriate to the complexity of the process and shall identify, evaluate, and control the hazards involved in the process. Employers shall determine and document the priority order for conducting process hazard analyses based on a rationale which includes such considerations as extent of the process hazards, number of potentially affected employees, age of the process, and operating history of the process. Consider that many hazard analysis techniques applied within system safety can be appropriate in conducting analyses.

Analysis Team   A team shall perform the process hazard analysis with expertise in engineering and process operations, and the team shall include at least one employee who has experience and knowledge specific to the process being evaluated. Also, one member of the team must be knowledgeable in the specific process hazard analysis methodology being used.

Risk Control   An output of the analysis process should be engineering and administrative controls applicable to the hazards and their interrelationships such as appropriate application of detection methodologies to provide early warning of releases. (Acceptable detection methods might include process monitoring and control instrumentation with alarms; and detection hardware, such as hydrocarbon sensors, process automation, instrumentation, and controls, is to be implemented.)

Hazard Tracking and Risk Resolution   The employer shall also establish a system to promptly address the team's findings and recommendations and include the following: (1) assure that the recommendations are resolved in a timely manner and that the resolution is documented; (2) document what actions are to be taken; (3) complete actions as soon as possible; (4) develop a written schedule of when these actions are to be completed; and (5) communicate the actions to operating maintenance and other employees whose work assignments are in the process and who may be affected by the recommendations or actions.

Employers shall retain process hazards analyses and updates or revalidations for each process covered by the standard, as well as the documented resolution of recommendations for the life of the process.

The employer shall develop and implement written operating procedures that provide clear instructions for safely conducting activities involved in each covered process consistent with the process safety information and shall address at least the following elements: initial startup, normal operations, temporary operations, emergency shutdown, procedure deviation, operating limits, hazardous chemicals, health hazards, and engineering and administrative controls.

Operating procedures shall be readily accessible to employees who work in or maintain a process. The operating procedures shall be reviewed as often as necessary to assure that they reflect current operating practice, including changes that result from changes in process chemicals, technology, and equipment, and changes to facilities. The employer shall certify annually that these operating procedures are current and accurate.

The employer shall also develop and implement safe work practices to provide for the control of hazards during operations such as lockout/tag-out; confined space entry; opening process equipment or piping; and control over entrance into a facility by maintenance, contractor, laboratory, or other support personnel. These safe work practices shall apply to employees and contractor employees.

Safety Training　Each employee presently involved in operating a process and each employee before being involved in operating a newly assigned process shall be trained in an overview of the process and in the operating procedures as specified in the standard. The training shall include emphasis on the specific safety and health hazards, emergency operations including shutdown, and safe work practices applicable to the employee's job tasks.

Refresher training shall be provided at least every 3 years, and more often if necessary, to each employee involved in operating a process to assure that the employee understands and adheres to the current operating procedures of the process. The employer, in consultation with the employees involved in operating the process, shall determine the appropriate frequency of refresher training.

Contractor Participation　The standard applies to contractors performing maintenance or repair, turnaround, major renovation, or specialty work on or adjacent to a covered process. It does not apply to contractors providing incidental services, which do not influence process safety, such as janitorial work, food and drink services, laundry, delivery, or other supply services.

Process Safety Information　The standard also addresses process safety information. In accordance with the schedule set forth in the standard, the employer shall complete a compilation of written process safety information before conducting any process hazard analysis required by the standard. The compilation of written process safety information is to enable the employer and the employees involved in operating the process to identify and understand the hazards posed by those processes involving highly hazardous chemicals. This process safety information shall include information pertaining to the hazards of the highly hazardous chemicals used or produced by the process, information pertaining to the technology of the process, and information pertaining to the equipment in the process.

Safety Review　The employer shall perform a pre-startup safety review for new facilities and for modified facilities when the modification is significant enough to require a change in the process safety information.

Maintenance Training Training for process maintenance activities is required. The employer shall train each employee involved in maintaining the ongoing integrity of process equipment in an overview of that process and its hazards and in the procedures applicable to the employee's job tasks to assure that the employee can perform the job tasks in a safe manner.

Inspections and Testing Inspections and tests shall be performed on process equipment. Inspection and testing procedures shall follow recognized and generally accepted good engineering practices. The frequency of inspections and tests of process equipment shall be consistent with applicable manufacturers' recommendations and good engineering practices, and more frequently if determined to be necessary by prior operating experience.

The employer shall document each inspection and test that has been performed on process equipment. The documentation shall identify the date of the inspection or test, the name of the person who performed the inspection or test, the serial number or other identifier of the equipment on which the inspection or test was performed, a description of the inspection or test performed, and the results of the inspection or test.

The employer shall correct deficiencies in equipment that are outside acceptable limits (defined by the process safety information in this standard) before further use or in a safe and timely manner when necessary means are taken to assure safe operation.

Quality Assurance In the construction of new plants and equipment, the employer shall assure that equipment as it is fabricated is suitable for the process application for which it will be used. Appropriate checks and inspections shall be performed to assure that equipment is installed properly and consistent with design specifications and the manufacturer's instructions. The employer shall assure those maintenance materials, spare parts, and equipment are suitable for the process application for which they will be used.

Hot Work Permit The employer shall issue a hot work permit for hot work operations conducted on or near a covered process. The permit shall document that the fire prevention and protection requirements in 29 CFR 1910.252(a) have been implemented prior to beginning the hot work operations; it shall indicate the date(s) authorized for hot work; and identify the object on which hot work is to be performed. The permit shall be kept on file until completion of the hot work operations.

Change Management The employer shall establish and implement written procedures to manage changes (except for "replacements in kind") to process chemicals, technology, equipment, and procedures; and changes to facilities that affect a covered process.

Incident Investigation The employer shall investigate each incident, which resulted in or could reasonably have resulted in a catastrophic release of highly hazardous chemical in the workplace. An incident investigation shall be initiated as promptly as possible, but not later than 48 hours following the incident. An incident investigation team shall be established and consist of at least one person knowledgeable in the process involved, including a contract employee if the incident involved work of the contractor, and other persons with appropriate knowledge and experience to thoroughly investigate and analyze the incident.

Emergency Planning and Response The employer shall establish and implement an emergency action plan for the entire plant in accordance with the provisions of 29 CFR 1910.38(a). In addition, the emergency action plan shall include procedures for handling small releases. Employers covered under this standard may also be subject to the hazardous waste and emergency response provisions contained in 29 CFR 1910.120(a), (p), and (q).

Compliance Audits Employers shall certify that they have evaluated compliance with the provisions of this section at least every 3 years to verify that the procedures and practices developed under the standard are adequate and are being followed. The compliance audit shall be conducted by at least one person knowledgeable in the process. A report of the findings of the audit shall be developed. The employer shall promptly determine and document an appropriate response to each of the findings of the compliance audit and document that deficiencies have been corrected.

Trade Secrets Employers shall make all information necessary to comply with the section available to those persons responsible for compiling the process safety information, those assisting in the development of the process hazard analysis, those responsible for developing the operating procedures, and those involved in incident investigations emergency planning and response and compliance audits without regard to possible trade secret status of such information.

11.2.3 System Safety Management

The tasks and activities of system safety management and engineering are defined in the System Safety Program Plan (SSPP).[3] An Integrated System Safety Program Plan (ISSPP) is modeled on the elements of a SSPP, which is outlined in Mil-Std 882C.[3] An ISSPP is required when there are large projects or large systems; the system safety activities should be logically integrated. Other participants, tasks, operations, or subsystems within a complex project should also be incorporated.

The first step is to develop a plan that is specifically designed to suit the particular project, process, operation, or system. A plan should be developed for each unique complex entity such as a particular line-of-business, project, system, development,

---

[3]Military Standard 882C explains and defines System Safety Program Requirements. Military Standard 882D is a current update as of 1999. This version no longer provides the details that version C had provided. At this writing, version E was just released. Version E is a movement toward providing more detail and discussion of system safety tasks, and system risks.

research task, or test. Consider a complex entity that is comprised of many parts, tasks, subsystems, operations, or functions and all of these subparts should be combined logically. This is the process of integration. All the major elements of the plan should be integrated. How this is accomplished is explained in the following paragraphs.

Integrated Plan   The Program Manager, Prime Contractor, Construction Manager, or Integrator develops the Integrated System Safety Program Plan. The plan includes appropriate integrated system safety tasks and activities to be conducted within the project. It includes integrated efforts of management, team members, subcontractors, and all other participants.

Program Scope and Objectives   The extent of the project, program, and system safety efforts is defined under scope. The system safety efforts should be in line with the project or program. Boundaries are defined as to what may be excluded or included within the plan.

The objective is to establish a management integrator to assure that coordination occurs between the many entities that are involved in system safety. The tasks and activities associated with integration management are defined in the document. The plan becomes a model for all other programs within the effort. Other participants, partners, and subcontractors are to submit plans, which are to be approved and accepted by the integrator. The plans then become part of the integrated plan.

Inputs to the Plan   The external inputs to the system safety process are the design concepts of the system, formal documents, engineering notebooks, and design discussions during formal meetings and informal communications. The ongoing output of the system safety process is hazard analysis, risk assessment, risk mitigation, risk management, and optimized safety.

System Safety Organization   The system safety organization is detailed within the plan. The duties and responsibilities are defined for the safety manager and staff. Each subentity such as a partner or subcontractor should appoint a manager or senior system safety engineer or lead safety engineer who will manage the entity's plan. All appropriate system safety participants are to be given specific responsibilities. The participants should have particular qualifications in system safety, safety engineering, and safety management, which should include a combination of experience and education.

System Safety Working Group   A System Safety Working Group (SSWG) is formed to help manage and conduct tasks associated with the program. The group specifically provides a consensus activity that enhances work performed. The SSWG is a major part of the program.

For large or complex efforts, where an integrated program has been established, activities of the Integrated System Safety Working Group (ISSWG) are defined. The ISSWG includes responsive personnel who are involved in the system safety process. The plan specifically indicates that, for example, reliability, maintainability, quality, logistics, human factors, software performance, and system effectiveness personnel are active participants in the ISSWG. The integrator may act as the chair of the ISSWG with key system safety participants from each subentity. The group may meet formally on a particular schedule. Activities are documented in meeting minutes. Participants are assigned actions.

Program Milestones   The system safety and program schedule is defined. The schedule indicates specific events and activities along with program milestones, to accomplish the specific tasks evolved. One example is the use of Program Evaluation Review Technique (PERT) [1]. It is essentially the presentation of tasks, events, and activities on a network in sequential and dependency format showing independence and task duration and completion time estimates. Critical paths are easily identifiable. Its advantage is the greater control provided over complex development and production programs as well as the capacity for distilling large amounts of scheduling data in brief, order reliability, maintainability, quality, logistics, human factors, software performance, and system effectiveness. Management decisions are implemented. Needed actions may be more clearly seen, such as steps to conduct a specific test.

System Safety Requirements   The engineering and administrative requirements for system safety are described. As the design and analysis matures, specific safety standards and system specifications are to be developed and the plan is to be updated. Initially, generic requirements are defined for the design, implementation, and application of system safety within the specific project or process. The integrator defines the requirements needed to accomplish the objectives. Specific are defined for the system safety products to be produced, the risk assessment code matrix, risk acceptability criteria, and residual risk acceptance procedures. This effort should also include guidelines for establishing project phases, review points, and levels of review and approval [2].

System Safety Objectives   The following system safety objectives are to be provided:

- Eliminate system risks early via design, material selection, and/or substitution.
- When hazardous components must be used, select those with the least risk throughout the life cycle of the integrated system.
- Design software-controlled or software-monitored functions to minimize initiation of hazardous events, through software controls such as error source detection, modular design, firewalls, command control response, and failure detection.
- Design to minimize risk created by human error in the operation and support of the system.
- Consider alternate approaches to minimize risk throughout the system, such as interlocks, fail-safe design, and redundancy.

- Provide design protection from uncontrolled energy sources (e.g., physical protection, shielding, grounding, and bonding).
- Locate equipment so that access during operations, servicing, maintenance, installation, and repair minimizes exposure to associated system risks.
- Categorize risks associated with the system and its performance.
- Identify the allocation of requirements to specification traceability and assess the effectiveness of this allocation on risk control.

Design for Safety: System Safety Precedence The order of precedence for satisfying system safety requirements and resolving identified risks is as follows:

DESIGN FOR SAFETY: SYSTEM SAFETY PRECEDENCE

Description	Priority	Definition
Design for minimum risk	1	From the first design to eliminate risks. If the identified risk cannot be eliminated, reduce it to an acceptable level through design selection.
Incorporate safety devices	2	If identified risks cannot be eliminated through design selection, reduce the risk via the use of fixed, automatic, or other safety design features or devices. Provisions shall be made for periodic functional checks of safety devices.
Provide warning devices	3	When neither design nor safety devices can effectively eliminate identified risks or adequately reduce risk, devices shall be used to detect the condition and to produce an adequate warning signal. Warning signals and their application shall be designed to minimize the likelihood of inappropriate human reaction and response.
Develop procedures and training	4	Where it is impractical to eliminate risks through design selection or specific safety and warning devices, procedures and training are used. However, concurrence of authority is usually required when procedures and training are applied to reduce risks of catastrophic or critical severity.

Risk/Hazard Tracking and Risk Resolution Risk/hazard tracking and risk resolution are described within the plan. This is a procedure to document and track contributory system risks and their associated controls by providing an audit trail of risk resolution. The controls are to be formally verified and validated and the associated contributory hazard is to be closed. This activity is conducted and/or reviewed during ISSWG meetings or formal safety reviews.

TABLE 11.1   Example of Customized Likelihood Definitions

Descriptor	Occurrence	Definition
A. Frequent	$X \geq$ EE-3	Continuously experienced
B. Probable	EE-3 $> X \geq$ EE-5	Expected to occur frequently
C. Remote	EE-5 $> X \geq$ EE-7	Expected to occur several times
D. Extremely Remote	EE-7 $> X \geq$ EE-9	Unlikely, but can reasonably be expected to occur
E. Extremely Improbable	$X <$ EE-9	Unlikely to occur, but possible

Risk Assessment   Risk is associated with a specific potential accident. It may be expressions of the worst, middle, and best case severity and likelihood of a specific events occurring at a particular time. The definitions of event likelihood and risk severity (Tables 11.1 and 11.2) are to be appropriate to support system hazard analysis activities, for those events that can occur at any time, considering any possible exposure within the system. Also consider events occurring at any time throughout the life cycle of the system. Furthermore, system hazard analysis considers interfaces and interactions of humans, hardware, software, firmware, and/or the environment.

Hazard Analysis Concepts   The goal is to optimize safety by the identification of safety-related risks, eliminating or controlling them via design and/or procedures, based on acceptable system safety precedence. Hazard analysis is the process of examining a system throughout its life cycle to identify inherent safety-related risks.

In conducting hazard analysis, the analyst should be concerned with machine–environment interactions resulting from change/deviation stresses as they occur in time/space; physical harm to persons; functional damage; and system degradation. The interactions between the human, the machine, and the environment are the elements of a system. The human parameter relates to appropriate human factors engineering and associated elements: biomechanics, ergonomics, and human performance variables. The machine equates to the physical hardware,

TABLE 11.2   Example of Customized Severity Definitions

Area Affected	Estimated Harm			
	Minor 4	Major 3	Catastrophic 1	Catastrophic 1+
General public	Minor recoverable injuries (TBD)	Major recoverable injuries (TBD)	Fatal injuries (1 through TBD)	Many fatalities (TBD)
Property	Minor property damage ($ TBD)	Major property damage ($ TBD)	Extensive property damage ($ TBD)	Extreme property damage ($ TBD)
Environment	Minor environmental damage ($ TBD)	Major environmental damage ($ TBD)	Extensive environmental damage ($ TBD)	Extreme environmental damage ($ TBD)

firmware, and software. The human and machine are within a specific environment. Adverse effects due to the environment are to be studied.

Specific integrated analyses are appropriate at a minimum to evaluate interactions:

- Human—human interface analysis
- Machine—abnormal energy exchange, software hazard analysis, fault hazard analysis
- Environment—abnormal energy exchange, fault hazard analysis

The interactions and interfaces between the human, machine, and environment can be evaluated by application of the above techniques, also with the inclusion of hazard control analysis or safety validation; the possibility of insufficient control of the system is analyzed.

Integrated Approach   An integrated approach is not simple; one does not simply combine many different techniques or methods in a single report and expect a logical evaluation of system risks and hazards. The logical combining of hazard analyses is called integrated system hazard analysis. To accomplish integrated system hazard analysis, many related concepts about system risks should be understood.

System Safety Data   Pertinent historical system safety-related data and specific lessons-learned information are to be used to enhance analysis efforts; specific knowledge concerning past contingencies, incidents, and accidents will also refine analysis activities.

Safety Verification and Validation   Specific verification techniques are discussed within the plan. Safety verification is needed to assure that system safety is adequately demonstrated and all identified system risks that have not been eliminated are controlled. Risk controls (mitigation) must be formally verified as being implemented. Safety verification is accomplished by the following methods:

- Inspection
- Analysis
- Demonstration
- Test
- Implicit verification

Audit Program   The plan should call for the quality assurance function to audit the program. All activities in support of system safety are to be audited; this includes contractor internal efforts. All external activities that support closed-loop hazard tracking and risk resolution must also be audited.

Training   Participants are to receive specific training in system safety in order to conduct analysis, hazard tracking, and risk resolution. Additional training is to be provided for SSWG members and program auditors to assure awareness of the system safety concepts discussed herein. Specific training is to be conducted for system users, systems engineers, and technicians. Training considers normal operations with standard operating procedures, maintenance with appropriate precautions, test and simulation training, and contingency response. Specific hazard control procedures will be recommended as a result of analysis efforts.

Incident Reporting and Investigation Any incident, accident, malfunction, or failure affecting system safety is to be investigated to determine causes and to enhance analysis efforts. As a result of investigation, causes are to be determined and eliminated. Testing and certification activities are also to be monitored; anomalies, malfunctions, and failures that affect system safety are to be corrected.

Concepts of system safety integration are also applied systematically through formal accident investigation techniques. Many systematic techniques have been successfully applied, for example [3], scenario analysis (SA), root cause analysis (RCA), energy trace barrier analysis (ETBA), management oversight and risk tree (MORT), and project evaluation tree (PET) [2]. For further details consult the references provided. Consider that hazard analysis is the inverse of accident investigation and similar techniques are applied in the application of inductive and deductive processes of hazard analysis and accident investigation.

System Safety Interfaces System safety interfaces with other applicable disciplines both internally to systems engineering and externally. System safety is involved with other program disciplines, for example, reliability, maintainability, quality, logistics, human factors, software performance, and system effectiveness. These disciplines should be involved in the hazard analysis, hazard control, hazard tracking, and risk resolution activities.

11.3 RESOURCE ALLOCATION AND COST ANALYSIS IN SAFETY MANAGEMENT

Knowledge is required in resource allocation and cost analysis to successfully manage a safety program. One of the most important decisions a safety manager has to make deals with the allocation of resources and the cost associated with those resources. Usually resource decisions are related to the size and complexity of the project, the nature of the contract, and the statement of work. All these factors dictate how large or small a safety management program should be. Program objectives can be broken down into specific tasks and activities. Estimates can be established as to who should conduct the work and how long it should take. Generally, budgets are established as a result of following such a process. However, there is the performance paradox to consider. Generally, when things are going along well, systems are functioning, and no accidents are occurring, resources associated with safety decrease—until something goes wrong. When catastrophic accidents take place, the resource expenditure in safety increases, and in

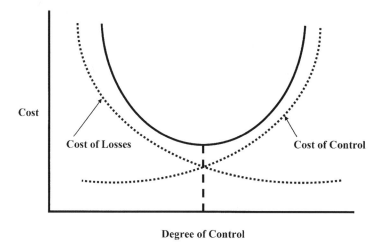

Cost

Cost of Losses   Cost of Control

Degree of Control

Figure 11.1   Plot of cost curves: cost verses degree of control.

some cases, it appears that resources and costs have increased exponentially. It is apparent that it would be beneficial to try to maintain a proactive view on accident prevention.

11.3.1   Cost of Losses Versus Cost of Control

In Fig. 11.1, an ideal situation is presented: the costs associated with prevention (control) versus costs associated with losses are exponentially symmetrical. The assumption indicates that losses would be uncontrolled initially with minimal prevention (control) applied. Eventually losses would decrease as prevention (control) is applied. The optimum point is indicated on the total cost curve, where its slope is level. (Cost curves are also discussed in Chapters 1, Fig. 1.3 and Chapter 4, Fig. 4.9.).

11.4   TOPICS FOR STUDENT PROJECTS AND THESES

1. Explore the differences between product, process, and system safety programs.
2. Choose and define a complex hazardous product and write a specific product safety program.
3. Define how reliability, maintainability, quality, logistics, human factors, software performance, and system effectiveness may compete with objectives of system safety.
4. Develop a process safety program for a hazardous process.
5. Explain how hazard analyses may differ between a product, process, and system.

6. Suggest your approach on conducting integrated hazard analysis of a complex system, considering the human, hardware, software, and environment.

REFERENCES

1. J. V. Grimaldi and R. H. Simonds, *Safety Management*, 3rd ed., Richard D. Irwin, Inc., Homewood, IL, 1975.
2. J. Stephenson, *System Safety 2000, A Practical Guide for Planning, Managing, and Conducting System Safety Programs*, Van Nostrand Reinhold, New York, 1991.
3. System Safety Society, *System Safety Analysis Handbook*, Unionville, Virginia, July 1997.

FURTHER READING

Allocco, M., Automation, System Risks and System Accidents. In: *Proceedings of the 17th International System Safety Conference*, System Safety Society, August 1999.

Allocco, M., Computer and Software Safety Considerations in Support of System Hazard Analysis. In: *Proceedings of the 21st International System Safety Conference*, System Safety Society, August 2003.

Allocco, M., W. E. Rice, and R. P. Thornburgh. System Hazard Analysis Utilizing a Scenario-Driven Technique. In: *Proceedings of the 20th International System Safety Conference*, System Safety Society, September 2001.

Anderson, D. R., D. J. Sweeney, and T. A. Williams, *An Introduction to Management Science Quantitative Approaches to Decision Making*, West Publishing, Belmont, CA, 1976.

American Society for Testing Materials, *Fire Risk Assessment*, ASTM STP762, 1980.

Bahr, N. J., *System Safety Engineering and Risk Assessment: A Practical Approach*, Taylor and Francis, Philadelphia, PA, 1997.

Chapanis, A., *Human Factors in Systems Engineering*, John Wiley & Son, Hoboken, NJ, 1996.

Clemens, P. L., A Compendium of Hazard Identification and Evaluation Techniques for System Safety Application, *Hazard Prevention*, March/April, 1982.

Cooper, J. A., Fuzzy-Algebra Uncertainty Analysis, *Journal of Intelligent and Fuzzy Systems*, vol. 2, no. 4, 1994.

DEF(AUST) 5679, Army Standardization (ASA), The Procurement of Computer-Based Safety Critical Systems, May 1999.

Department of Labor, *OSHA Regulations for General Industry*, 29 CFR 1910, July 1992.

Department of Labor, *Process Safety Management of Highly Hazardous Chemicals*, 29 CFR 1910.119, *Federal Register*, 24 February 1992.

Department of Labor, *OSHA Regulations for Construction Industry*, 29 CFR 1926, July 1992.

Department of Labor, *Process Safety Management Guidelines for Compliance*, OSHA 3133, 1992.

Ebeling, C. E., *An Introduction to Reliability and Maintainability Engineering*, McGraw-Hill, New York, 1997.

Electronic Industries Association, EIA-6B, G-48, *System Safety Engineering in Software Electrical/Electronic/Programmable Electronic Safety-Related Systems*, Draft 61508-2 Ed 1.0, 1998.

Endsley, M. R., Situation Awareness and Human Error: Designing to Support Human Performance. In: *Proceedings of the High Consequence Systems Surety Conference*, Albuquerque, NM, 1999.

Environmental Protection Agency, *Exposure Factors Handbook*, EPA/600/8-89/043, Office of Health and Environmental Assessment, Washington DC, 1989.

Environmental Protection Agency, *Guidance for Data Usability in Risk Assessment*, EPA/540/G-90/008, Office of Emergency and Remedial Response, Washington DC, 1990.

Everdij, M. H. C., and H. A. P. Blom, Safety Assessment Techniques Database, Version 0.3, March 2005.

FAA Safety Risk Management, FAA Order 8040.4.

FAA, *System Safety Handbook: Practices and Guidelines for Conducting System Safety Engineering and Management*, Revised: September 30, 2001.

Hammer, W., *Occupational Safety Management and Engineering*, 2nd ed., Prentice-Hall, Englewood Cliffs, NJ, 1981.

Heinrich, H. W., D. Petersen, and N. Roos, *Industrial Accident Prevention: A Safety Management Approach*, 5th ed., McGraw-Hill, New York, 1980.

Institute for Electrical and Electronic Engineers, *Standard for Software Safety Plans*, IEEE Std 1228, 1994.

International System Safety Conference, Proceedings through 23rd.

Johnson, W. G., MORT—The Management Oversight and Risk Tree, SAN 821-2, U.S. Atomic Energy Commission, 12 February 1973.

Joint Services Computer Resources Management Group, *Software System Safety Handbook: A Technical and Managerial Team Approach*, Published on Compact Disc, December 1999.

Leveson, N., *SAFEWARE: System Safety and Computers, A Guide to Preventing Accidents and Losses Caused by Technology*, Addison Wesley, Boston, 1995.

Martin, L., and J. Orasanu, Errors in Aviation Decision Making: A Factor in Accidents and Incidents, HESSD, 1998.

Modarres, M., *What Every Engineer Should Know About Reliability and Risk Analysis*, Marcel Dekker, New York, 1993.

Moriarty, B., and H. E. Roland, *System Safety Engineering and Management*, 2nd ed., John Wiley & Sons, Hoboken, NJ, 1990.

NASA Dryden Policy Directive, DPD-8740.1A, Range Safety Policy for Dryden Flight Research Center (DFRC), July 2000.

NASA NSTS 22254, *Methodology for Conduct of NSTS Hazard Analyses*, May 1987.

Nguyen, L., Unmanned Aircraft System Safety and Operational Safety Assessments Considerations, White Paper, RTCA SC-203 Safety Assessment Subgroup, March 15, 2005.

Nuclear Regulatory Commission (NRC), Safety/Risk Analysis Methodology, April 12, 1993.

Process Safety Management, 29 CFR 1910.119, U.S. Government Printing Office, Washington DC, July 1992.

Radio Technical Commission for Aeronautics (RTCA) DO-264, Operational Safety Assessment for the Approval of the Provision and Use of Air Traffic Services Supported by Data Communications.

Raheja, D. G., *Assurance Technologies: Principles and Practices*, McGraw-Hill, New York, 1991.

RTCA-DO 178B, *Software Considerations in Airborne Systems and Equipment Certification*, December 1, 1992; COMDTINST M411502D, *System Acquisition Manual*, December 27, 1994; DODD 5000.1, *Defense Acquisition*, March 15, 1996.

Sheridan, T. B., *Humans and Automation: System Design and Research Issues*, John Wiley & Sons, Hoboken, NJ, 2002.

Society of Automotive Engineers, *Aerospace Recommended Practice 4754: Certification Considerations for Highly Integrated or Complex Aircraft Systems*, November 1996.

Society of Automotive Engineers, *Aerospace Recommended Practice 4761: Guidelines and Methods for Conducting the Safety Assessment Process on Civil Airborne Systems and Equipment*, December 1996.

System Safety Program Requirements, Mil-Std 882D, February 10, 2000.

Tarrants, W. E., *The Measurement of Safety Performance*, Garland STPM Press, New York, 1980.

U.K. Ministry of Defense, *Defense Standard 00-55: Requirements for Safety Related Software in Defense Equipment*, Issue 2, 1997.

U.K. Ministry of Defense, *Defense Standard 00-56: Safety Management Requirements for Defense Systems*, Issue 2, 1996.

U.K. Ministry of Defense, *Interim DEF STAN 00-54: Requirements for Safety Related Electronic Hardware in Defense Equipment*, April 1999.

CHAPTER 12

STATISTICAL CONCEPTS, LOSS ANALYSES, AND SAFETY-RELATED APPLICATIONS

12.1 USE OF DISTRIBUTIONS AND STATISTICAL APPLICATIONS ASSOCIATED WITH SAFETY

Statistical information can be used and applied in numerous ways in support of safety analysis. There are many related objectives when statistics are used in analysis. An analyst must have an understanding of what has occurred in the past, what is currently happening, and what is possible in the future, concerning past, current, and future risks associated with the system under evaluation. Also, it is appropriate to have an understanding of the risks in regard to the engineering, the physics of failure, the dynamics related to human performance and error, the state of the system, the environment, and human health.

12.2 STATISTICAL ANALYSIS TECHNIQUES USED WITHIN SAFETY ANALYSIS

A brief description of statistical analysis techniques that are used within safety analysis is provided below. This is not an all-inclusive listing of all techniques used within safety, however. Almost any engineering, technical, or scientific statistical application can be applied toward solving a safety-related problem.

Assurance Technologies Principles and Practices: A Product, Process, and System Safety Perspective, Second Edition, by Dev G. Raheja and Michael Allocco
Copyright © 2006 John Wiley & Sons, Inc.

Loss Analysis This technique is a safety analysis-based process to semiquantitatively analyze, measure, and evaluate planned or actual loss outcomes resulting from the action of equipment, procedures, and personnel during emergencies or accidents. Any safety-related operation should have an emergency contingency plan to handle unexpected events.

This approach defines the organized data needed to assess the objectives, progress, and outcome of an emergency response; to identify response problems; to find and assess options to eliminate or reduce response problems and risks; to monitor future performance; to investigate accidents; and to be used for safety planning purposes.

Probabilistic Design Analysis This method is used to assess hardware, software, and human reliability for given failure modes, errors, and hazards. Random variables are characterized by probability density functions (PDFs) or cumulative distribution functions (CDFs). A random variable (RV) is a variable that is a numerical value within some probability distribution. A RV can be either continuous such as real numbers or discrete such as nonnegative integer values. The probability distribution defines a probability for each value of a discrete RV or considers a probability for an interval of values of a continuous RV [1, pp. 16–17].

Probabilistic Risk Analysis This is a number of methods that are used to evaluate reliability where the probability of failure can be equated directly to a hazard. The probability of failure is determined by other supportive analysis such as Bayesian networks, fault-tree analysis, and event trees. Probabilistic data is estimated from prior history, current observation, and testing. Prior, current, and future PDFs are weighted using Bayesian techniques. These techniques rely on conditional probabilities.

Sensitivity Analysis This is the study of how changes of some input affect some output [2, pp. 92–93]. Complex mathematical models can be used to examine the effects of changes such as on human error probabilities that may affect system unavailability.

Scatter Diagram Raw data is plotted to determine if there is any relationship between two variables. The graphs displayed on the scatter diagram can help the analyst determine possible causes of problems, even when the connection between two variables is unexpected. The direction and compactness of the cluster of points provides a clue as to the relationship between the variable causes and effects [3, pp. 195–199].

Control Chart/Trending Analysis Control charts are graphical plots where sampling events such as observations of unsafe acts, accidents, incidents, or deviations, are plotted along the vertical and time is indicated along the horizontal.

Bar Chart   This chart shows a comparison of quantities of data to help identify quantity changes. The lengths of the bars, which can represent percentage or frequency of events, depict the quantities of data, such as types of accidents. Bars may be horizontal or vertical. Bar charts can be shown in double or triple bars to compare different information.

Stratification Chart   Stratification involves sorting the data into different groups that share a common characteristic. Comparisons of different groups, units, or other types of strata can lead to suggesting an improvement or mitigation or recognizing a hazard.

Pareto Chart   When there is a need to know the relative importance of data or variables (problems, hazards, causes, conditions), a Pareto chart can be used. The chart can help highlight data or variables that may be vital. The chart can be an illustration of the data as of a specific time period. The data are arranged in descending order with the most important to the left. The Pareto chart is based on the "Pareto Principle," which states that a few of the causes often account for most of the effects.

Histograms   Another form of bar chart is called a histogram, which shows a spreading of data over a specified range. This spread of data makes presentations easier to interpret. When data are plotted on histograms, many items tend to fall toward the center of the data distribution. Fewer items fall on either side of the center. The bars are proportional in height to the frequency of the group represented. Since group intervals are equal in size, the bars are equal width. Each bar within the graph can be referred to as a class. The thickness of the bar is the class interval. The numerical values corresponding to the borders of the bars are the class boundaries; the central value of the class is called the representative value or min-value.

Student-t Analysis   The Student-*t* compares the sample statistic *t*, which is based on the sample mean and standard deviation, to the *t*-distribution for the same sample size and a desired significance (probability of error). The *t*-distribution is similar to the normal distribution in that, with an infinite sample size, the *t*-distribution is equivalent to the standard normal distribution. At sample sizes lower than infinity, the *t*-distribution becomes "lower and flatter" than the normal distribution.

Analysis of Variance (ANOVA)   This technique is used in the design of experiments (reliability testing) to compare sample statistics to determine if the variation of the mean and variance between two or more populations are attributable to sources other than random variation.

Correlation Analysis   Correlation is a measure of the relation between two or more variables. The measurement scales used are interval scales, but other correlation coefficients are available to handle other types of data. Correlation coefficients can range from -1.00 to $+1.00$. The value of -1.00 represents a perfect negative correlation, while a value of $+1.00$ represents a perfect positive correlation. A value of 0.00 represents a lack of correlation.

Confidence Analysis This analysis compares sample values, means, or standard deviations with population standard deviations to obtain a confidence interval, with a chosen significance. Confidence analysis is used to determine the interval of values, with a chosen probability of being within that interval. Confidence analysis can be used with individual points, means, standard deviations, regression lines, or reliability measurements such as mean time between failures.

Regression Analysis This analysis is a form of curve-fitting to find a mathematical relationship for a group of data. A typical regression analysis approach is called a least squares curve-fitting. This evolves using a probability plot and fitting a linear regression line. The transformation will depend on the distribution used. A goodness-of-fit test is often performed to determine how the generated relationship fits the data. Usually, reliability data such as failure or repair time are plotted. The least squares method can fit exponential, Weibull, normal, and log normal distributions.

Critical Incident Technique This is a method of identifying errors and unsafe conditions that contribute to both potential and actual accidents and incidents within a given population by means of a stratified random sample of participant-observers selected from within the population [4, pp. 41–43]. Operational personnel can collect information on potential or past errors or unsafe conditions. Hazard controls are then developed to minimize the potential error or unsafe condition.

Delphi Technique This technique involves an iterative process that results in a consensus by a group of subject matter experts. The issue is presented to the experts. Without discussion, the experts communicate their comments to a facilitator or judge [2, p. 141]. They may use numerical or rank-order evaluation, stating reasons why conclusions have been made. The facilitator or judge reviews the comments and eliminates those not applicable to the topic. Then, the comments are redistributed to the experts for further review. The iteration is repeated until a consensus is reached.

12.3 USING STATISTICAL CONTROL IN DECISION-MAKING FOR SAFETY

Consider the concept that managers can monitor a complex system from a safety point of view and be able to detect anomalies when they occur, by using statistical control charts [5, pp. 61–85]. The anomalies are imbalances in the system and if they are unabated accidents can occur. This is the basis of safety observation where observers identify unsafe acts and unsafe conditions.

System Monitoring Using Control Charts   Thinking in terms of system safety, it is applicable to include the monitoring of deviations within a system since these deviations can be hazards. Trained collectors acquire data and this data is plotted over time. Statistically, it is possible to identify deviations with trends in a complex system. Once the trends are identified, corrections could be made to offset the system imbalance before harm occurs. Not only are unsafe acts and unsafe conditions corrected but any system deviation that could affect safety. By applying the statistical control, it then becomes possible to decrease the risks associated with very complex system accidents.

Control Charts   Control charts are graphical plots where sampling events such as observations of unsafe acts, accidents, incidents, or deviations are plotted along the vertical axis and time is indicated along the horizontal axis. A control chart is useful to discover how much variability there is in a stable process due to random variation or due to unique events or individual actions in order to determine whether a process is in statistical control—that is, the process is consistent.

Types of Control Charts   There are various types of control charts and they vary according to the data they contain. Certain data are based on measurements, such as the measurement of unit parts or yields of a chemical process. These are known as indiscrete values or continuous data. Data that are based on counting, such as observations, are known as discrete values or enumerated data. Control charts can also be divided into types according to their usage, such as how data is influenced by various factors, material, human factors, or methods, or if two or more different factors are exerting an influence. Data may have to be stratified and separate charts developed to determine the source of influence. It would be logical to assume that a "system" of control charts would be needed to monitor a complex system. An example control chart is illustrated in Fig. 12.1.

Safety Observations and the Use of the Attributes Control Chart   An attributes control chart or "P chart" is used when the samples collected are considered qualitative characteristics such as safety observations. To develop a P chart requires the following calculations [6, pp. 573–575]:

$$\text{UCL} = p + 1.96\sqrt{p(1-p)/n} \qquad \text{LCL} = p - 1.96\sqrt{p(1-p)/n}$$

where p is the mean proportion of observed behaviors that are unsafe or safe for all observation periods, and n is the number of observation periods. The number of readings N required for a certain level of accuracy at a 95% confidence is

$$N = \frac{4(1-p)}{(s^2)p}$$

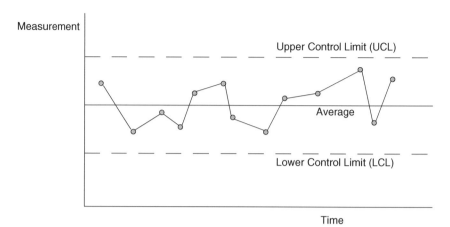

Figure 12.1   Example statistical control chart.

where p is the proportion of safe or unsafe acts observed during the study, and S is the desired accuracy (percent per 100 readings). The accuracy of reading is the proportion observed plus or minus some percent accuracy.

Attributes Control Chart Procedure   It is assumed that the appropriate control chart has been selected by analysis. A decision has been make to conduct behavioral sampling and the data that is to be acquired may be subjective and qualitative; consequently, the attributes control chart or p chart is to be used. The general procedure for constructing a p chart is indicated below.

1. Select trained observers to collect the data.
2. Select the desired accuracy S.
3. Collect the data randomly including the number of observations and the number of safe or unsafe acts observed for the study.
4. Compute p, the fraction defective for each subgroup and enter it on a datasheet:

$$p = \frac{pn}{n} = \frac{\text{number of rejects in subgroup}}{\text{number observed in subgroup}}$$

5. Compute the central line, that is, the average fraction defective or p average:

$$p \text{ average} = \frac{pn + pn + pn + \cdots}{n + n + n + \cdots} = \frac{\text{total defectives}}{\text{total observations}}$$

6. Compute upper and lower control limits (UCL and LCL).
7. Construct the control chart and plot p.

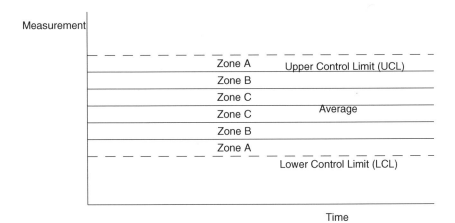

Figure 12.2 Statistical control chart with zone indications.

Interpreting Control Charts [7, p. 55] The process is said to be "out of control" if one or more points fall outside the control limits. Divide the control chart into zones (see Fig. 12.2), and examine what has changed and possibly make an adjustment if:

- Two points out of three successive points are on the same side of the centerline in zone A or beyond.
- Four points out of five successive points are on the same side of the centerline in zone B or beyond.
- Nine successive points are on one side of the centerline.
- There are six consecutive points, increasing or decreasing.
- There are fourteen points, in a row alternating up and down.
- There are fifteen points in a row within zone C (above and below centerline).

12.4 BEHAVIOR SAMPLING

Behavior sampling [4, pp. 283–298] is based on the principle of random sampling. By observing a portion of the total, a prediction can be made concerning the makeup of the whole. Behavior sampling can be defined as the evaluation of an activity performed in an area using a statistical measurement technique based on a series of instantaneous random observations or samples.

The behavior sampling method is based on the laws of probability, the concept of the normal distribution, and randomness. To determine the confidence level accuracy and the number of observations are required.

Consider that the "probability" [8, pp. 1153–1158] of an event A in an experiment is supposed to measure how frequently A is about to occur if there are many trials. If a coin is flipped, then heads H and tails T will appear about equally often or are "equally likely." When there are a large number of trials, it is possible to obtain an approximation of an unknown probability by dividing the number of times A occurs by the number of trials (observations divided by trials).

Behavior sampling is based on the principle that the readings taken will conform to the characteristics of the normal distribution or Gauss distribution. The characteristic of a normal distribution is that the curve is symmetrical around the mean, which is also equal to the median and mode. The larger the number of samples or the number of observations made, the closer the plot of them will approach the normal curve, and the more confidence one can have that the sample readings are representative of the population.

It is necessary to take only the number of observations required to satisfy the desired confidence level. The confidence level equates to a particular area under the curve. A confidence level of 95% represents an area of 95%, or $\pm 2\sigma$, and is considered adequate for most behavior sampling. A confidence level of 95% means that the conclusions will be representative of the true population 95% of the time and 5% of the time they will not. Accuracy is the tolerance limit of the readings that fall within a desired confidence level. Accuracy is a function of the number of readings taken. The tolerance becomes smaller as the number of observations increases.

In applying behavior sampling, it is necessary to determine the percent of time a person is behaving safely and the percent of time the person is behaving in an unsafe manner. The analyst can observe the person throughout the total shift or observe the person at certain times. A record is made of whether the person is working safely or unsafely. The total number of observations is noted and safe and unsafe behavior observations are prorated.

12.5 CALCULATING HAZARDOUS EXPOSURES TO THE HUMAN SYSTEM

In conducting hazard analysis, consideration must be given to the identification of health-related risks. Specific techniques can be applied, such as operating and support hazard analysis, health hazard assessment, or job safety analysis. These efforts should identify hazards associated with the human system. The human may be exposed to hazardous materials, hazardous substances, and biological pathogens. Statistical studies are used to determine exposure effects and limiting values for which a human can be exposed. Guidelines and standards are then developed. It is important to understand the statistics and science that support such guidelines and standards.

Hazardous Materials A system under evaluation may contain hazardous materials, which present health-related risks. Within the United States, a material is considered hazardous if it is:

- Specifically listed in the law, 29 CFR Part 1910, Subpart Z, Toxic and Hazardous Substances (the Z list)
- Assigned a threshold limit value (TLV) by the American Conference of Governmental Industrial Hygienists
- Determined to be cancer causing, corrosive, toxic, an irritant, a sensitizer, or has damaging effects on specific body organs [9]

Sources of Information Related to Hazardous Materials, Hazardous Substances, and Biological Pathogens When conducting evaluations associated with hazardous exposures information can be acquired from the following:

- Material Safety Data Sheets (MSDS). They are sources of information on hazardous substances within the United States. MSDS provide information on the material's physical properties or fast-acting health effects that make the material dangerous to handle. They also provide information on personal protective equipment required, first aid treatment if exposed, and the preplanning needed for safely handling spills, fires, and day-to-day operations.
- National Fire Protection Association 704M publication provides a code for indicating the storage of hazardous materials and related hazards (*NAPA Hazard Rating*).
- OSHA Flammable/Combustible Liquid Classification, 29 CFR 1910.106 lists standard classifications used to identify the risks of fire or explosion associated with a liquid.

A partial list of sources for identifying hazards is presented below for reference purposes. When evaluating environmental health-related exposures (within the United States), it is recommended that a Certified Industrial Hygienist (CIH) be consulted for the most current ACGIH guidelines and standards.

Inadvertent Overexposure to Hazardous Substances: The ACGIH lists 26 pages of substances from acetaldehyde to zirconium [10, pp. 12–37]. Exposures to these substances may occur during chemical processing, assembly, manufacturing, disassembly, operation, and disposal of the system. The analyst should consider any possible exposure during the system life cycle.

Inadvertent Exposure to Carcinogens: Appendix A of the *ACGIH Handbook* discusses classification of substances that are categorized as human carcinogens. For A1 carcinogens with Threshold Limit Value (TLV)[1] and for A2 and A3 carcinogens, worker exposure by all routes should be carefully controlled to levels as low as possible below the TLV.

A1–Confirmed Human Carcinogen: The agent is carcinogenic to humans based on the weight of evidence from epidemiological studies of, or convincing clinical evidence in, exposed humans.

A2–Suspected Human Carcinogen: The agent is carcinogenic in experimental animals at dose levels, by route(s) of administration, at site(s), of histology type(s), or by mechanism (s) that are considered relevant to worker exposure. Available epidemiological studies[2] are conflicting or insufficient to confirm an increased risk of cancer in exposed humans.

A3–Animal Carcinogen: The agent is carcinogenic in experimental animals at a relatively high dose, by route(s) of administration, at site(s), of histology type(s), or by mechanism(s) that are not considered relevant to worker exposure. Available epidemiological studies do not confirm an increased risk of cancer in exposed humans. Available evidence suggests that the agent is not likely to cause cancer in humans except under uncommon or unlikely route or levels of exposure.

A4–Not Classifiable as a Human Carcinogen: There are inadequate data on which to classify the agent in terms of its carcinogenicity in humans and/or animals.

A5–Not suspected as a Human Carcinogen: The agent is not suspected to be a human carcinogen on the basis of properly conducted epidemiological studies in humans. These studies have sufficiently long follow-up, reliable exposure histories, sufficiently high dose, and adequate statistical power to conclude that exposure to the agent does not convey a significant risk of cancer to humans.

---

[1]Consult the most current *ACGIH Handbook* for a complete discussion on the following:

Threshold Limit Values (TLVs) refer to airborne concentrations of substances and represent conditions under which it is believed that nearly all workers may be repeatedly exposed day after day without adverse health effects.

Threshold Limit Value–Time-Weighted Average (TLV-TWA) is the time-weighted average concentration for a normal 8-hour workday and a 40-hour workweek, to which nearly all workers may be repeatedly exposed, day after day, without adverse effect.

Threshold Limit Value–Short-Term Exposure Limit (TLV-STEL) is the concentration to which workers can be exposed continuously for a short period of time without suffering from (1) irritation, (2) chronic or irreversible tissue damage, or (3) narcosis of sufficient degree to increase the likelihood of accidental injury, impair self-rescue, or materially reduce work efficiency, and provided that the daily TLV-TWA is not exceeded.

A STEL is defined as a 15-minute TWA exposure, which should not be exceeded at any time during a workday even if the 8-hour TWA is within the TLV-TWA.

Threshold Limit Value–Ceiling (TLV-C) is the concentration that should not be exceeded during any part of the working exposure.

[2]Epidemiology is the study of disease in a general population. Determination of the incidence and distribution of a particular disease may provide information about the causes of the disease.

Evidence suggesting a lack of carcinogenicity in experimental animals will be considered if it is supported by other relevant data.

Inadvertent Exposure to Biological Hazards: Bloodborne pathogens, insects, molds, fungi, and bacteria can present hazards to humans. Safety analysis can be conducted to evaluate any system including facilities where biological research is being conducted. Analysts must be aware of any potential biological risk.

Inadvertent Exposure to Bloodborne Pathogens (such as HIV) Will Be Fatal: Bloodborne pathogens are pathogenic microorganisms that are present in human blood and can cause disease in humans. These pathogens include, but are not limited to, hepatitis B virus (HBV) and human immunodeficiency virus (HIV). OSHA issued the Bloodborne Pathogens Standard (29 CFR 1910.1030) in 1991 to prevent needle sticks and other exposures to blood and other body fluids that contain blood at work.

Insects Can Present Health Risks: Mosquitoes, flies, ticks, and other insects can cause illness such as Lyme disease and other adverse allergic responses.

Biological Agents Can Cause Dermatitis Due to Exposure from Bacteria, Fungi, or Parasites: Anthrax infection can occur from handling hides, tularemia can result from handling skins, erysipeloid can occur from handling animal products, boils and folliculitis can be caused by staphylococci and streptococci, and general infections can result from wounds.

Severe allergic Reactions Can Cause Illness and as a Result Contribute to Accidents: Allergies are the result of a response by the human's immune system to agents perceived as possibly dangerous. Allergens are present in many forms. They may be in medications, parts of foliage and plants, dust, animal dander, molds, fungi, foods, insect venom, and as a result of exposure to hazardous substances.

12.6 TOPICS FOR STUDENT PROJECTS AND THESES

1. Select a specific operation, facility, set of tasks, or function, and compile a list of unsafe acts and unsafe conditions (hazards) and conduct safety observations. Apply statistical analysis techniques and present your findings.

2. Select a statistical analysis method and explain how such an approach may be applied to solve a particular safety problem.

3. Acquire reported accident/loss information and data. Analyze the information and data using a least three different methods. Explain your results. Provide recommendations, and discuss how the use of the different methods enhanced or limited your work.

4. Select a specific operation, facility, set of tasks, or function, and conduct a health hazard assessment and provide safety requirements, recommendations, and precautions. Reference the appropriate standards, references, and studies to support your recommendations.

REFERENCES

1. C. E. Ebeling, *An Introduction to Reliability and Maintainability Engineering*, McGraw-Hill, New York, 1997.
2. T. B. Sheridan, *Humans and Automation: System Design and Research Issues*, John Wiley & Sons, Hoboken, NJ, 2002.
3. K. Ishikawa, *Guide to Quality Control*, Asian Productivity Organization, 1982.
4. W. E. Tarrants, *The Measurement of Safety Performance*, Garland Publishing, 1980.
5. K. Ishikawa, *Guide to Quality Control*, JUSE Press Ltd., Tokyo, 1994.
6. R. L. Brauer, *Safety and Health for Engineers*, Van Nostrand Reinhold, New York, 1990.
7. M. Brassard, *The Memory Jogger, A Pocket Guide of Tools for Continuous Improvement*, 2nd ed., GOAL/QPC, Methuen, MA, 1988.
8. E. Kreyszig, *Advanced Engineering Mathematics*, 7th ed., John Wiley & Sons, Hoboken, NJ, 1993.
9. J. O. Accrocco, *The MSDS Pocket Dictionary*, Genium Publishing, Schenectady, NY, 1991.
10. *American Conference of Governmental Industrial Hygienists* (ACGIH) *Handbook, current version of: Documentation of the TLVs and BEIs with Worldwide Occupational Exposure Values CD-ROM-2005*, Single User Version, 2005. ACGIH, 1330 Kemper Meadow Drive, Cincinnati, Ohio.

FURTHER READING

29 CFR Part 1910.106 *OSHA Flammable/Combustible Liquid Classification.*

29 CFR Part 1910, Subpart Z, *Toxic and Hazardous Substances* (the Z list).

29 CFR Part 1910.1030, *OSHA Blood borne Pathogens Standard*, 1991.

CHAPTER 13

MODELS, CONCEPTS, AND EXAMPLES: APPLYING SCENARIO-DRIVEN HAZARD ANALYSIS

13.1 ADVERSE SEQUENCES

In order to think in terms of accidents in conducting accident analysis or scenario-driven hazard analysis, many analysis techniques can be combined and used. Either accident sequences are reconstructed or potential future accidents are hypothesized. A model of the accident or potential accident is to be constructed. An adverse sequence is to be defined. This sequence is initiated and it progresses to an outcome—harm; then recovery is to be conducted.

13.1.1 Scenarios Within Safety Analysis

Scenario thinking has been in existence since the inception of formal safety analysis. The analyst must be able to picture what can happen or picture how events have occurred. The picture is a scenario—a snapshot in time that depicts an accident that has occurred or can occur. The scenario presents a process, an adverse event flow. There are many abstractions that can be used to depict scenarios; they are discussed in this chapter.

13.1.2 Modeling Within Safety Analysis

Modeling of sequences can be conducted in many ways, in some cases with specially designed models, automated models, and manual depictions. There are dynamic and

Assurance Technologies Principles and Practices: A Product, Process, and System Safety Perspective, Second Edition, by Dev G. Raheja and Michael Allocco
Copyright © 2006 John Wiley & Sons, Inc.

static models. A typical technique evolves the use of event trees that show sequences with branches or success or failure states. Nodes of event trees are fault or success trees. Dynamic models—such as digraphs, Petri nets, Markov models, differential equations, and state-transition diagrams—have been applied within system safety analysis. Dynamic modeling can be used in simulation in the attempt to determine what can happen under changing conditions or how events could have occurred. Simulations can provide information on physical configurations, visualizations, and physics of failure.

13.1.2.1 *Overviews and Models*   It is also appropriate to get an overall view of the system by accessing or developing an overview diagram, which shows a top-level detail of the concept of the system. Later, functional breakdown diagrams can be made for the subsystems that are a part of the total system that was described in the top-level conceptual approved diagram.

There are many types of models/diagrams that can be useful. Examples of models and diagrams are listed here:

- Process flowcharts
- Production flowcharts
- Program evaluation review technique (PERT) chart
- System interface diagrams
- Math models
- Dispersion models
- Strategic concept diagram
- Reliability block diagrams
- Event trees
- Success trees
- Fault trees
- Computer-aided drawings
- As-built drawings
- Photo records
- Digital records
- Videos
- Exploded views
- Wiring diagrams
- Software threads
- Software flowcharts
- Truth tables
- Logic trees
- Fishbone charts
- Statistical analysis charts

- Carrier diagrams
- Markovian models
- Simulation models
- Master logic diagrams
- Network diagrams
- Petri network diagrams

13.1.2.2 *Visualization*

Acquiring visual data can be most helpful in gaining knowledge of a complex system or system accident. Visual study of existing or similar systems will provide additional insight into how a system is operating in a real-time situation, in a test condition, or during a simulation. Visual records are acquired and systematically reviewed. Visual information can be acquired from many sources:

- Videos of manufacturing, installation, assembly, or disassembly operations
- Abstractions and animated redesigns of complex assemblies
- Site photographs and videos
- Animation
- Pictures and video mock-ups
- Computer simulations of operations, assembly tasks, or specific sequences
- Virtual simulations

13.1.2.3 *Scenarios, Reality, and Benefits*

There are many apparent benefits derived from scenario thinking. The more vigorous the process, the more benefits obtained. Caution should be exercised, however; models and simulations must reflect reality as close as possible. Deltas (differences) between models and simulations, that have not been identified, could introduce additional risk. Here are the benefits of scenario thinking:

Holistic View   Models can show a more holistic picture of the potential accident within the proposed system. There should be no missing logic in the constructed potential accident. The scenario is based on physical reality, the physics of the system, an understanding of all possible energy interactions, the physical transfer of energy, and even the physiological considerations of the human. A logical accident progression must be defined.

Negative Events   The goal is to provide a more comprehensive picture of the hypothesis, the details that comprise the sequence. In the abstract, the analyst uses a negative camera to take pictures of future negative events that are possible, given all the parameters of the system, including boundaries and assumptions. The negative or failure state is to be defined in these pictures.

Prototyping   System models could be developed, which, in actuality, prototype or breadboard the system and system accident. Accident models can be integrated into the overall system model and simulations can be conducted.[1]

Logical Subsets   Any logical subset within a system can be modeled. Consider modeling newly developed maintenance approaches, assembly tasks, safe operating procedures, hazardous tasks, contingencies, or operational sequences. Hazard analysis could be conducted based on any logical subset developed.

Additional Knowledge   Very large complex systems could be modeled and bounded for systematic hazard analysis. The more knowledge acquired for model development and simulation, the more the analysis could be enhanced.

System States   Complex system state conditions can be identified via modeling and simulation. These complex system states could present additional system risks. Once complex system states have been defined, enhanced hazard analysis could be conducted to identify any additional risk.

Reliability and Availability   System reliability and availability could be tested via prototype simulation and testing. If system reliability and availability are considered hazard controls, simulation and testing provide part of the hazard control validation and verification process. In cases where systems are automated, there are associated risks that may be controlled via reliability or availability.

Integrating Contingency Data   Safety-related data and safe operating and contingency procedures could be modeled, enabling a more detailed hazard analysis. Once systems are modeled, data, procedures, and processes could be displayed for quick access during an actual contingency. Rather than providing safety data in various manuals, handbooks, plans, or charts, this information could be automated and displayed on multifunction displays or monitors. Such a computer–human interface must also be evaluated from a system safety view.

Personnel Exposures   With access to large models, training, when properly conducted, could also be enhanced. Hazardous operations, processes, and tasks could be simulated and the simulations used for training purposes. Personnel who would normally be exposed to a high-risk situation will no longer be

---

[1]Model development of the Space Station Freedom was begun in the early 1990s. Stage 6 of the Station was modeled in digraph and actual FMEA data was integrated. Simulations were then conducted via computer to test system redundancy.

exposed during all phases of training. Again, such uses must also be evaluated from a system safety view.

Iterative Analysis Scenario-driven hazard analysis is an iterative process and developing models of the system and potential accidents enhances the hazard analysis efforts. The analyst can work from system or accident models or worksheets. The scenario model and worksheet forms are comprised of hazards, initiators, contributors, and primary hazards. The analyst jogs between models and worksheets. This process enables the analyst to look at problems differently, and as a result additional details are identified for further model and worksheet development. This is the concept of "memory jogging." The additional advantage of combining the scenario-driven technique with the tabular worksheet format is that it allows the entire sequence of the scenario to be conceptualized, visualized, and presented in a holistic yet nonlinear format. The scenario—the event logic—is laid out in a format that allows for each scenario's initiators, contributors, and subsequent primary hazards to be developed, along with possible effects, appropriate controls, and recorded associated risk and risk rack codes. The scenario-driven technique with tabular worksheet format also lends itself to more efficient hazard tracking and risk resolution.

Concurrent Engineering Integrating evaluation activities can enhance the concurrent engineering efforts. Specialists within assurance technologies can access system and accident models to conduct coordinated analyses concurrently. During these efforts, communication between the specialists can be increased and this can enhance the integration of the system.

Cost Savings Appropriate model simulations can also decrease the costs associated with testing. Accurate simulations can provide similar information that prototype testing provides. Through simulation and testing, more complex system risks can be identified earlier by the analysis process, which can enable the early identification of safety-related risks to assure these risks can be designed out early in the system life cycle. Reengineering the system to eliminate hazards and their associated risks at later stages is costly.

Documentation Scenarios could provide extensive documentation about safety-related risks. Good documentation could enhance liability claims defense, for example, by providing documentation of best practices and by showing great care for the protection of the general public.

Accident Investigation In conducting accident investigation and accident reconstruction, scenario models could be invaluable in documenting possible accident scenarios. Failure state modeling and simulation could document appropriate data and information at a particular accident site. Simulations can help prove theories related to complex system accidents.

13.1.3 Integration and Presentation of Analysis Information

In order to conduct scenario-driven hazard analysis, safety-related information must be integrated to present scenarios. The first step in this integration of analysis information is to design tables, spreadsheets, or worksheets, which will contain information in a logical way so that the analysis can progress systematically and be presented in an understandable fashion. Analyses should contain a scenario theme, initiators, contributory hazards, and primary hazards. Depending on the stage of the analysis—preliminary or system hazard analysis—risk parameters are indicated. Initial risk and residual risk may also be listed. In some cases, risk assessment code (RAC) is provided. Other specific risk parameters can also be indicated, defining and indicating criteria for risk ranking.

13.1.4 Narrative Reports Versus Tabular Formats

When extensive details are required, narrative formats are used. Narrative reports provide details needed to document specialized safety analysis. Formal reports, such as a Safety Engineering Report (SER) or a Safety Assessment Report (SAR), are used to document the analysis efforts, findings, results, and recommendations, but for scenario development, tabular formats are most appropriate. Here are the benefits of using tabular formats in conducting scenario-driven hazard analysis.

Iterative Process and Information Scanning By listing appropriate data in short narrative form, the analyst can scan pages and pages of data, which list scenarios and related information. An iterative process takes place and the analyst is able to develop additional similar scenarios during a review of previous work. Slightly changing the logic associated with the scenario will result in the development of additional scenarios. For example, altering the outcome or primary hazard, or changing the system state will modify a potential accident.

Each scenario in a specific tabular sheet provides appropriate information about the potential accident under study. A descriptive picture or snapshot can be provided so one can get a mental picture of how this particular accident can occur.

Enhancing Cross-checking Capabilities The tabular format enhances cross-checking capabilities; for example, grouping similar controls and altering wording to change the control logic may develop additional hazard controls.

Cross-referencing of information can be accomplished by using a tabular form. Common hazard controls are identified for determination of hazard control validation and verification purposes.

When the analyst develops accident models using event trees, dependency diagrams, or logic trees, additional cross-referencing of information can be accomplished by using a tabular form. The analysis efforts are enhanced by the concurrent activities of developing models and integration information into the tabular worksheet. Tables can easily be cross-linked with other tables or narratives to enable extensive review of information. For example, a requirements cross-check analysis, which addresses requirements associated with procedures, could be linked with an operating and support hazard analysis.

Scenario Sequencing Tabular analysis forms allow for the scenario illustration. Scenario logic progression, theme, initiator, contributors, and outcome can show scenario sequencing.

Use of Short Concise Statements Within tabular worksheets, short concise statements are used rather than long verbose statements that may be hard to review and comprehend. Long narrative statements can cause confusion especially during extensive safety reviews.

Presentation Enhancement Tabular forms are also good for the presentation of information to groups of reviewers during safety reviews. A fairly comprehensive scenario can be presented on a single slide within a power point presentation.

Customization of Analysis The analysis and the way in which information is presented or recorded in tabular form can be customized to suite the analysis type, system, review, or documentation.

Sorting and Segregating Analysis Data Sorting within the analysis is easily accomplished when the material is in tabular form. Sorting and segregating analysis data can enhance analysis efforts. Presenting scenarios information by type or hazard control or by risk will allow the analyst to evaluate the risk in a different context. In doing so, there is the potential for additional enhancement in applying the iteration concept.

Quality Review Being able to review extensive information at a glance is also an asset when using a table. The analysis will be improved as a result of extensive review and re-review. The overall quality of the analysis can be improved upon by review. The analyst checks for inconsistent logic, errors in assumptions, typos, errors in control logic, redundancy in scenarios, and verification of controls.

13.2 DESIGNING FORMATS FOR CONDUCTING ANALYSIS AND REPORTING RESULTS

Analyses can be developed and presented in many ways depending on the objective, the types of methods used, the techniques applied, and the documentation requirement. Consideration is given to whether or not the analysis is at a preliminary or at a subsystem or system level. Further consideration must be given to the type

S #	Scenario Description	Initial Contributors	Subsequent Contributors	Phase	Possible Effect	Recommendations, Precautions, and Controls
System State and Exposure						

Figure 13.1 Example of a basic tabular format heading for scenario-driven hazard analysis.

S #	Scenario Description	Initial and Subsequent Contributors	Possible Effect	Phase	Initial Risk/ Residual Risk	Recommendations, Precautions, and Controls
System State and Exposure						
System or Subsystem Evaluated						

Figure 13.2 Additional example of a tabular format heading with more detail.

and amount of information needed to support the analysis. A tabular format should be designed to accommodate very concise information, which is presented in an integrated fashion. The scenario "picture" has to be defined. Note that it will not be possible to present all pertinent information and data; referencing has to be made to other supportive information. Below are examples of format headings that have been developed and used within the process.

A typical tabular format heading is displayed in Fig. 13.1. The Scenario Number in indicated in the first column. The second column contains the Scenario Description or Scenario Theme. The Initiators are indicated in the third column. All other Contributors are listed in the fourth column. The Mission Phase or Operational Phase in indicated in column 5. The Possible Effect or Primary Hazard is presented in column 6. The Recommendations, Precautions, and Controls are provided in the last column. The next row contains a description of the System State and Exposure.

As stated previously, the analysis and the way in which information is presented or recorded in tabular form can be customized to suite the analysis type, system, review, or documentation. There are many ways to customize worksheets and Fig. 13.2 shows an additional example.

Figure 13.2 illustrates a slightly more complicated format. The Scenario Number is provided in column 1. Column 2 provides the Scenario Description. In column 3, both the Initial and Subsequent Contributors are listed. The Possible Effect in provided in column 4. The Phase or Life Cycle in which the accident can occur is indicated in column 5. Initial Risk and Residual Risk are listed in column 6. Recommendations, Precautions, and Controls for residual risk are provided in the last column. The System State and Exposure information is to be provided in the second row. Within the third row, the System or Subsystem under analysis is indicated.

Figure 13.3 displays a more detailed worksheet format. Note that Baseline Recommendations, Precautions, and Controls are provided in the first row, column 6. The baseline controls are defined in order to identify initial risk within a complex existing system. The Final Recommendations, Precautions, and Controls

S#	Scenario Description	Initial and Subsequent Contributors	Possible Effect	Phase	Baseline Recommendations, Precautions, and Controls	Final Recommendations, Precautions, and Controls
S T C	System State and Exposure				Initial Risk/RAC	Residual Risk/RAC
	System or Subsystem Evaluated					

Figure 13.3 Example of a tabular format heading with additional details and risk assessment information.

S #	Scenario Description	Initial Hazard	Contributory Hazards	Primary Hazards	Initial Risk/ Residual Risk	Recommendations, Precautions, and Controls
	System State and Exposure		Comments			Additional References
	System or Subsystem Evaluated		Tracking and Risk Resolution			

Figure 13.4 Example of a tabular format heading with additional details and conventional terms.

are provided in the first row, column 7. Risk Assessment Codes have also been added to the analysis. In column 1 row 2, the Scenario Type Code (STC) has been indicated. Scenarios are separated by STC, which is an indicator of the initiators associated with the scenario.

Figure 13.4 provides an example of a hazard analysis format using more conventional terms addressed by Hammer [1]. In row 1, column 1 indicates the Scenario Number, column 2 provides the Scenario Description, which is the potential accident theme. Columns 3, 4, and 5 indicate the hazard sequence from the Initial, Contributory, and Primary Hazards. In column 6, both the Initial Risk and Residual Risk are defined. Recommendations, Precautions, and Controls are provided in column 7. Within the second row, The System State and Exposure is indicated, along with relevant Comments and References. Here reference is made to other support work such as other analyses, trade studies, or design documentation. The System or Subsystem under evaluation is indicated in the third row and status associated with the risk is also indicated under Tracking and Risk Resolution. The status providing the details involving Hazard Control Validation and Verification is provided. The Scenario (risk) could be in the Open, Closed, or Monitoring State.

13.3 DOCUMENTATION REPORTS

For documentation purposes, there are many types of system safety-related reports. For example, there are Safety Engineering Reports, Safety Assessment Reports, Safety Action Records, Hazard Action Reports, Accident Analysis Reports, and Specialized Safety-Related Reports documenting the state of design, investigation results, studies, and special analyses results.

13.3.1   Reporting Analysis Results

Hazard analysis efforts are documented in Safety Engineering Reports or Safety Assessment Reports. These reports can be used to present analysis results or to report current system safety efforts. Depending on the extensiveness of the project, a single report may be developed or many reports may be written. The purposes of these reports are to document the analysis results. Report content will vary depending on the purpose of the communication and the communication objectives. Consider the audience who will be reading the report. As an analyst, consider what you want to communicate. Extensive details may be provided to engineers, or high-level summary information may be communicated to upper management. Extensive detail may be needed to adequately document your analysis.

13.4   CONCEPTUAL MODELS

To further enhance and illustrate the discussion of the scenario thinking, a number of models are presented for consideration.

13.4.1   Hammer Model

The first and most appropriate model addressed is what is considered the "Hammer model." The scenario concept first came to mind after study of Hammer's books and material on system safety and in later discussions with Hammer. Hammer

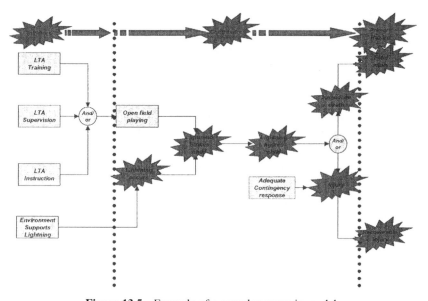

Figure 13.5   Example of a complex scenario model.

[1, pp. 63–64] initially discussed concepts of initiators, contributors, and primary hazards in the context of hazard analysis. Hammer noted that determining exactly which hazard is or has been directly responsible for an accident is not as simple as it seems.

Figure 13.5 presents an oversimplified example of a potential accident involving a child playing in an open field during lightning conditions, with possible outcomes. The initiating hazards are: less than adequate (LTA) training, supervision, and instruction, including the existing lightning conditions. The subsequent contributors involve the exposure of the child to lightning because of the open field playing, the actual lightning strike, and the intensity of the strike. The primary hazards include possible fatal or nonfatal injury.

13.4.2 Complex Scenario Models

A more complex scenario (a system accident) involving an operator and an automated process is illustrated in Fig. 13.6. The illustration represents initiating events (I) and contributory events (C). The adverse flow is from left to right. Arrows indicate directional flow. An oval represents a node, an important event in the sequence. There are branches leading into nodes, which define underlying events.

An initial malfunction, failure, or anomaly occurs due to software coding error, specification error, or calculation error. This situation leads to a malfunction; the

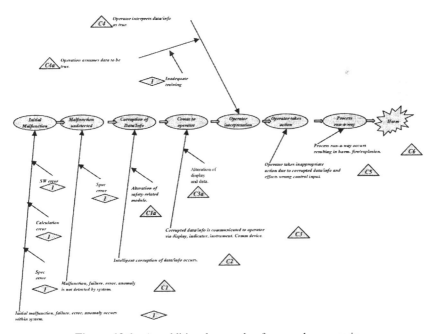

Figure 13.6 An additional example of a complex scenario.

malfunction is undetected due to specification error, consequently resulting in the alteration or corruption on a safety-related module. The corrupted data/information is communicated to an operator via a display. The operator does not recognize the hazardous misleading data/information. The operator incorrectly determines that the corrupted data/information is true. The operator then takes inappropriate action and inputs an incorrect control command. Consequently, a process run-a-way occurs resulting in a fire and explosion.

13.4.3 Fishbone Diagrams

Scenarios can also be depicted via fishbone diagrams (Fig. 13.7). Adverse event flows are shown by arrow, which flow from left to right. The scenario theme can be defined within the arrow, with inputs from branches indicating initiating and contributory hazards. Each branch defines the initiating and contributory hazards associated with hardware, human, software, and environment. The integrated scenario is to be presented. The symbol key below the diagram defines the elements within the diagram. Such diagrams can be designed to show various levels of decomposition and branches can be added or excluded to suite the scenario or system under study.

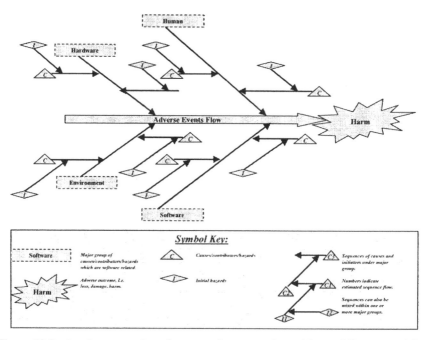

Figure 13.7 Another example of a complex scenario, with a different modeling convention.

13.5 LIFE CYCLE OF A SYSTEM ACCIDENT

An additional model discussed here addresses and illustrates the concept that a system accident has a life cycle associated with it (Fig. 13.8). Accidents are initiated, they progress, and harm can result. In conducting scenario-driven system hazard analysis, the analyst should consider this concept. A system is in dynamic equilibrium when it is appropriately designed. The system is operating within specification and within design parameters. The system is operating within the envelope. However, when something goes wrong and an initiator occurs, the system is no longer in balance. The adverse sequence progresses and the imbalance worsens until a point-of-no-return, and harm results. In conducting analysis, consider the accident life cycle and how these adverse sequences progress. By the application of hazard control, the adverse flow can be stopped. It is important that any imbalance is detected and the system is stabilized and brought back to a stable state. Furthermore, should the adverse sequence progress past the point-of-no-return, the resultant harm may be minimized or decreased by the application of hazard control. Should harm result, the system should be brought back to a normal stable state. Consider that additional harm can occur during casualty, contingency, or attempted recovery. During hazard control application, not only are all the initiators, contributors, and primary hazards controlled or eliminated, but also contingency, recovery, and damage and loss control are applied toward the system life cycle. The system must be restabilized.

13.5.1 Complex Interactions

System accidents can be complex interactions between system elements: humans, hardware, software, and the environment. Consequently, in constructing scenarios,

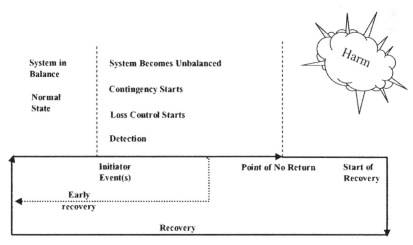

Figure 13.8 Life cycle of a system accident revisited.

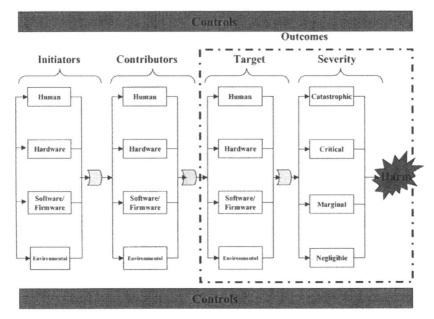

Figure 13.9 Universal scenario-driven model. (Illustration by Robert Thornburgh.)

the analyst must consider such interactions. The elements of potential accidents are made up of success and failure states—positive and negative elements. Consider initiating and contributory hazards as negative elements and positive elements as appropriate system state situations. The model in Fig. 13.9 illustrates this concept.

13.6 OPERATING AND SUPPORT HAZARD ANALYSIS EXAMPLE

Table 13.1 shows a sample part of an operating and support hazard analysis involving electronic maintenance of a complex system. An actual analysis may consist of hundreds of scenarios with related initiating and contributory hazards. The worksheet is typical, there are nine columns: Type, Scenario #, Scenario Description, Initial Contributors, Subsequent Contributors, Phase, Possible Effect, Recommendations for Precautions, Controls, and Mitigations, and finally Risk. Columns can be provided if more detail is needed (e.g., initial or current risk, system states, existing requirements, or additional references or analyses). Further studies or analyses may be applicable to acquire more details on hardware failure, human error, or software malfunction. Cross-referencing to specific failure modes and effects analysis, human interface analysis, and software hazard analysis may be appropriate.

TABLE 13.1 A Sample Part of an Operating and Support Hazard Analysis

Type	Scenario #	Scenario Description	Initial Contributors	Subsequent Contributors	Phase	Possible Effect	Recommendations for Precautions, Controls & Mitigations	Risk
E	X	Technician may be inadvertently exposed to core high voltage when maintaining monitor (e.g., MONITOR) on the workbench. While accessing core, a technician inadvertently contacts high voltage. This can result in possible fatality.	While accessing core, a technician inadvertently contacts high voltage due to: Inadequate design (core is accessible) Human error Distraction Inadequate procedure Inadequate warnings provided Inadequate training Moisture Inadequate illumination Inappropriate use of tools or equipment	Technician contacts core. Technician provides conductive source to ground. Electrical shock occurs. Unable to let go in time. Unable to recover, apply first aid, or enable rescue in time. Energy level is fatal.	Installation/ Maintenance/ Removal/Disposal	Fatal injury	2—Ensure qualified personnel are trained in the maintenance procedures. 3—Stored energy within the core must be removed via grounding prior to initiating work. 13—Provide recurring training for affected employees in appropriate safe operating procedures associated with electrical hazards, i.e., Lockout/Tag out procedure (LOTO). 14—Design consoles/racks/electrical equipment such that all power can be removed from a single device prior to performing maintenance and develop and document the procedures to accomplish this. If it is not possible to deenergize all power within consoles/racks/electrical equipment, such power must be isolated, guarded, and identified to prevent accidental contact.	XX

(continued)

TABLE 13.1 *Continued*

Type	Scenario #	Scenario Description	Initial Contributors	Subsequent Contributors	Phase	Possible Effect	Recommendations for Precautions, Controls & Mitigations	Risk
							31—Design equipment, devices, subsystems, and systems to enable zero energy states in order to preclude injury or damage as a result of uncontrolled energy. 35—Clearly identify stored power sources to preclude injury or damage due to uncontrolled energy. 89—Design equipment, devices, subsystems, and systems to provide lighting and task illumination to preclude injury or damage, in accordance with Human Factors requirements.	
E	XX	Technician may be inadvertently exposed to core high voltage when maintaining monitor (e.g., MONITOR) on the workbench.	While accessing core, a technician inadvertently contacts high voltage due to: Inadequate design (core is accessible)	Technician contacts core. Technician provides conductive source to ground.	Installation/ Maintenance/ Removal/Disposal	Major injury	2—Ensure qualified personnel are trained in the maintenance procedures. 3—Stored energy within the core must be removed via grounding prior to initiating work.	XX

While accessing core, a technician inadvertently contacts high voltage. This can result in possible major injury.	Human error Distraction Inadequate procedure Inadequate warnings provided Inadequate training Moisture Inadequate illumination Inappropriate use of tools or equipment	Electrical shock occurs. Able to let go in time. Able to recover, apply first aid, or enable rescue in time. Energy exposure is not fatal.	13—Provide recurring training for affected employees in appropriate safe operating procedures associated with electrical hazards, i.e., LOTO. 14—Design consoles/racks/ electrical equipment such that all power can be removed from a single device prior to performing maintenance and develop and document the procedures to accomplish this. If it is not possible to deenergize all power within consoles/racks/electrical equipment, such power must be isolated, guarded, and identified to prevent accidental contact. 31—Design equipment, devices, subsystems, systems to enable zero energy states in order to preclude injury or damage as a result of uncontrolled energy. 35—Clearly identify stored power sources to preclude injury or damage due to uncontrolled energy. 89—Design equipment, devices, subsystems, and systems to provide lighting and task illumination to preclude injury or damage, in accordance with Human Factors requirements.

(*continued*)

TABLE 13.1 *Continued*

Type	Scenario #	Scenario Description	Initial Contributors	Subsequent Contributors	Phase	Possible Effect	Recommendations for Precautions, Controls & Mitigations	Risk
E	XXX	Technician may be inadvertently exposed to core high voltage when maintaining monitor (e.g., MONITOR) on the workbench. While accessing core, a technician inadvertently contacts high voltage. This can result in possible minor injury.	While accessing core, a technician inadvertently contacts high voltage due to: Inadequate design (core is accessible) Human error Distraction Inadequate procedure Inadequate warnings provided Inadequate training Moisture Inadequate illumination Inappropriate use of tools or equipment	Technician contacts core. Technician provides conductive source to ground. Minor electrical shock occurs. Able to let go in time. Able to recover, apply first aid, or enable rescue in time. Energy exposure is minimal.	Installation/ Maintenance/ Removal/Disposal	Minor injury	2—Ensure qualified personnel are trained in the maintenance procedures. 3—Stored energy within the core must be removed via grounding prior to initiating work. 13—Provide recurring training for affected employees in appropriate safe operating procedures associated with electrical hazards, i.e., LOTO. 14—Design consoles/racks/ electrical equipment such that all power can be removed from a single device prior to performing maintenance and develop and document the procedures to accomplish this. If it is not possible to deenergize all power within consoles/racks/electrical equipment, such power must be isolated, guarded, and identified to prevent accidental contact.	XX

E	XXXX	Technician could be inadvertently exposed to electrical power during removal and replacement of line replacement units (LRUs). This can result in possible fatality.	Technician is inadvertently exposed to electrical power due to: Inadequate design (power is accessible) Human error Distraction Inadequate procedure Inadequate warnings provided Inadequate training Moisture Inadequate illumination Inappropriate use of tools or equipment	Technician contacts core. Technician provides conductive source to ground. Electrical shock occurs. Unable to let go in time. Unable to recover, apply first aid, or enable rescue in time. Energy level is fatal.	Installation/ Maintenance/ Removal/Disposal	Fatal injury
						2—Ensure qualified personnel are trained in the maintenance procedures. 9—Provide guarding for each line replacement unit (LRU) associated equipment (e.g., relays, switches, bus bars) such that inadvertent contact with energized components cannot occur during installation, replacement, and/or removal of other LRUs, devices, or equipment. 10—Lockout and Tag out (LOTO) procedures must be followed and enforced prior to any replacement,	XX
						31—Design equipment, devices, systems to enable zero energy states in order to preclude injury or damage as a result of uncontrolled energy. 35—Clearly identify stored power sources to preclude injury or damage due to uncontrolled energy. 89—Design equipment, devices, subsystems, and systems to provide lighting and task illumination to preclude injury or damage, in accordance with Human Factors requirements.	XX

(continued)

TABLE 13.1 *Continued*

Type	Scenario #	Scenario Description	Initial Contributors	Subsequent Contributors	Phase	Possible Effect	Recommendations for Precautions, Controls & Mitigations	Risk
							or during any access to powered equipment, devices, or subsystems. 13—Provide recurring training for affected employees in appropriate safe operating procedures associated with electrical hazards, i.e., LOTO. 14—Design consoles/racks/ electrical equipment such that all power can be removed from a single device prior to performing maintenance and develop and document the procedures to accomplish this. If it is not possible to deenergize all power within consoles/racks/electrical equipment, such power must be isolated, guarded, and identified to prevent accidental contact. 89—Design equipment, devices, subsystems, and systems to provide lighting and task illumination to preclude injury or damage, in accordance with Human Factors requirements.	

| E | XXXXX | Technician could be inadvertently exposed to electrical power during removal and replacement of LRUs. This can result in possible major injury. | While accessing core, a technician inadvertently contacts high voltage due to: Inadequate design (power is accessible) Human error Distraction Inadequate procedure Inadequate warnings provided Inadequate training Moisture Inadequate illumination Inappropriate use of tools or equipment | Technician contacts core. Technician provides conductive source to ground. Electrical shock occurs. Able to let go in time. Able to recover, apply first aid, or enable rescue in time. Energy exposure is not fatal. | Installation/ Maintenance/ Removal/Disposal | Major injury | 2—Ensure qualified personnel are trained in the maintenance procedures. 9—Provide guarding for each LRU associated equipment (e.g., relays, switches, bus bars) such that inadvertent contact with energized components cannot occur during installation, replacement, and/or removal of other LRUs, devices, or equipment. 10—Lockout and Tag out (LOTO) procedures must be followed and enforced prior to any replacement, or during any access to powered equipment, devices, or subsystems. 13—Provide recurring training for affected employees in appropriate safe operating procedures associated with electrical hazards, i.e., LOTO. 14—Design consoles/racks/ electrical equipment such that all power can be removed from a single device prior to performing maintenance and develop and document the procedures to accomplish this. If it is not possible to | XX |

(continued)

TABLE 13.1 *Continued*

Type	Scenario #	Scenario Description	Initial Contributors	Subsequent Contributors	Phase	Possible Effect	Recommendations for Precautions, Controls & Mitigations	Risk
							deenergize all power within consoles/racks/electrical equipment, such power must be isolated, guarded, and identified to prevent accidental contact. 89—Design equipment, devices, subsystems, and systems to provide lighting and task illumination to preclude injury or damage, in accordance with Human Factors requirements.	
E	XXXXXX	Technician could be inadvertently exposed to electrical power during removal and replacement of LRUs. This can result in possible minor injury.	While accessing core, a technician inadvertently contacts high voltage due to: Inadequate design (power is accessible) Human error Distraction Inadequate procedure Inadequate warnings provided Inadequate training Moisture Inadequate illumination Inappropriate use of tools or equipment	Technician contacts core. Technician provides conductive source to ground. Minor electrical shock occurs. Able to let go in time. Able to recover, apply first aid, or enable rescue in time. Energy exposure is minimal.	Installation/ Maintenance/ Removal/Disposal	Minor injury	2—Ensure qualified personnel are trained in the maintenance procedures. 9—Provide guarding for each LRU associated equipment (e.g., relays, switches, bus bars) such that inadvertent contact with energized components cannot occur during installation, replacement, and/or removal of other LRUs, devices, or equipment. 10—Lockout and Tag out (LOTO) procedures must be followed and enforced	XX

prior to any replacement, or
during any access to
powered equipment, devices,
or subsystems.

13—Provide recurring training
for affected employees in
appropriate safe operating
procedures associated with
electrical hazards, i.e.,
LOTO.

14—Design consoles/racks/
electrical equipment such that
all power can be removed
from a single device prior to
performing maintenance and
develop and document the
procedures to accomplish
this. If it is not possible to
deenergize all power within
consoles/racks/electrical
equipment, such power must
be isolated, guarded, and
identified to prevent
accidental contact.

89—Design equipment, devices,
subsystems, and systems to
provide lighting and task
illumination to preclude
injury or damage, in
accordance with Human
Factors requirements

13.7 TOPICS FOR STUDENT PROJECTS AND THESES

1. Conduct research on a past catastrophic accident and discuss the published findings. Determine any shortcomings, biases, oversights, omissions, or errors that may have been made in conducting the accident investigation.
2. Investigate a number of accidents that have occurred within a particular facility, industry, or company, and conduct a loss analysis. Determine what were the initiators and contributors within the past accidents and make recommendations to assure acceptable risk.
3. Acquire loss data and apply statistical analysis and explain your findings.
4. Define the benefits of conducting scenario-driven hazard analysis.
5. Select a complex system and conduct a scenario-driven hazard analysis and explain your findings.
6. Select a hazardous system and define the scope, purpose, and objective of conducting system safety engineering and management. Define the criteria and information needed to document system safety efforts within a safety engineering report.
7. Evaluate a particular high-risk operation and determine appropriate risk assessment criteria: risk acceptance, exposure, severity, and likelihood scales. Define what particular factors will be associated with risk acceptance.
8. Assume you have been selected to be a Chair of a formal safety review board. The board is to evaluate the development and design of a highly hazardous complex system. Design appropriate processes to conduct successful safety reviews and document the processes in a formal plan.
9. Choose a complex system and apply five modeling techniques and explain why the techniques were selected. What are your observations after application?

REFERENCE

1. W. Hammer, *Handbook of System and Product Safety*, Prentice-Hall, Englewood Cliffs, NJ, 1972.

FURTHER READING

Allocco, M., Hazards in Context with System Risks. In: *Proceedings of the 23rd International System Safety Conference*, System Safety Society, April 2005.

Allocco, M., and J. F. Shortle, Applying Qualitative Hazard Analysis to Support Quantitative Safety Analysis for Proposed Reduce Wake Separation CONOPS. In: *Proceedings of the NASA/FAA ATM (Air Traffic Management) Workshop*, Baltimore, MD, June 2005.

Allocco, M., Key Concepts and Observations Associated with a Safety Management System. In: *Proceedings of the 22nd International System Safety Conference*, System Safety Society, August 2004.

Allocco, M., Computer and Software Safety Considerations in Support of System Hazard Analysis. In: *Proceedings of the 21st International System Safety Conference*, System Safety Society, August 2003.

Allocco, M., and R. P. Thornburgh, A Systemized Approach Toward System Safety Training with Recommended Learning Objectives. In: *Proceedings of the 20th International System Safety Conference*, System Safety Society, August 2002.

Allocco, M., W. E. Rice, and R. P. Thornburgh, System Hazard Analysis Utilizing a Scenario Driven Technique. In: *Proceedings of the 20th International System Safety Conference*, System Safety Society, August 2002.

Allocco, M., Consideration of the Psychology of a System Accident and the Use of Fuzzy Logic in the Determination of System Risk Ranking. In: *Proceedings of the 19th International System Safety Conference*, System Safety Society, September 2001.

Allocco, M., Appropriate Applications Within System Reliability Which Are in Concert with System Safety; The Consideration of Complex Reliability and Safety-Related Risks Within Risk Assessment. In: *Proceedings of the 17th International System Safety Conference*, System Safety Society, August 1999.

Allocco, M., Automation, System Risks and System Accidents. In: *Proceedings of the 17th International System Safety Conference*, System Safety Society, August 1999.

Allocco, M., Development and Applications of the Comprehensive Safety Analysis Technique, *Professional Safety*, December 1997.

Allocco, M., Hazard Control Considerations in Computer Complex Designs, *Professional Safety*, March 1990.

Allocco, M., Focus on Risk, *Safety & Health*, September 1988.

Allocco, M., Chemical Risk Assessment and Hazard Control Techniques, *National Safety News*, April 1985.

Allocco, M., Hazard Recognition for the Newcomer, *National Safety News*, December 1983.

CHAPTER 14

AUTOMATION, COMPUTER, AND SOFTWARE COMPLEXITIES

14.1 COMPLEX SYSTEM ANALYSIS

Software is becoming an extensive part of a complex automated system[1] (Fig. 14.1). In conducting system hazard analysis, the entire system must be evaluated considering the complex human, sophisticated hardware designs, and microelectronics providing the firmware interface between hardware and software machine instruction. Because of these complexities, special knowledge and experience are required in human engineering, hardware and material engineering, and software engineering. Specialty engineers in human and software reliability, computer engineering, quality assurance, maintainability, and logistics are also needed to concurrently develop a successful complex system design. To keep up with these disciplines, system safety specialties have also evolved, specializing in hardware, software, and system analysis.

---

[1] The concepts involving the system context that are discussed within this section have been presented in the following: M. Allocco, Computer and Software Safety Considerations in Support of System Hazard Analysis. In: *Proceedings of the 21st International System Safety Conference*, System Safety Society, August 2003.

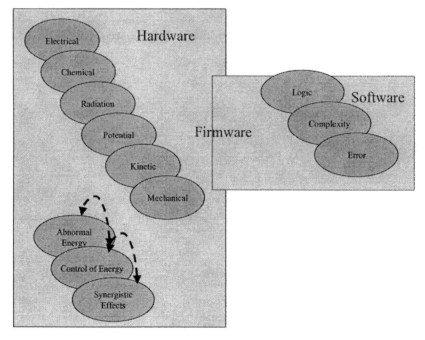

Figure 14.1 Complex automated system.

14.2 SYSTEM CONTEXT

It is up to the "system" integrator (the system safety engineer) to ensure that all safety-related risks are identified, eliminated, or controlled to an acceptable level. Consider that system accidents are the result of many hazards. Accidents can be driven by hardware failures and code errors, which affect computer subsystems. Generally, human error contributes to the hardware and software problem that can result in a system accident.

Thinking logically, the system is comprised of the human, machine, and environment. Furthermore, the machine can be decomposed into the computer subsystem. The computer subsystem is also broken down into elements involving computer concept development, logic design, coding/writing software, compiling the higher-level code into machine language, the consequent loading of the instruction into the computer, and the designing of the appropriate computer architecture initially.

Almost any entity can be considered a system meeting the definition: composite, at any level of complexity, of personnel, procedures, materials, tools, equipment, facilities, and software. The elements of this composite entity are used together in the intended operational or support environment to perform a given task or achieve a specific production support requirement: a set or arrangement of components so related or connected as to form a unity or organic whole.

Currently, there is a concept of a hierarchy where the term "systems of systems" is in use. Eventually, rather than using the term subsystem, or component, or part, any entity can be considered a system within a system. Consider the microworld of subatomic particles and the macroworld of a modern aircraft carrier. Furthermore, the aircraft carrier "system" is integrated into the world, where it interfaces with the sea and oceans. In conducting system safety for computers and software, the analyst must be able to understand how a latent problem in the microworld (e.g., microelectronics) can ultimately initiate an adverse event sequence in the macroworld.

14.3 UNDERSTANDING THE ADVERSE SEQUENCE

The system safety analyst must understand the potential adverse sequence of the system. Consider a system accident as a potential adverse integration of success and failure states that will eventually result in harm. The analyst must be able to decompose this potential adverse integration of success and failure states.

Errors can occur anywhere along the sequence, which can contribute to latent or real-time hazards. Errors can occur during computer concept development, when poor assumptions are made and objectives are ill defined; errors can occur during coding and the writing of software; and errors can happen during the compiling of the code. When compiling is conducted automatically, not only can errors be introduced, but malfunctions can also take place within the complier. Furthermore, errors, malfunctions, and failures could occur within the end of the adverse sequence.

14.3.1 Malfunction and Failure Modes

Once the code is compiled, it becomes firmware, which is an electromagnetic state maintained by microelectronics—hardware. Following this logic, the computer system will be affected by an abnormal energy exchange. Stress, heat, vibration, power, electromagnetism, chemical reactions, and consequent synergistic effects—the physics of failure—affect microelectronics. Any applicable energy form appropriate to the computer, if uncontrolled, will adversely affect the system. This possibility must always be a consideration in safety analysis.

14.3.2 Understanding System Functions

If the analyst is thinking in terms of system functions by conducting functional hazard analysis, the output of such an analysis will generate high-level functional hazard controls. To understand the system risk, the analyst must define what conditions drive the function. The analyst must determine how the functions (or objects) will work and malfunction in the real world. Latent hazards can be introduced anywhere within the creation, concept development, and initial design phases. These hazards are initiators because they were introduced into the system.

If code is being evaluated, consider the potential for human error, logic errors, errors of assumption, and errors of omission.

14.3.3   Understanding Conceptual Processes

An analyst must be able to decompose the system by abstractions and by conceptual processes—keeping in mind that those abstractions and conceptual processes must reflect the real world. In evaluating computers and software modeling, a form of abstraction is applied. Consequently, the model must truly reflect the system, including mathematical abstractions and calculations.

An example of a conceptual process is the object-oriented methodology (OOM), which is a process that is independent of any programming language until the final stage [1, pp. 267–270]. It is not the intention here to fully describe the OOM process; however, the analyst must have an understanding of various abstractions and concepts to be able to conduct the analysis. It is therefore appropriate to plan on working with a subject matter expert.

OOM is a formal method of creating a software structure and abstraction. The concept involves the use of "objects," which are represented as instances of a specific class in programming languages. A class represents a template for several objects and describes how these objects are structured internally. An object retains certain information on how to perform certain operations, using input provided by the data and the method imbedded in the object. An object contains data structures (attributes) and behavior (operations or methods) that are grouped into a class. There are two types of object diagrams: class and instance diagrams. A class diagram can be a schematic, pattern, or template that describes many possible instances of data. An instance diagram describes how objects relate to each other. A class diagram describes object classes. An instance diagram describes object instances.

14.4   ADDITIONAL SOFTWARE SAFETY ANALYSIS TECHNIQUES

There are many specific software safety techniques used to accomplish specific detailed analysis. A review of these is provided in Table 14.1.

14.4.1   Software Malfunction

Software does not fail; hardware and firmware can fail. Humans can make software-related errors. Design requirements can be inappropriate. Humans can make errors in coding. The complexity or extensive software design could add to the error potential. There could be other design anomalies, sneak paths, and inappropriate do-loops. There are specific software analysis and control methods that can be applied successfully to initiating and contributory hazards, which are related to software.

TABLE 14.1 Example Software Safety Analysis Techniques

Technique	Summary	Applicability and Use
Hardware/software safety analysis	The analysis evaluates the interface between hardware and software to identify hazards within the interface.	Any complex system with hardware and software.
Interface analysis	The analysis is used to identify hazards due to interface incompatibilities. The methodology entails seeking those physical and functional incompatibilities between adjacent, interconnected, or interacting elements of a system, which, if allowed to persist under all conditions of operation, would generate risks.	Interface analysis is applicable to all systems. All interfaces should be investigated; machine–software, environment–human, environment–machine, human–human, machine–machine, and so on.
Modeling and simulation	There are many modeling technique forms that are used in system engineering. Failures, events, flows, functions, energy forms, random variables, hardware configuration, accident sequences, and operational tasks—all can be modeled.	Modeling is appropriate for any system or system safety analysis.
Network logic analysis	Network logic analysis is a method to examine a system in terms of mathematical representation in order to gain insight into a system which might not ordinarily be achieved.	The technique is universally appropriate to complex systems.
Petri net analysis	Petri net analysis is a method to model unique states of a complex system. Petri nets can be used to model system components, or subsystems at a wide range of abstraction levels; for example, conceptual, top–down, detail design, or actual implementations of hardware, software, or combinations.	The technique is universally appropriate to complex systems.

(continued)

TABLE 14.1 *Continued*

Technique	Summary	Applicability and Use
Software failure modes and effects analysis	This technique identifies software-related design deficiencies through analysis of process flowcharting. It also identifies areas for verification/validation and test evaluation.	Software is embedded into vital and critical systems of current as well as future aircraft, facilities, and equipment. This methodology can be used for any software process; however, application to software-controlled hardware systems is the predominant application. It can be used to analyze control, sequencing, timing monitoring, and the ability to take a system from an unsafe to a safe condition.
Software fault-tree analysis	This technique is employed to identify the root cause(s) of a "top" undesired event, to assure adequate protection of safety-critical functions by inhibits, interlocks, and/or hardware.	Any software process at any level of development or change can be analyzed deductively.
Software hazard analysis	The purpose of this technique is to identify, evaluate, and eliminate or mitigate software hazards by means of a structured analytical approach that is integrated into the software development process.	This practice is universally appropriate to software systems.
Software sneak circuit analysis	Software sneak circuit analysis (SSCA) is designed to discover program logic that could cause undesired program outputs or inhibits or incorrect sequencing/timing.	The technique is universally appropriate for any software program.

14.4.2   Manifestation of Software-Related Risks

Because systems are becoming more and more complex, more involved with millions of lines of code, it may not be possible to test every safety-related function. This being the case, there is always the situation where something could have been overlooked. This is why many system accidents involving software do not manifest immediately following a new design or modification. It may take years for the latent logical error to initiate. As discussed, system accidents could be the result of many initiators, contributors, and particular system states. To assure acceptable risk, it may be prudent to assume that such complex computer systems will malfunction, resulting in system accidents. The analyst should address the life cycle of the accident and consider all means of controlling risk.

System monitoring and detection are vital. Contingency should also be addressed. Both physical and software redundant hazard controls may be required to control abnormal energy reactions associated with automated processes and functions. Misdirection, inappropriate indication, and miscommunication between the human and automated subsystem should also be evaluated. If the human is a vital link in the safety-critical function, the human can become the contingency control. Human factors controls should, for example, address hazardous misleading information that can occur in the event of computer malfunction. It is important for the human to recognize a safety-related deviation and react appropriately if possible. Concepts of human reliability become paramount.

14.4.3   Understanding Anomalies

Unfortunately, it may not be possible to determine what the initiator was in all situations when complex computer system anomalies occur. It may not be possible to replicate the anomaly. Also consider security requirements, privileged information, trade secrets, and copyright privilege or inappropriate reliability data as limiting factors on acquiring appropriate design information. We may never know what the problem was after the crash or when the computer hangs or stops or corrupts data. The analyst should not automatically have confidence in such complex computer systems and should assume the potential for malfunction.

14.4.4   Complexity, Understanding Risks, and System States

Because of the complexity of large automated systems, the safety analyst is faced with a resource problem. It may not be possible to drill down into particular software/computer logic threads within very complex convoluted systems. The analyst must consider possible hazards associated with the human, machine, and environment. Each of these efforts can be very extensive. It is important to use resources prudently. First off, it is appropriate to understand the potential system accidents throughout the life cycle of the system. This can be accomplished by conducting system hazard analysis. As an output of such analysis, risks are to be identified, eliminated, or controlled. These risks can then be ranked relatively and resources can be expended logically.

14.4.5 System States

Considering the computer subsystem, it is very important to expend resources to identify all unsafe system states and to define the robust data flow and state transition paths to the unsafe states, including the unsafe transitions. The dynamics of such system states in conjunction with timing of transition are also to be evaluated. Thought should also be given to the evaluation of all safety-critical functional logic, code logic, and physical architecture design.

14.4.6 Complexity of Initiators, Contributors, and System Accidents

System accidents can be simplistic or complex. Accidents can occur during automated operations due to errors in coding or logic errors or incomplete specifications. Single anomalies can have catastrophic outcomes. Consider a weapon system inadvertently operating or a medical device malfunctioning at a critical time due to an apparent single error. A simple logic error could have a direct adverse outcome. Therefore, it is prudent to perform hazard analysis at the functional level, logic design level, and code level, where inputs and outputs could be erroneous.

In constructing potential accidents, consider the apparent simplistic initiators as well as very complex potential accidents with many initiators and contributory hazards. Thinking should not be totally confined to so-called software and hardware hazards only. It should include EMI, RFI, and system noise. Separate analyses for software and hardware may or may not be appropriate depending on the situation. Analysis efforts should be conducted to address integrated risks as well as separate software and hardware hazards.

14.4.7 Functional Abstractions and Domains

There are limitations when analysis work is confined only to abstract software logic or system functions. A characteristic assumption is to consider the failure of, or loss of, a particular function as a hazard. Consequently, a safety criticality is assigned to the "hazard" and mitigation is then provided to assure that the function does not fail.

Consider that there can be many domains that can be used to define the real world. For example, within the design world there are four domains: the customer domain, functional domain, physical domain, and process domain [1, p. 10]. The customer domain is characterized by the needs or attributes that the customer is looking for in a product, process, or system. Within the functional domain, needs are transformed into functional requirements and constraints. Design parameters are within the physical domain; they satisfy the functional requirements. To produce the product in terms of the design parameters, process variables are applied within a process domain. Latent hazards and errors can be introduced within any of the four domains.

14.4.8 Latent Hazards Throughout the Life Cycle

Thinking more out of the box may help an analyst address potential complex system accidents. The human, hardware, software, or environment can introduce latent hazards into a system in many ways. The introduction of latent hazards, or accidents, can occur at any time within the life cycle of the system—during creation, concept development and initial design, assessment of the concept, demonstration of the prototype, manufacture, deployment, and disposal.

14.4.9 Errors in Model Use and Development

Errors can also be introduced when inappropriate abstracted models are used to conduct analysis, do bread boarding, and conduct calculations or simulations. Functional models are used to reflect system operations. Fault trees or event trees are constructed to reflect failure states of the system. Latent hazards can be introduced when models deviate from design reality, or when they are in error or inappropriately constructed or used. Automated tools used for calculations, model development, and graphical presentation can also be inappropriate or in error.

14.4.10 Understanding Safety Criticality

Autonomous automated systems could be considered safety-critical in that if they malfunction a catastrophic accident results. *Anything that interfaces with such a system and that can introduce hazards should also be considered safety-critical.* When models and automated support tools are used, they too should be considered safety-critical when they can adversely affect the safety-critical system. Errors are made when there is a failure to consider this relationship and exposure. This logic is also applicable when evaluating a backup subsystem, standby safety device, or any hazard control associated with a safety-critical system. The backups, standby safety devices, and controls all should then be considered safety-critical. An inappropriate assumption is made when backups, standby safety devices, and controls are not considered as important because they are only backups. Should these backups, standby safety devices, and controls fail when needed, a catastrophic outcome can occur.

14.4.11 Understanding Transfer and Switching Complications

There is an additional complication to consider with autonomous automated systems. These systems may rely on automated transfer or switching capabilities in order for a redundant backup to work. They may also rely on automated detection of the malfunction prior to initiating the transfer or switching. The detection, transfer, or switching operation should also be considered safety-critical if the initial lag or thread is considered safety-critical. Should automated detection, transfer, or switching malfunction when needed, a catastrophic outcome will result. In many situations, there are single-point failures/malfunctions that have been overlooked, as common causes. In these situations, the hardware, software, firmware, environment, and human must also be evaluated for single events and common causes.

14.5   TRUE REDUNDANCY

The lack of understanding of what is true redundancy can introduce latent hazards. A truly redundant subsystem must be as independent as possible from the initial subsystem, lag, or thread. Utilizing two diverse architectures with diverse computers and hardware with common software defeats the redundancy concept in that there may be common software errors between the seemingly diverse architectures. The concept of N-Version Software is applied for true diversity. This software is developed and tested to fulfill a set of requirements where multiple versions of software are intentionally made independent and different. Differences can be in some or all of specifications, design, use of language, algorithms, or data structures.

In order for true redundancy to exist, the state of the detection, transfer, or switching and the redundant lag must be known. A system should not be considered redundant unless it is known that the redundancy will work.

14.5.1   System Redundancy

Redundancy can also be applied or satisfied in many ways. Hardware, software, firmware, environment, and human redundancy can be applied. It may not be necessary to develop a similar redundant function or lag. It may be possible to back up an automated function by the human operator via a safety procedure. A hardware device can back up a human or automated function. Many redundant complexities have been utilized from a systems view.

14.6   COMPLEXITIES AND HAZARDS WITHIN COMPUTER HARDWARE

Electronics, avionics, and microelectronics can all be affected by abnormal energy; heat, cold, and extreme temperature change; vibration; excessive current flow; microcracking of silica; delaminating of substrates; electromagnetic environmental effects; and neutron and gamma radiation. External hazards can also cause harm: water damage, sprinkler discharge, earthquake, physical damage, connector mismating, bent pins within connectors, electrical shorts, chafing of wires or cables, the introduction of foreign objects, smoke damage, burning of composite materials, incompatibility of materials, corrosion, ionization of materials, and rodent or animal damage.

14.6.1   Controls, Mitigations, and Added Complexities

Providing additional redundancy, adding built-in testing capabilities, automating an operation, and designing a standby or switching capability obviously add complexity and may introduce more safety-related risk. All controls, mitigations, and added complexities must be evaluated from a system safety view. A technique called hazard control analysis should be applied to evaluate these added extras.

The analyst must not do additional harm or increase safety-related risks when recommending additional controls to be designed. The analyst must determine the adequacy, effectiveness, and efficiency in the application of a hazard control. The control must be designed to work when required.

14.7 INITIATORS AND CONTRIBUTORS: THE ERRORS ASSOCIATED WITH SOFTWARE

Errors associated with software are initiators and contributors within the potential accident sequence. Table 14.2 lists example initiators and contributors that can occur within inputs, processing, and output software logic. Consider the software logic involving processing as a white box with software logic inputs and outputs.

TABLE 14.2   Initiators, Contributors, and Errors Associated with Software

Factor Name[a]	Errors Resulting in the Inappropriate Implementation of Software
Computational error	Coded equations, algorithms, and/or models may be in error.
Initiator	Assumptions, logic, knowledge, and experience are less than adequate (LTA); initiates errors in math
Initiator	Assumptions, logic, knowledge, and experience are less than adequate (LTA); initiates errors in algorithm.
Initiator	Assumptions, logic, knowledge, and experience are less than adequate (LTA) initiates errors in models.
Subsequent Initiator	LTA system specification adversely affects software specification.
Subsequent Initiator	LTA software specification affects math equation application.
Subsequent Initiator	LTA software specification affects algorithm.
Subsequent Initiator	LTA software specification affects model development and use.
Subsequent Initiator	LTA system specification adversely affects hardware/ computer specification.
Subsequent Initiator	LTA hardware specification affects math equation application.
Subsequent Initiator	LTA hardware specification affects algorithm.
Subsequent Initiator	LTA hardware specification affects model development and use.
Contributor	Math equation is in error.
Contributor	Algorithm is in error.
Contributor	Models used in analysis are in error.
Contributor	Models used in code design are in error.

(*continued*)

TABLE 14.2 *Continued*

Factor Name[a]	Errors Resulting in the Inappropriate Implementation of Software
Configuration error	Configuration error occurs when incompatibly occurs between versions or types of software, that is, operating and application software, updates, patches, and conversions.
Initiator	Assumptions, logic, knowledge, and experience are less than adequate (LTA); initiates errors in configuration design.
Subsequent Initiator	LTA software specification affects configuration design.
Subsequent Initiator	LTA system specification adversely affects hardware/ computer specification, involving configuration.
Subsequent Initiator	LTA hardware specification affects configuration design.
Contributor	Error occurs within the logic associated with the interface between operating and application software.
Contributor	Error occurs within the logic associated with the compiler.
Contributor	Error occurs within the logic associated with the interfaces between updates, patches, and conversions.
Contributor	Error occurs within the logic associated with the interface timing between operating and application software.
Contributor	Error occurs within the logic associated with timing and compiler application.
Contributor	Error occurs within the logic associated with the interface timing between updates, patches, and conversions.
Data handling errors	Data handling errors are made during reading, writing, moving, storing, and modifying data via code instruction.
Initiator	Assumptions, logic, knowledge, and experience are less than adequate (LTA); initiates errors in data handling logic.
Subsequent Initiator	LTA software specification affects data handling design.
Subsequent Initiator	LTA system specification adversely affects hardware/ computer specification, involving data handling.
Subsequent Initiator	LTA hardware specification affects data handling design.
Contributor	Data handling errors occur during reading of data.
Contributor	Data handling errors occur during writing of data.
Contributor	Data handling errors occur during data movement.
Contributor	Data handling errors occur associated with data storage and retrieval.
Contributor	Data handling errors occur during data modification.
Contributor	Data handling errors occur associated with timing.
Database errors	Errors within database design, selection, addressing, sequencing, inappropriate data, data storage, and intelligent corruption of data.
Initiator	Assumptions, logic, knowledge, and experience are less than adequate (LTA); initiates errors in database logic.

(*continued*)

TABLE 14.2 *Continued*

Factor Name[a]	Errors Resulting in the Inappropriate Implementation of Software
Subsequent Initiator	LTA software specification affects database design.
Subsequent Initiator	LTA system specification adversely affects hardware/ computer specification, involving database.
Subsequent Initiator	LTA hardware specification affects database design.
Contributor	Database errors occur within design.
Contributor	Database errors occur within selection of data.
Contributor	Database errors occur within addressing of data.
Contributor	Database errors occur within sequencing of data.
Contributor	Database errors occur due to inappropriate data use.
Contributor	Database errors occur due to inappropriate data storage.
Contributor	Database errors occur due to intelligent corruption of data.
Definition errors	Errors occur due to the inappropriate definition of concepts, functions, variables, parameters, or constants.
Initiator	Assumptions, logic, knowledge, and experience are less than adequate (LTA); initiates errors in definition logic.
Subsequent Initiator	LTA software specification affects definition logic.
Subsequent Initiator	LTA system specification adversely affects hardware/ computer specification, involving definition logic.
Subsequent Initiator	LTA hardware specification affects definition logic.
Contributor	Definition error occurs within concept.
Contributor	Definition error occurs associated with functions.
Contributor	Definition error occurs within variables.
Contributor	Definition error occurs within parameters.
Contributor	Definition error occurs within constants.
Intermittent errors	Temporary states that are not appropriate and that appear to have occurred as a result of inappropriate software logic; there is no apparent hardware contributor.
Initiator	Assumptions, logic, knowledge, and experience are less than adequate (LTA); initiates errors in logic.
Subsequent Initiator	LTA software specification affects logic.
Subsequent Initiator	LTA system specification adversely affects hardware/ computer specification, involving logic.
Subsequent Initiator	LTA hardware specification affects logic.
Contributor	Temporary state occurs with adverse affects.
Recurrent errors	Recurrent errors that are not appropriate and that appear to have occurred as a result of inappropriate software logic; there is no apparent hardware contributor.
Initiator	Assumptions, logic, knowledge, and experience are less than adequate (LTA); initiates errors in logic.
Subsequent Initiator	LTA software specification affects logic.

(*continued*)

TABLE 14.2 *Continued*

Factor Name[a]	Errors Resulting in the Inappropriate Implementation of Software
Subsequent Initiator	LTA system specification adversely affects hardware/computer specification, involving logic.
Subsequent Initiator	LTA hardware specification affects logic.
Subsequent Initiator	Recurrent errors occur with adverse effects.
Input errors	Input errors arise when any deviation occurs from expected inputs.
Initiator	Assumptions, logic, knowledge, and experience are less than adequate (LTA); initiates errors in logic.
Subsequent Initiator	LTA software specification affects logic.
Subsequent Initiator	LTA system specification adversely affects hardware/computer specification, involving logic.
Subsequent Initiator	LTA hardware specification affects logic.
Contributor	Input error occurs within concept.
Contributor	Input error occurs associated with functions.
Contributor	Input error occurs within variables.
Contributor	Input error occurs within parameters.
Contributor	Input error occurs within constants.
Input failures/ malfunctions/ anomaly	Input failures/malfunctions/anomalies occur due to abnormal energy exchange affecting hardware and analog input. The input deviation is not detected and it is accepted as reasonable.
Initiator	Assumptions, logic, knowledge, and experience are less than adequate (LTA); initiates errors in logic.
Subsequent Initiator	LTA software specification affects logic.
Subsequent Initiator	LTA system specification adversely affects hardware/computer specification, involving logic.
Subsequent Initiator	LTA hardware specification affects logic.
Subsequent Initiator	Input error occurs within concept.
Contributor	Input error occurs associated with functions.
Contributor	Input error occurs within variables.
Contributor	Input error occurs within parameters.
Contributor	Input error occurs within constants.
Contributor	Failures are due to electromagnetic environmental effects.
Contributor	Failures are due to single events, BIT flips, or neutron event.
Contributor	Failures are due to microelectronics problem, electron migration, fracture cracking, delaminating, and synergistic environmental effects.
Contributor	Failures are due to physical damage.
Contributor	Failures occur due to process error.
Contributor	Failures are due to external hazard.

(*continued*)

TABLE 14.2 *Continued*

Factor Name[a]	Errors Resulting in the Inappropriate Implementation of Software
Factor name	Output Errors and Harm
Computational error: Subsequent Contributor	Computational error adversely affects computer output.
Configuration error: Subsequent Contributor	Configuration error adversely affects computer output.
Data handling error: Subsequent Contributor	Data handling error adversely affects computer output.
Database error: Subsequent Contributor	Database error adversely affects computer output.
Definition error: Subsequent Contributor	Definition error adversely affects computer output.
Intermittent error: Subsequent Contributor	Intermittent error adversely affects computer output.
Recurrent error: Subsequent Contributor	Recurrent error adversely affects computer output.
Input error: Subsequent Contributor	Input error adversely affects computer and computer output.
Subsequent Contributor(s) to Harm	Computer output deviation contributes to: Hazardous misleading information (HMI) Hazardous automated function (HAF) Loss of safety-critical system Loss of safety-critical information at required time Loss of safety-critical output Changing/altering of safety-critical information Undetected loss of safety-critical function
Subsequent Contributor(s) to Harm	Computer output deviation contributes to harm; inappropriate information is displayed resulting in improper human response.
Subsequent Contributor(s) to Harm	Computer output deviation contributes to harm; automated function occurs inappropriately and inadvertently.
Subsequent Contributor(s) to Harm	Computer output deviation contributes to harm; loss of safety-critical system occurs at the most inappropriate time.
Subsequent Contributor(s) to Harm	Computer output deviation contributes to harm; loss of safety-critical information occurs at the most inappropriate time.
Subsequent Contributor(s) to Harm	Computer output deviation contributes to harm; loss of safety-critical output occurs at the most inappropriate time.
Subsequent Contributor(s) to Harm	Computer output deviation contributes to harm; safety-critical information is changed/altered and situation occurs at the most inappropriate time.
Subsequent Contributor(s) to Harm	Computer output deviation contributes to harm; undetected loss of safety-critical function occurs.

[a]Coding error types; see Ref. 2, pp. 288, 301.

14.8 OTHER SPECIALIZED TECHNIQUES, ANALYSIS METHODS, AND TOOLS FOR EVALUATING SOFTWARE AND COMPUTER SYSTEMS

14.8.1 Software Reliability

Over the past 30 years many techniques, methods, and tools have been developed and used to conduct software complexity analysis and software reliability. The output of this effort provides quantitative information on code to the designing, testing, and maintenance organizations involved in the software life cycle. Information on code structure, critical components, risk, and testing deficiencies can be obtained as a result of software complexity analysis, which is considered part of the software reliability function. From a strict reliability view, software reliability provides estimation and prediction on the readiness of software prior to operation implementation [3]. Caution should be applied when these techniques, methods, and tools are used and applied in support of system safety. The analyst must be aware of how these analyses were conducted, the assumptions involved in the processes, and variances that have occurred in data collection and analysis. Reliable software does not automatically equate to the safety of the system.

14.8.2 Static Complexity Analysis

To determine extensive involvement of code, static complexity analysis is conducted. Complexity metrics are defined to measure software structure, size, and interfaces. Matrices can range from measures of number of lines of code to complicated metrics that measure abstract characteristics of software. Metrics can be collected on the number of requirements, sequential timing parameters, number of errors, number of statements, total number of operands, total number of operations, number of inputs, number of outputs, and number of direct calls.

14.8.3 Dynamic Analysis

To evaluate the efficiency and effectiveness of software testing, the software system is monitored during its execution phase. Data is collected during execution to evaluate the thoroughness of the testing; the efficiency of software testing is evaluated by measuring the amount of actual code executed during a software test run. A dynamic analysis will track the execution path through the code and determine what logic paths are executed. The analysis considers the portions of code executed, the frequency of execution, and the amount of time spent in each application.

14.8.4 Test Coverage Monitoring

To evaluate the test case coverage, there are methods of monitoring the execution status of the software component during the execution of the code. The methods used will depend on the tool used for monitoring as well as the importance of the component being monitored. These methods require a technique referred to as instrumentation, which is a software program that involves adding additional code, which records the execution status of various parts of code. Status flags are set to indicate execution sequence through the source code. Types of monitoring are:

- Entry point monitoring involves testing to determine that a given component is used by a test case but indicates nothing about which parts of the component were executed.
- Segment monitoring determines whether all the statements in the code have been executed but does not provide information as to the logic paths.
- Transfer monitoring measures the logical branches from one segment to another. Executing all transfers equates to the execution of all the statements as well as all the logic paths in the code.
- Path monitoring attempts to monitor the execution of all the possible paths through the code. Path monitoring can be attempted on very small routines or modules.
- Profile monitoring measures the frequency and execution times of individual components. Profiling is monitoring a program's execution by outputting a detailed procedure-by-procedure analysis of the execution time, including how many times a procedure is called, what procedure called it, how much time was spent in the procedure, and what other procedures it called during its execution.

14.9 EXISTING LEGACY SYSTEMS, REUSABLE SOFTWARE, COMMERCIAL OFF THE SHELF (COTS) SOFTWARE, AND NONDEVELOPMENT ITEMS (NDIs)

The most interesting challenges from a system safety view involve the integration of new technologies into existing systems, which include legacy systems, reusable software, COTS software and NDIs. Usually there is very little data and information known about such systems, including software, firmware, and hardware. There is an old adage—if you don't know the risks associated with a system when you acquire it, you may as well assume the risks. Unfortunately, there are concerns with technology transfer, trade secrets, and security. There is a need for upfront contractual work (from a risk management point of view) or the buyer, integrator, construction management, or prime contractor will assume risk.

There is also an additional complexity involving the concept of intended use. For example, consider the intended use of a typical laptop computer. Generally, such a device is not used in safety-critical applications. Consequently, minimal mitigations were designed into the device to prevent undetectable altering of information displayed. Furthermore, assume that such a device was not to be used for a safety-critical operation—like providing details on a complex hazardous process or medical procedure. Consequently, the laptop was dropped and firmware was affected, which inadvertently altered critical data. Expect a complex system accident. There was no initial need for additional mitigation since the original intended use was not for a safety-critical operation.

The following precautions are recommended, this list is not all-inclusive:

- If the seller is not willing to provide the details—data, information, and safety documentation requested—contractually the seller should be made to assume all risks.
- If appropriate data, information, and safety documentation are not acquired, assume the worst and provide isolation, barriers, and segregation, to prevent failure propagation or hazard initiation.
- Consider such systems as white or black boxes and seek to control inputs and outputs.
- Provide maximum safety requirements to be met by the seller, for example, built-in test (BIT) capabilities, hardware test verification, reliability, availability, and other system requirements.
- Provide for system-level controls, end-to-end testing, and continuous BIT.
- Provide for contingencies, backups, and recovery since the worst is expected.
- Based on known information, conduct the appropriate hazard analyses and risk assessments.
- Have operators validate and verify from the acquired system: operation, data, and information.
- Inform all participants of the risks and controls.
- Provide appropriate cautions, warnings, and alerts.
- Initially provide upfront safety requirements to be met by the seller.
- Provide for latency and data transfer requirements.
- Provide for physical protection of the acquired system.
- Have the seller modify the system to meet safety requirements.
- Externally prevent an inappropriate operation that may occur from the acquired system.
- Include service requirements that support system safety.

14.10 TOPICS FOR STUDENT PROJECTS AND THESES

1. Select a complex automated system and develop a software safety program plan.
2. Assume that you are evaluating a safety-critical automated system and you are concerned with errors within the system. Explain and design mitigations for error handling.
3. Define the attributes and requirements for the development of a fail-safe recovery capability within a safety-critical automated system.

4. Select a complex automated system with extensive human interaction. Define the hazards associated with the computer–human interface and provide design controls.

5. Assume that you are evaluating a complex computer program and you have been informed that there are problems with latency and control flow. Develop a plan and check list that will enable trouble shooting and mitigating the problem.

REFERENCES

1. Suh Pyo Nam, *Axiomatic Design: Advances and Applications*, Oxford University Press, New York, 2001.
2. Raheja, D.G., assurance Technologies Principles and Practices, First Edition, McGraw-Hill, Inc., New York, 1991.
3. A. T. Lee, T. Gunn, T. Pham, and R. Ricaldi. NASA Technical Memorandum 104799, Software Analysis Handbook: Software Complexity Analysis and Software Reliability Estimation and Prediction, August 1994.

FURTHER READING

Adrian, W. R., M. A. Brans, and J. C. Cherniavaky, *Validation, Verification, and Testing of Computer Software*, Special Publication 500-75, National Bureau of Standards, Gaithersburg, MD, February 1981.

Allocco, M., Computer and Software Safety Considerations in Support of System Hazard Analysis. In: *Proceedings of the 21st International System Safety Conference*, System Safety Society, August 2003.

ANSI/IEEE STD 1002-1992, *IEEE Standard Taxonomy for Software Engineering Standards.*

Chikofsky, E. J., *Computer-Aided Software Engineering*, IEEE Computer Society Press, New York, 1988.

Chow, T. S., *Software Quality Assurance: A Practical Approach*, IEEE Computer Society Press, New York, 1985.

Curtis, B., *Human Factors in Software Development*, IEEE Computer Society Press, New York, 1986.

Defense System Software Development Handbook, DoD-Hdbk-287, Naval Publications and Forms Centers, Philadelphia, February 29, 1988.

Department of Defense, *Military Standard Defense System Software Development*, DoD-Std 2167 A, February 29, 1988.

Donahoo, J. D., and D. Swearinger. A Review of Software Maintenance Technology, Report RADC-TR-80-13, Rome Air Development Center, Rome, NY, 1980.

Dunn, R., and R. Ullman. *Quality Assurance for Computer Software*, McGraw-Hill, New York, 1981.

Electronic Industries Association, EIA-6B, G-48, System Safety Engineering in Software Development, 1990.

Freeman, P., *Software Reusability*, IEEE Computer Society Press, New York, 1987.

Glass, R. L., *Checkout Techniques, Software Reliability Guidebook*, Prentice-Hall, Englewood Cliffs, NJ, 1979.

Handbook for Software in Safety-Critical Applications, English edition, Swedish Armed Forces, Defense Material Administration, March 15, 2005.

Herrmann, D. S., *Software Safety and Reliability*, IEEE Computer Society Press, Los Alamitos, 1999.

IEEE Standard Dictionary of Measures to Product Reliable Software, IEEE Standard 982.1-1988, IEEE, New York, 1988.

IEEE Standard for Software Reviews and Audits, IEEE Standard 1028-1988, IEEE, New York, 1988.

IEEE Standard for Software Safety Plans, IEEE Standard 1228, IEEE, New York, 1994.

IEEE Standard for Software Test Documentation, IEEE Standard 829, IEEE New York, 1983.

IEEE Guide to Software Requirements Specification, IEEE Standard 830, IEEE, New York, 1984.

Leveson, N., *Safeware: System Safety and Computers*, Addison-Wesley, Boston, 1995.

Musa, J. D., A. Lannine, and K. Okumotoko, *Software Reliability: Measurement, Prediction, Application*, McGraw-Hill, New York, 1987.

Myers, G. J., *The Art of Software Testing*, John Wiley & Sons, Hoboken, NJ, 1979.

NASA-Std-8719.13A, *Software Safety*, September 1997.

Reviews and Audits for Systems, Equipment, and Computer Programs, Mil-Std-1521B, Naval Publications and Forms Center, Philadelphia.

RTCA-DO 178B, *Software Considerations in Airborne Systems and Equipment Certification*, December 1, 1992.

RTCA-DO 278B, *Guidelines foe Communication, Surveillance, Air Traffic Management (CNS/ATM) System Software Integrity Assurance*, 2004.

Shooman, M., *Software Engineering—Design, Reliability, Management*, McGraw-Hill, New York, 1983.

Software Qualify Evaluation, DoD-Std-2168, Naval Publications and Forms Center, Philadelphia.

Storey, N., *Safety-Critical Computer Systems*. Addison-Wesley, Boston, 1996.

System Safety Program Requirements, Mil-Std-882, Naval Publications and Forms Center, Philadelphia, 1984.

APPENDIX A

REFERENCE TABLES

TABLE A   Standard Normal Distributions

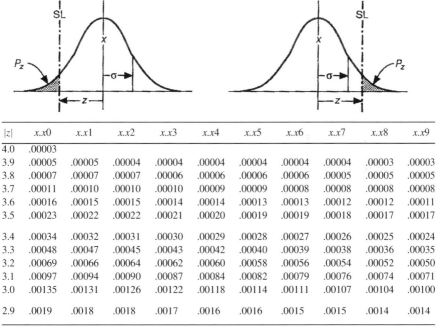

|z|	x.x0	x.x1	x.x2	x.x3	x.x4	x.x5	x.x6	x.x7	x.x8	x.x9
4.0	.00003									
3.9	.00005	.00005	.00004	.00004	.00004	.00004	.00004	.00004	.00003	.00003
3.8	.00007	.00007	.00007	.00006	.00006	.00006	.00006	.00005	.00005	.00005
3.7	.00011	.00010	.00010	.00010	.00009	.00009	.00008	.00008	.00008	.00008
3.6	.00016	.00015	.00015	.00014	.00014	.00013	.00013	.00012	.00012	.00011
3.5	.00023	.00022	.00022	.00021	.00020	.00019	.00019	.00018	.00017	.00017
3.4	.00034	.00032	.00031	.00030	.00029	.00028	.00027	.00026	.00025	.00024
3.3	.00048	.00047	.00045	.00043	.00042	.00040	.00039	.00038	.00036	.00035
3.2	.00069	.00066	.00064	.00062	.00060	.00058	.00056	.00054	.00052	.00050
3.1	.00097	.00094	.00090	.00087	.00084	.00082	.00079	.00076	.00074	.00071
3.0	.00135	.00131	.00126	.00122	.00118	.00114	.00111	.00107	.00104	.00100
2.9	.0019	.0018	.0018	.0017	.0016	.0016	.0015	.0015	.0014	.0014

(Continued)

Assurance Technologies Principles and Practices: A Product, Process, and System Safety Perspective,
Second Edition, by Dev G. Raheja and Michael Allocco
Copyright © 2006 John Wiley & Sons, Inc.

TABLE A *Continued*

$|z|$	$x.x0$	$x.x1$	$x.x2$	$x.x3$	$x.x4$	$x.x5$	$x.x6$	$x.x7$	$x.x8$	$x.x9$
1.9	.0287	.0281	.0274	.0268	.0262	.0256	.0250	.0244	.0239	.0233
1.8	.0359	.0351	.0344	.0336	.0329	.0322	.0314	.0307	.0301	.0294
1.7	.0446	.0436	.0427	.0418	.0409	.0401	.0392	.0384	.0375	.0367
1.6	.0548	.0537	.0526	.0516	.0505	.0495	.0485	.0475	.0464	.0455
1.5	.0668	.0655	.0643	.0630	.0618	.0606	.0594	.0582	.0571	.0559
1.4	.0808	.0793	.0778	.0764	.0749	.0735	.0721	.0708	.0694	.0681
1.3	.0968	.0951	.0934	.0918	.0901	.0885	.0869	.0853	.0838	.0823
1.2	.1151	.1131	.1112	.1093	.1075	.1056	.1038	.1020	.1003	.0985
1.1	.1357	.1335	.1314	.1292	.1271	.1251	.1230	.1210	.1190	.1170
1.0	.1587	.1562	.1539	.1515	.1492	.1469	.1446	.1423	.1401	.1379
0.9	.1841	.1814	.1788	.1762	.1736	.1711	.1685	.1660	.1635	.1611
0.8	.2119	.2090	.2061	.2033	.2005	.1977	.1949	.1922	.1894	.1867
0.7	.2420	.2389	.2358	.2327	.2297	.2266	.2236	.2206	.2177	.2148
0.6	.2743	.2709	.2676	.2643	.2611	.2578	.2546	.2514	.2483	.2451
0.5	.3085	.3050	.3015	.2981	.2946	.2912	.2877	.2843	.2810	.2776
0.4	.3446	.3409	.3372	.3336	.3300	.3264	.3228	.3192	.3156	.3121
0.3	.3821	.3783	.3745	.3707	.3669	.3632	.3594	.3557	.3520	.3483
0.2	.4207	.4168	.4129	.4090	.4052	.4013	.3974	.3936	.3897	.3859
0.1	.4602	.4562	.4522	.4483	.4443	.4404	.4364	.4325	.4286	.4247
0.0	.5000	.4960	.4920	.4880	.4840	.4801	.4761	.4721	.4681	.4641

P_z = the proportion of process output beyond a particular value of interest (such as a specification limit) that is z standard deviation units away from the process average (for a process that is in statistical control and is normally distributed). For example, if $z = 2.17$, $P_z = .0150$ or 1.5%. In any actual situation, this proportion is only approximate.
(*Courtesy*: FORD MOTOR CO.)

TABLE B Constants and Formulas for Control Charts

Sub-Group Size n	$\overline{X}$ and R Charts*					$\overline{X}$ and s Charts*			
	Chart for Averages ($\overline{X}$)	Chart for Ranges (R)				Chart for Averages ($\overline{X}$)	Chart for Standard Deviations s		
	Factors for Control Limits	Divisors for Estimate of Standard Deviation	Factors for Control Limits			Factor for Control Limits	Divisors for Estimate of Standard Deviation	Factors for Control Limits	
	A_2	d_2	D_3	D_4		A_3	c_4	B_3	B_4
2	1.880	1.128	—	3.267		2.659	0.7979	—	3.267
3	1.023	1.693	—	2.574		1.954	0.8862	—	2.568
4	0.729	2.059	—	2.282		1.628	0.9213	—	2.266
5	0.577	2.326	—	2.114		1.427	0.9400	—	2.089
6	0.483	2.534	—	2.004		1.287	0.9515	0.030	1.970
7	0.419	2.704	0.076	1.924		1.182	0.9594	0.118	1.882
8	0.373	2.847	0.136	1.864		1.099	0.9650	0.185	1.815
9	0.337	2.970	0.184	1.816		1.032	0.9693	0.239	1.761
10	0.308	3.078	0.223	1.777		0.975	0.9727	0.284	1.716
11	0.285	3.173	0.256	1.744		0.927	0.9754	0.321	1.679
12	0.266	3.258	0.283	1.717		0.886	0.9776	0.354	1.646
13	0.249	3.336	0.307	1.693		0.850	0.9794	0.382	1.618
14	0.235	3.407	0.328	1.672		0.817	0.9810	0.406	1.594
15	0.223	3.472	0.347	1.653		0.789	0.9823	0.428	1.572
16	0.212	3.532	0.363	1.637		0.763	0.9835	0.448	1.552
17	0.203	3.588	0.378	1.622		0.739	0.9845	0.466	1.534
18	0.194	3.640	0.391	1.608		0.718	0.9854	0.482	1.518
19	0.187	3.689	0.403	1.597		0.698	0.9862	0.497	1.503
20	0.180	3.735	0.415	1.585		0.680	0.9869	0.510	1.490

(Continued)

TABLE B *Continued*

Sub-Group Size n	$\overline{X}$ and R Charts*				$\overline{X}$ and s Charts*			
	Chart for Averages ($\overline{X}$)	Chart for Ranges (R)			Chart for Averages ($\overline{X}$)	Chart for Standard Deviations s		
	Factors for Control Limits	Divisors for Estimate of Standard Deviation	Factors for Control Limits		Factor for Control Limits	Divisors for Estimate of Standard Deviation	Factors for Control Limits	
	A_2	d_2	D_3	D_4	A_3	c_4	B_3	B_4
21	0.173	3.778	0.425	1.575	0.663	0.9876	0.523	1.477
22	0.167	3.819	0.434	1.566	0.647	0.9882	0.534	1.466
23	0.162	3.858	0.443	1.557	0.633	0.9887	0.545	1.455
24	0.157	3.895	0.451	1.548	0.619	0.9892	0.555	1.445
25	0.153	3.931	0.459	1.541	0.606	0.9896	0.565	1.435

$$\mathrm{UCL}_{\overline{X}}, \mathrm{LCL}_{\overline{X}} = \overline{\overline{X}} + A_2\overline{R}$$
$$\mathrm{UCL}_{\overline{R}} = D_4\overline{R}$$
$$\mathrm{LCL}_{\overline{R}} = D_3\overline{R}$$
$$\hat{\sigma} = \overline{R}/d_2$$

$$\mathrm{UCL}_{\overline{X}}, \mathrm{LCL}_{\overline{X}} = \overline{\overline{X}} \pm A_3\overline{s}$$
$$\mathrm{UCL}_{s} = B_4\overline{s}$$
$$\mathrm{LCL}_{s} = B_3\overline{s}$$
$$\hat{\sigma} = \overline{s}/c_4$$

*From ASTM publication STP-15D, *Manual on the Presentation of Data and Control Chart Analysis*, 1976; pp. 134-136. Copyright ASTM, 1916 Race Street, Philadelphia, Pa. 19103. Reprinted with permission.

434

TABLE C Percentage Points of the χ^2 Distribution

Table of $\chi_{\alpha;\nu}$—the $100\,\alpha$ percentage point of the χ^2 distribution for ν degrees of freedom

ν \ α	.995	.99	.98	.975	.95	.90	.80	.75	.70	.50
1	$.0^3393$	$.0^3157$	$.0^3628$	$.0^3982$	.00393	.0158	.0642	.102	.148	.455
2	.0100	.0201	.0404	.0506	.103	.211	.446	.575	.713	1.386
3	.0717	.115	.185	.216	.352	.584	1.005	1.213	1.424	2.366
4	.207	.297	.429	.484	.711	1.064	1.649	1.923	2.195	3.357
5	.412	.554	.752	.831	1.145	1.610	2.343	2.675	3.000	4.351
6	.676	.872	1.134	1.237	1.635	2.204	3.070	3.455	3.828	5.348
7	.989	1.239	1.564	1.690	2.167	2.833	3.822	4.255	4.671	6.346
8	1.344	1.646	2.032	2.180	2.733	3.490	4.594	5.071	5.527	7.344
9	1.735	2.088	2.532	2.700	3.325	4.168	5.380	5.899	6.393	8.343
10	2.156	2.558	3.059	3.247	3.940	4.865	6.179	6.737	7.267	9.342
11	2.603	3.053	3.609	3.816	4.575	5.578	6.989	7.584	8.148	10.341
12	3.074	3.571	4.178	4.404	5.226	6.304	7.807	8.438	9.034	11.340
13	3.565	4.107	4.765	5.009	5.892	7.042	8.634	9.299	9.926	12.340
14	4.075	4.660	5.368	5.629	6.571	7.790	9.467	10.165	10.821	13.339
15	4.601	5.229	5.985	6.262	7.261	8.547	10.307	11.036	11.721	14.339
16	5.142	5.812	6.614	6.908	7.962	9.312	11.152	11.912	12.624	15.338
17	5.697	6.408	7.255	7.564	8.672	10.085	12.002	12.792	13.531	16.338
18	6.265	7.015	7.906	8.231	9.390	10.865	12.857	13.675	14.440	17.338
19	6.844	7.633	8.567	8.907	10.117	11.651	13.716	14.562	15.352	18.338

(Continued)

TABLE C *Continued*

v α	.995	.99	.98	.975	.95	.90	.80	.75	.70	.50
20	7.434	8.260	9.237	9.591	10.851	12.443	14.578	15.452	16.266	19.337
21	8.034	8.897	9.915	10.283	11.591	13.240	15.445	16.344	17.182	20.337
22	8.643	9.542	10.600	10.982	12.338	14.041	16.314	17.240	18.101	21.337
23	9.260	10.196	11.293	11.668	13.091	14.848	17.187	18.137	19.021	22.337
24	9.886	10.856	11.992	12.401	13.848	15.659	18.062	19.037	19.943	23.337
25	10.520	11.524	12.697	13.120	14.611	16.473	18.940	19.939	20.867	24.337
26	11.160	12.198	13.409	13.844	15.379	17.292	19.820	20.843	21.792	25.336
27	11.808	12.879	14.125	14.573	16.151	18.114	20.703	21.749	22.719	26.336
28	12.461	13.565	14.847	15.308	16.928	18.939	21.588	22.657	23.647	27.336
29	13.121	14.256	15.574	16.047	17.708	19.768	22.475	23.567	24.577	28.336
30	13.787	14.953	16.306	16.791	18.493	20.599	23.364	24.478	25.508	29.336

.30	.25	.20	.10	.05	.025	.02	.01	.005	.001	α v
1.074	1.323	1.642	2.706	3.841	5.024	5.412	6.636	7.879	10.827	1
2.408	2.773	3.219	4.605	5.991	7.378	7.824	9.210	10.597	13.815	2
3.665	4.108	4.642	6.251	7.815	9.348	9.837	11.345	12.838	16.268	3
4.878	5.385	5.989	7.779	9.488	11.143	11.668	13.277	14.860	18.465	4
6.064	6.626	7.289	9.236	11.070	12.832	13.388	15.086	16.750	20.517	5
7.231	7.841	8.558	10.645	12.592	14.449	15.033	16.812	18.548	22.457	6
8.383	9.037	9.803	12.017	14.067	16.013	16.622	18.475	20.278	24.322	7
9.524	10.219	11.030	13.362	15.507	17.535	18.168	20.090	21.955	26.125	8
10.656	11.389	12.242	14.684	16.919	19.023	19.679	21.666	23.589	27.877	9
11.781	12.549	13.442	15.987	18.307	20.483	21.161	23.209	25.188	29.588	10

12.899	13.701	14.631	17.275	19.675	21.920	22.618	24.725	26.757	31.264	11
14.011	14.845	15.812	18.549	21.026	23.337	24.054	26.217	28.300	32.909	12
15.119	15.984	16.985	19.812	22.362	24.736	25.472	27.688	29.819	34.528	13
16.222	17.117	18.151	21.064	23.685	26.119	26.873	29.141	31.319	36.123	14
17.322	18.245	19.311	22.307	24.996	27.488	28.259	30.578	32.801	37.697	15
18.418	19.369	20.465	23.542	26.296	28.845	29.633	32.000	34.267	39.252	16
19.511	20.489	21.615	24.769	27.587	30.191	30.995	33.409	35.718	40.790	17
20.601	21.605	22.760	25.989	28.869	31.526	32.346	34.805	37.156	42.312	18
21.689	22.718	23.900	27.204	30.144	32.852	33.687	36.191	38.582	43.820	19
22.775	23.828	25.038	28.412	31.410	34.170	35.020	37.566	39.997	45.315	20
23.858	24.935	26.171	29.615	32.671	35.479	36.343	38.932	41.401	46.797	21
24.939	26.039	27.301	30.813	33.924	36.781	37.659	40.289	42.796	48.268	22
26.018	27.141	28.429	32.007	35.172	38.076	38.968	41.638	44.181	49.728	23
27.096	28.241	29.553	33.196	36.415	39.364	40.270	42.980	45.558	51.179	24
28.172	29.339	30.675	34.382	37.652	40.646	41.566	44.314	46.928	52.620	25
29.246	30.434	31.795	35.563	38.885	41.923	42.856	45.642	48.290	54.052	26
30.319	31.528	32.912	36.741	40.113	43.194	44.140	46.963	49.645	55.476	27
31.391	32.620	34.027	37.916	41.337	44.461	45.419	48.278	50.993	56.893	28
32.961	33.711	35.139	39.087	42.557	45.722	46.693	49.588	52.336	58.302	29
33.530	34.800	36.250	40.256	43.773	46.979	47.962	50.892	53.672	59.703	30

Source: REALIABILITY HANDBOOK, AMCP 702-3, HEADQUARTERS, U. S. ARMY MATERIAL COMMAND, WASHINGTON D.C., 1968.

TABLE D 5, Median, and 95 Percent Ranks for Various Sample Sizes

5 Percent Ranks

Sample Size, n

j	1	2	3	4	5	6	7	8	9	10	11	12	13	14	15	16	17	18	19	20
1	5.000	2.532	1.695	1.274	1.021	0.851	0.730	0.639	0.568	0.512	0.465	0.426	0.394	0.366	0.341	0.320	0.301	0.285	0.270	0.256
2		22.361	13.535	9.761	7.644	6.285	5.337	4.639	4.102	3.677	3.332	3.046	2.805	2.600	2.423	2.268	2.132	2.011	1.903	1.806
3			36.840	24.860	18.925	15.316	12.876	11.111	9.775	8.726	7.882	7.187	6.605	6.110	5.685	5.315	4.990	4.702	4.446	4.217
4				47.237	34.259	27.134	22.532	19.290	16.875	15.003	13.507	12.285	11.267	10.405	9.666	9.025	8.464	7.969	7.529	7.135
5					54.928	41.820	34.126	28.924	25.137	22.244	19.958	18.102	16.566	15.272	14.166	13.211	12.377	11.643	10.991	10.408
6						60.696	47.930	40.031	34.494	30.354	27.125	24.530	22.395	20.607	19.086	17.777	16.636	15.634	14.747	13.955
7							65.184	52.932	45.036	39.338	34.981	31.524	28.705	26.358	24.373	22.669	21.191	19.895	18.750	17.731
8								68.766	57.086	49.310	43.563	39.086	35.480	32.503	29.999	27.860	26.011	24.396	22.972	21.707
9									71.687	60.584	52.991	47.267	42.738	39.041	35.956	33.337	31.083	29.120	27.395	25.865
10										74.113	63.564	56.189	50.535	45.999	42.256	39.101	36.401	34.060	32.009	30.195
11											76.160	66.132	58.990	53.434	48.925	45.165	41.970	39.215	36.811	34.693
12												77.908	68.366	61.461	56.022	51.560	47.808	44.585	41.806	39.358
13													79.418	70.327	63.656	58.343	53.945	50.217	47.003	44.197
14														80.736	72.060	65.617	60.436	56.112	52.420	49.218
15															81.896	73.604	67.381	62.332	58.088	54.442
16																82.925	74.988	68.974	64.057	59.897
17																	83.843	76.234	70.420	65.634
18																		84.668	77.363	71.738
19																			85.413	78.389
20																				86.089

Courtesy: GENERAL MOTORS.

95 Percent Ranks

Sample Size

j	1	2	3	4	5	6	7	8	9	10	11	12	13	14	15	16	17	18	19	20
1	95.000	77.639	63.160	52.713	45.072	39.304	34.816	31.234	28.313	25.887	23.840	22.092	20.582	19.264	18.104	17.075	16.157	15.332	14.587	13.911
2		97.468	86.468	75.139	65.741	58.180	52.070	47.068	42.914	39.416	36.436	33.868	31.634	29.673	27.940	26.396	25.012	23.766	22.637	21.611
3			98.305	90.239	81.075	72.866	65.874	59.969	54.964	50.690	47.009	43.811	41.010	38.539	36.344	34.383	32.619	31.026	29.580	28.262
4				98.726	92.356	84.684	77.468	71.076	65.506	60.662	56.437	52.733	49.465	46.566	43.978	41.657	39.564	37.668	35.943	34.366
5					98.979	93.715	87.124	80.710	74.863	69.646	65.019	60.914	57.262	54.000	51.075	48.440	46.055	43.888	41.912	40.103
6						99.149	94.662	88.889	83.125	77.756	72.875	68.476	64.520	60.928	57.744	54.835	52.192	49.783	47.580	45.558
7							99.270	95.361	90.225	84.997	80.042	75.470	71.295	67.497	64.043	60.899	58.029	55.404	52.997	50.782
8								99.361	95.898	91.274	86.492	81.898	77.604	73.641	70.001	66.663	63.599	60.784	58.194	55.803
9									99.432	96.323	92.118	87.715	83.434	79.393	75.627	72.140	68.917	65.940	63.188	60.641
10										99.488	96.668	92.813	88.733	84.728	80.913	77.331	73.989	70.880	67.991	65.307
11											99.535	96.954	93.395	89.595	85.834	82.223	78.809	75.604	72.605	69.805
12												99.573	97.195	93.890	90.334	86.789	83.364	80.105	77.028	74.135
13													99.606	97.400	94.315	90.975	87.623	84.366	81.250	78.293
14														99.634	97.577	94.685	91.535	88.357	85.253	82.269
15															99.659	97.732	95.010	92.030	89.009	86.045
16																99.680	97.868	95.297	92.471	89.592
17																	99.699	97.989	95.553	92.865
18																		99.715	98.097	95.783
19																			99.730	98.193
20																				99.744

(Continued)

TABLE D Continued

Median Ranks

j	\multicolumn{20}{c}{Sample Size, n}

j	1	2	3	4	5	6	7	8	9	10	11	12	13	14	15	16	17	18	19	20
1	50.000	29.289	20.630	15.910	12.945	10.910	9.428	8.300	7.412	6.697	6.107	5.613	5.192	4.830	4.516	4.240	3.995	3.778	3.582	3.406
2		70.711	50.000	38.573	31.381	26.445	22.849	20.113	17.962	16.226	14.796	13.598	12.579	11.702	10.940	10.270	9.678	9.151	8.677	8.251
3			79.370	61.427	50.000	42.141	36.412	32.052	28.624	25.857	23.578	21.669	20.045	18.647	17.432	16.365	15.422	14.581	13.827	13.147
4				84.090	68.619	57.859	50.000	44.015	39.308	35.510	32.380	29.758	27.528	25.608	23.939	22.474	21.178	20.024	18.988	18.055
5					87.055	73.555	63.588	55.984	50.000	45.169	41.189	37.853	35.016	32.575	30.452	28.589	26.940	25.471	24.154	22.967
6						89.090	77.151	67.948	60.691	54.831	50.000	45.951	42.508	39.544	36.967	34.705	32.704	30.921	29.322	27.880
7							90.572	79.887	71.376	64.490	58.811	54.049	50.000	46.515	43.483	40.823	38.469	36.371	34.491	32.795
8								91.700	82.038	74.142	67.620	62.147	57.492	53.485	50.000	46.941	44.234	41.823	39.660	37.710
9									92.588	83.774	76.421	70.242	64.984	60.456	56.517	53.059	50.000	47.274	44.830	42.626
10										93.303	85.204	78.331	72.472	67.425	63.033	59.177	55.766	52.726	50.000	47.542
11											93.893	86.402	79.955	74.392	69.548	65.295	61.531	58.177	55.170	52.458
12												94.387	87.421	81.353	76.061	71.411	67.296	63.629	60.340	57.374
13													94.808	88.298	82.568	77.526	73.060	69.079	65.509	62.289
14														95.169	89.060	83.635	78.821	74.529	70.678	67.205
15															95.484	89.730	84.578	79.976	75.846	72.119
16																95.760	90.322	85.419	81.011	77.033
17																	96.005	90.849	86.173	81.945
18																		96.222	91.322	86.853
19																			96.418	91.749
20																				96.594

APPENDIX B

AN OUTSTANDING APPLICATION OF ASSURANCE TECHNOLOGIES

EATON TRUCK TRANSMISSIONS

At Eaton Truck Components, design assurance technologies are embedded early into the design, manufacturing, and service process. The company, a Kalamazoo, Michigan-based division of Eaton Corporation, designs and manufactures transmissions for the heavy-duty, medium-duty, and light-duty trucks, including hybrid electric vehicles (HEV).

The process begins at the Production Definition phase, where the key goals for each product are specified:

- System Reliability Goal—No mission-stopping failure during the expected life.
- System Safety Goal—No fatalities any time even beyond expected life.
- Maintainability Goal—Eliminate as much downtime as possible, by design.
- Logistics Engineering Goal—Improve serviceability and fault isolation capabilities.
- Human Factors Engineering Goal—Design to reduce the impact of driver errors.
- System Integration Goal—Aim for high Availability and smart diagnostics.

Each of these goals is communicated to customers to ensure quality in the design and manufacturing processes. Managers believe that high reliability reduces the product cost, because eliminating problems is cheaper than fixing them in the field. That is

Assurance Technologies Principles and Practices: A Product, Process, and System Safety Perspective, Second Edition, by Dev G. Raheja and Michael Allocco
Copyright © 2006 John Wiley & Sons, Inc.

why Eaton managers have developed a continuous innovation approach to problem solving. Many small innovations add up to radical improvements in the product reliability, durability, safety, and serviceability. To encourage innovations, the managers require 500% Return on Investment (ROI) for every problem eliminated.

[The ROI is computed as the ratio of life cycle savings and the investment required. For an example of life cycle costs, see Chapter 3.]

Implementing assurance technologies for Eaton is a three-step process:

1. Mitigating known risks
2. Mitigating unknown risks
3. Using new paradigms for robustness

Mitigating Known Risks

The process inputs for known risks are taken from lessons learned, warranty reports, and customer feedback. Eaton engineers mitigate these risks by putting tougher requirements in the system specifications. They include reliability, durability, maintainability, serviceability, human factors, logistics support, and interface requirements in the specs. They pay close attention to the reliability of incoming and outgoing interfaces. This helps system engineers gain clarity. The following features are the primary requirements while specifying a system:

Application Environment
Functions
Active Safety
Reliability
Durability (Duty Cycles)
Serviceability/Maintainability (including prognostics)
Human Factors
Logistics
Produceability
Input Interface Requirements
Output Interface Requirements
Installation Requirements
Shipping / Handling Requirements

The specification not only contains what the product should do but also what it "should not" do, such as "no unintended motion of the vehicle." This is important for the validation of safety.

Mitigating Unknown Risks

These are unrecognized risks and they need to be uncovered. They are unknown until formal reliability, safety, use, and misuse analysis is performed. It is difficult to identify them because of complex interactions of components, software, interfacing systems, and variations in customer inputs. A thorough brainstorming is required to discover the unknown failure modes, system hazards, and user frustration. This work requires creative thinking by a good cross-functional team from System Engineering, Reliability Engineering, Service Engineering, Marketing, Manufacturing Engineering, and Supply Chain Engineering.

The bottom line is to identify the major risks and mitigate them when it is cheapest for Eaton and the customer. Three analyses are required before releasing the specification. These are:

- System Functions Failure Modes and Effects Analysis (FMEA)
- System Hazard Analysis
- Use/Misuse Analysis

Often these analyses result in numerous improvements, leading to a comprehensive and unambiguous system specification. Such a system specification also sets a single goal for all the engineering activities. The aim of this thorough work is to mitigate at least 80% of the risks in the system specification. Figure 1 shows the three stages of continuous innovation during the design phase. The operational changes in this figure include design improvements made to avoid manufacturing and service problems. Innovation is defined as designing out the failure or a customer complaint, with at least 500% ROI.

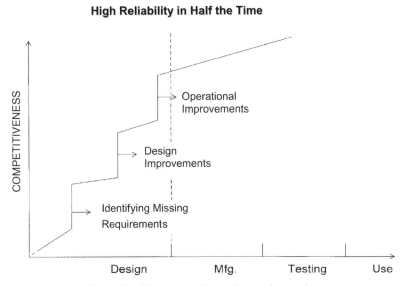

Figure 1   Three stages in continuous innovation.

System Functions FMEA   This FMEA is about the functions as seen by the user—the driver in this case. It is not about the failure of the components that are analyzed later during the detailed design. The driver is primarily concerned about completing the trip safely in order to deliver perishables or time-sensitive goods. This FMEA is done on all the functions of the system. A question is asked: "What can go wrong with this function at the system level?" The system is made up of the interactions among the driver, the road conditions, the environment, the vehicle software, hardware, and interfaces, the mission being "no mission-stopping failures." Usually several things can go wrong. Each is treated as a failure mode of the system.

The high risk items in a FMEA are defined as those having a "severity rating" of 8, 9, or 10 on a scale of 1 to 10. The rating 8 is for a mission-stopping failure, 9 for a major safety concern or potential recall, and 10 is where a fatality is possible. Risks in this range are not acceptable and must be mitigated. The mitigation criteria, in the order of precedence, are:

Eliminate the risk by changing the design.

Tolerate the fault, usually with redundancy or an alternate mode.

Design for failsafe and limp home mode.

Design early prognostics warning.

Note that every action listed above is mitigation by design. This eliminates the need for unnecessary testing. For example, the risk posed by a seal in a new product was mitigated by eliminating the joint. Since there was no way for the assembly to leak, a design qualification test for leaks was not required. This kind of failure mitigation increases the ROI more than 1000%.

As a result, no testing, no statistical quality control on mating components, and no production testing were required. No warranty costs had to be paid. That is just a fraction of the savings. The customer saved even more: no failures for life, no maintenance, no repairs, and no downtime! This is the business model for Eaton. A win–win model!

FUNCTION	FAILURE MODE	CAUSES	EFFECT	SYSTEM RESPONSE	RECOMMENDED ACTION
Provide High Voltage device interface for 340V DC devices	Loss of inter-face Inter-mittent	Connection degradation Heat Environmental Damage	Loss of Hybrid Function.	Current leakage fault	Change High Voltage connector design to add signal disconnect + VSE to route wiring and protect high voltage system with shield.

Figure 2   Partial table for system functions FMEA.

System Hazard Analysis   The system hazard analysis examines every possible accident scenario to identify unsafe conditions or hazards that could result in accidents. A hazard can be a component, an event, a set of components, a series of events, a breakdown of safety barriers, or can be caused by environmental factors. Hazards alone do not cause accidents. For example, an H-bridge circuit (similar to an H pattern circuit covered in the sneak circuit analysis in Chapter 5) can be a hazard because the current could potentially flow in the opposite direction. But this hazard does not result in an accident until something, such as a stuck-at-fault, occurs in the integrated circuit chip. This event is called a trigger event. The accident only takes place when the "hazard" and the "trigger" exist simultaneously. Therefore, if the design eliminates one of the two, an accident can be prevented. Eaton prevented this "hazard" from becoming an "accident" by introducing a built-in software check to make sure there was no stuck-at fault in the chip that supports the H-bridge.

Use/Misuse Analysis   Human factor engineering analysis helps to prevent frustrations of the user and safety related mishaps. The analysis has two parts. First an FMEA is performed on the process of using and misusing the product. The failure mode in this analysis is defined as anything that can go wrong when no component

1. **Product Over Speed Requirements**
   a. Need separate meeting to assess the risk of this issue
2. **Power Transmission Overdrive (PTO) Requirements**
   a. Need to assess PTO startup loads
   b. Need to address PTO operating modes (i.e., pump and roll, driving with the PTO engaged, etc.)
3. **Idle Requirements**
   a. Need to assess idling with the clutch closed or open.
   b. Efficiency, damper life, idle rattle, and spline wear are concerns.
4. **Shuttle Shifting**
   a. Does Traction Control System impact the ability to shuttle shift?
   b. Does the product define the requirements for shuttle shifting/programmable option?
   c. If shuttle shifting is required, what is the duty cycle?
5. **Clutch Protection Requirements**
   a. Does the product open the clutch to protect in the event of high thermal limit operation?
   b. If yes, is this a safety concern?
6. **Hill Start Aide Requirements**
   a. How does this work?
   b. How does this address faults and fallback modes?
7. **Maneuverability Requirements**
   a. Express in terms of speed in reverse?
   b. How does this product compare to current products (specifically DM)?
   c. Is there a switched input needed for high-speed reverse operation?

Figure 3   Partial tasks from use/misuse analysis.

or system has failed. Examples are: the ride may not be comfortable because of normal vibration, the vehicle is shifted into the wrong gear, or a driver inadvertently takes a wrong action. Second, an interview is conducted with the user to learn first-hand about the problem. The same risk mitigation guideline applies here as in the System Functions FMEA above.

USING THE NEW PARADIGMS FOR ROBUSTNESS

The design specification analyses described above are good starting points in most cases. When a more durable or more intelligent product is needed to satisfy the goal of "no mission-stopping failure," two new paradigms usually provide more choice. They are:

- Design for "twice-the-life" for hardware failures.
- Design for prognostics for soft failures from complex interactions.

Designing for Twice-the-Life

Why "twice-the-life"? The simple answer is that it is cheaper than designing for one life. It requires understanding the life cycle costs. When a Weibull Analysis (Chapter 3) is performed, the usual method is to use median ranks, which implies that 50% of the time the life will be less than predicted. In other words, either Eaton or the customer has to pay for 50% failures during the product cycle. This is expensive for both parties. Besides, there are many indirect costs to monitoring, production testing, and maintaining excessive inventories to replace failed parts. Eaton's requirement of twice-the-life at 500% ROI actually turns the situation into a positive cash flow. There is nothing to be monitored since the failures are going to be beyond the "first life." The 50% failure rate is now shifted to the "second life" when the product will be obsolete.

Another reason for a "twice-the-life" design is the need for basic engineering. A manager in Supply Chain at Eaton, explains: Imagine a bridge designed for 20-ton trucks. It may have no problems in the beginning. But the bridge degrades over time. After five years it may not be strong enough to take even a 15-ton truck and it is very likely to collapse. If it were designed for 40-ton trucks, it would be very safe. This is the same as the 100% safety margin we were taught in engineering schools a long time back. For the same reason, the components in the aerospace industry are de-rated 50%.

Since the Eaton attention is on continuous innovation, engineers design for "twice-the-life" creatively. They try to do it without increasing the size or weight of the components, the main cost drivers. Occasionally they may increase the size by a minor amount to expedite the solution. This is acceptable, as long as the ROI is at least 500%. There are numerous examples of "twice-the-life" design that were accomplished without changing size or weight. For example, the expected life of a transmission shift-key assembly was increased several-fold by using a different method of heat treating and by using a cheaper round key that had practically no stress concentration points. In the Brazilian operation of Eaton, "twice-the-life" design was achieved by molding two parts into a single piece, preventing stresses at the joint. The cost was lower because no assembly was required, there were fewer parts needed in the inventory, and there were no failures and no downtime for the customer. Such a robust mitigation of risks is always encouraged by Eaton managers.

Designing for Prognostics

As a result of the "twice-the-life" approach, hardware is becoming more reliable. The trend is towards "No Faults Found" from complex interactions. In such cases, understanding interactions among subsystems, interfaces, and software is difficult but necessary. Since a system behaves differently in the case of an anomaly, it could be possible to detect variation and inform the driver to take a proactive action. A good prognostics design allows the driver sufficient time to complete the trip. This analysis is done during the detailed design. Service Engineering predicts the possible mission-stopping failures and accidents, determines what unusual behaviors of the vehicle need to be tracked through built-in checks, and suggests design improvements. Hybrid vehicles contain many prognostics requirements, which require revising the system specification.

Function	Driver	AMT	Hydraulics	Clutch	X/Y Shifter	Trans. Harness	ECU	CAN Bus	OEM Interface
Engage Clutch	X	X	X	X	X	X	X	X	X
Engage Gear	X	X	X	X	X	X	X	X	X
Isolate Fault	X	X	X	X	X	X	X	X	X
Reduce No Fault Founds	X	X	X	X	X	X	X	X	X
Smooth shifting	X	X	X	X	X	X	X	X	X
Starting the engine	X	X			X	X	X		
Acceleration	X	X	X	X	X	X	X	X	X
No unintended Motion		X	X	X	X	X	X	X	X
Robust OEM communication	X	X	X	X	X	X	X	X	X
Rel. Centered Maintenance	X		X	X					
Speed control on grade	X	X	X	X	X	X	X	X	X
PTO Operation	X	X	X		X		X	X	
Maneuverability	X	X	X		X	X	X	X	
Low Temperature Operation		X	X			X	X	X	X
Speed Control (Cruise)	X	X		X	X		X	X	
Diagnostics		X		X	X	X	X	X	X

Figure 4   Functions and subsystems interaction matrix.

This is the stage when the system specification is considered complete and clear. It becomes the master document for all the assurance functions. Then a matrix of all subsystems and functions is constructed, as shown in Fig. 4, to identify all the interactions involved, including the communications with the customer's systems.

The matrix serves as the baseline for developing subsystem specifications, such as for the Clutch and the OEM Interface. For example, the Clutch specification will contain all the functions marked X in the Clutch column. The horizontal rows for each function show the interactions among subsystems. Such a matrix is a very useful tool for developing software specification because most of the interactions are visible. It helps engineers to be able to see the whole picture for system integration. They can then work together for the ultimate goal—"No Mission –Stopping Failures."

[*Note: The authors are thankful to the following Eaton executives for sharing their world class design assurance process in this book. Tim Morscheck, Vice President, Technology; George Nguyen, Vice President for Heavy Duty Trucks; Ken Davis, Vice President for Medium/Light Duty Trucks; Kevin Beaty, Business Manager, Hybrid Electric Vehicle; and Paul Kellberg, Manager for Advance Purchasing.*]

INDEX

*Assurance Technologies Principles and Practices: A Product, Process, and System Safety Perspective,
Second Edition*, by Dev G. Raheja and Michael Allocco
Copyright © 2006 John Wiley & Sons, Inc.

449